OPTIMIZATION TECHNIQUES IN OPERATIONS RESEARCH

PRENTICE-HALL INTERNATIONAL SERIES
IN INDUSTRIAL AND SYSTEMS ENGINEERING

W. J. Fabrycky and J. H. Mize, Editors

OPTIMIZATION TECHNIQUES IN OPERATIONS RESEARCH

B. D. SIVAZLIAN
University of Florida

L. E. STANFEL
University of Texas at Arlington

PRENTICE-HALL, INC., *Englewood Cliffs, New Jersey*

Library of Congress Cataloging in Publication Data

Sivazlian, B.D.
  Optimization techniques in operations research.

  (Prentice-Hall international series in
industrial and systems engineering)
  Companion volume to Analysis of systems in
operations research.
  Includes bibliographies.
  1. Operations research. 2. Mathematical
optimization. 3. Programming (Mathematics)
I. Stanfel, L. E.,   joint author.
II. Title.
T57.6.S59   001.4'24   73–18320
ISBN 0–13–638163–4

TO JANE

No dedication can match her own

© 1975 by Prentice-Hall, Inc.
Englewood Cliffs, New Jersey

10 9 8 7 6 5 4 3 2 1

Printed in the United States of America

PRENTICE-HALL INTERNATIONAL, INC., *London*
PRENTICE-HALL OF AUSTRALIA, PTY. LTD., *Sydney*
PRENTICE-HALL OF CANADA, LTD., *Toronto*
PRENTICE-HALL OF INDIA PRIVATE LIMITED, *New Delhi*
PRENTICE-HALL OF JAPAN, INC., *Tokyo*

CONTENTS

3 Linear Programming 130

4 PERT and Network Flows 220

5 Transportation and Assignment Problems 258

6 Search Techniques 344

7 Dynamic Programming 402

PREFACE

The aim of this book is to provide an introduction to the principal methods of optimization in present day operations research. Together with its companion volume, *Analysis of Systems in Operations Research,* it provides comprehensive coverage for the undergraduate engineering or mathematics student of junior or senior level whose mathematical background is assumed to include elementary calculus and a modicum of differential equations. In addition, the book will serve as an introductory text for qualified students in such fields as business and economics, and for freshly entering graduate students in certain programs who have not had exposure to the techniques of operations research. The practicing operations researcher will also find the book a helpful reference.

It is the case that the preparation mentioned is most often realized by students during their second undergraduate year, and portions of the book will comprise a cohesive, understandable introduction to certain areas in operations research for the student of late sophomore standing. The authors feel that an early introduction of this sort is most desirable; for example, the pre-engineering student rarely receives a hint of the nature of this area of study, whereas other fields of engineering are represented by required courses during the first two years.

The book is not an entirely theoretical treatment of the various topics; first, this is not possible in terms of the background presumed and the audience being addressed; secondly, feasibility not withstanding, that sort of exposition would not serve as the undergraduate-level introductory treatment we hope to provide—namely, to impart a problem-solving facility to the students.

On the other hand, the treatment of a given topic does not consist of the presentation of a few algorithms and a collection of solved problems. This would be, the authors feel, meaningless from a pedagogical point of view, with respect to both the material under consideration and the preparation and stimulation of the students for possible further pursuit of the topics. The understanding of the book's scope and content is a first step for the ambitious student who will be delving into the more complex and specialized literature.

We have attempted to strike a useful balance between these two extremes. Numerous pertinent results have been derived in terms meaningful to the group being addressed, the more rigorous portion of this book, as well as its companion book, corresponding to those topics whose underlying foundations rest more heavily upon the students' presumed minimum background. Thus, for example, the chapter dealing with classical optimization will be found to be more rigorously developed than, say, the chapter on linear programming.

When the presentation becomes, of necessity, less rigorous—and in fact, whenever possible—we have attempted to provide insight by relying upon illustrations of results and a sufficient development to make a final result intuitively appealing and meaningful. We have frequently utilized geometry in two and three dimensions for illustrative purposes, and numerous examples have been included. Various computational aspects of the application of the techniques have been discussed, since the utility of an application often hinges upon these.

It is believed that both the set of *optimization techniques* and the *fields of application* constitute operations research. Thus, dynamic programming is an optimization technique, whereas inventory systems is a field in operations research in which such optimization techniques are useful. The two books are separated within this conceptual framework. The present text treats some of the primary optimization techniques, including classical mathematics of optimization, linear programming, PERT and network flows, transportation and assignment problems, search techniques, dynamic programming, and geometric programming. Its initial chapter provides the elementary concepts of linear algebra and matrix manipulations which are utilized in subsequent chapters, while the essentials of probability theory are presented at the beginning of *Analysis of Systems in Operations Research*. Students with no exposure to these are in no way disadvantaged.

Optimization Techniques in Operations Research

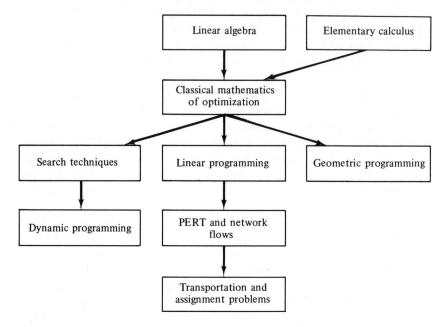

The book contains chapters not ordinarily found in texts of this kind. A chapter on geometric programming, a recent and—for certain problems—a most effective optimization technique is included, as well as one devoted to search techniques, a topic of much importance in any practical considerations of optimization problems.

In choosing notations we have attempted to select those that are fairly conventional within subject areas and which are, simultaneously, not unwieldy or cumbersome.

The problems at the end of each chapter will be found to include not only a number of the problem-solving variety, but also several requiring the proofs of statements relative to the development, extensions thereof, etc. The problems will assist the student in mastering certain fundamental techniques and the reader is urged to solve them, as they are believed to form an integral part of the text.

B. D. Sivazlian
L. E. Stanfel

OPTIMIZATION
TECHNIQUES IN
OPERATIONS
RESEARCH

LINEAR ALGEBRA

1.1. Introduction

It is the purpose of this chapter to provide the reader with a few fundamental concepts of linear algebra and matrices. The brief coverage should enable such a reader to undertake—without disadvantage—those portions of the book which require some familiarity with these concepts.

1.2. Vector Spaces

i. *Definitions*

Definition. A set of objects, called *vectors*, V, is said to be a vector space over the real numbers if the following conditions hold:

1. Sums of vectors from V form other vectors that belong to V.
2. Scalar multiples of vectors by real numbers form vectors that belong to V.

1

The Euclidean vector spaces E^n over the set of real numbers consist of all n-tuples $(x_1, x_2, \ldots, x_n)$ of real numbers. Here the vector addition is defined as

$$(x_1, x_2, \ldots, x_n) + (y_1, y_2, \ldots, y_n) = (x_1 + y_1, x_2 + y_2, \ldots, x_n + y_n)$$

Multiplication by a scalar c is defined as

$$c(x_1, x_2, \ldots, x_n) = (cx_1, cx_2, \ldots, cx_n)$$

Vectors will be denoted by block symbols; thus

$$\mathbf{x} = (x_1, x_2, \ldots, x_n)$$

and

$$\mathbf{y} = (y_1, y_2, \ldots, y_n)$$

Example 1.1

$$(1, 2, 3) + (4, 5, 6) = (5, 7, 9).$$

ii. *Vector Equalities and Inequalities in* E^n

Define in the vector space E^n vectors **x** and **y** where

$$\mathbf{x} = \begin{pmatrix} x_1 \\ x_2 \\ \cdot \\ \cdot \\ \cdot \\ x_n \end{pmatrix}, \qquad \mathbf{y} = \begin{pmatrix} y_1 \\ y_2 \\ \cdot \\ \cdot \\ \cdot \\ y_n \end{pmatrix}$$

Then 1. $\mathbf{x} = \mathbf{y}$ iff $x_j = y_j, j = 1, \ldots, n$.
2. $\mathbf{x} \geq \mathbf{y}$ iff $x_j \geq y_j, j = 1, \ldots, n$.
3. $\mathbf{x} > \mathbf{y}$ iff $x_j > y_j, j = 1, \ldots, n$.

iii. *A Word on Vector Products*

The reader may wonder since we have a well-defined vector addition that associates with any two vectors $\mathbf{v}_1, \mathbf{v}_2 \in V$ a vector $\mathbf{v}_1 + \mathbf{v}_2 = \mathbf{v}_3 \in V$, whether it is possible to define a *vector product* that associates with $\mathbf{v}_1, \mathbf{v}_2 \in V$ a vector $\mathbf{v}_1\mathbf{v}_2 \in V$. The reader may recall, for example, the *cross product* often used in physics, which does this. (The nonassociativity of the cross product is an undesirable feature of that particular "multiplication.") The answer is that this is sometimes possible, and if the multiplication has an appropriate set of properties, the vector space becomes a *linear algebra*, which we mention more to indicate an ambiguity of connotation than to introduce a new concept.

Of greater interest to us here is the "dot product" or "scalar product," which we prefer to define here as *inner product* on the vector space E^n.

Definition. If $\mathbf{v}_1, \mathbf{v}_2 \in E^n$, then the *inner product* of $\mathbf{v}_1, \mathbf{v}_2$, which we shall write $\mathbf{v}_1 \cdot \mathbf{v}_2$, or $\mathbf{v}_1\mathbf{v}_2$ is defined as the scalar

$$\mathbf{v}_1 \cdot \mathbf{v}_2 = \sum_{i=1}^{n} v_{1i}v_{2i} = \mathbf{v}_1\mathbf{v}_2$$

Example 1.2

$$\begin{pmatrix} -1 \\ 1 \\ 3 \\ 4 \end{pmatrix} \cdot \begin{pmatrix} -2 \\ 0 \\ 0 \\ \frac{1}{2} \end{pmatrix} = 2 + 0 + 0 + 2 = 4$$

iv. *Additional Concepts*

a. LINEAR COMBINATIONS

Let $\{v_1, v_2, \ldots, v_n, \ldots,\}$ be a collection of vectors. Then a sum of the form $\lambda_1 v_1 + \lambda_2 v_2 + \cdots + \lambda_n v_n + \cdots$, where $\lambda_1, \lambda_2, \ldots, \lambda_n, \ldots$ are arbitrary scalars, is called a *linear combination* of the vectors $v_1, \ldots, v_n, \ldots$. If the set of vectors is finite, we have a *finite linear combination*.

Theorem 1.1

Let W be a collection of vectors from a vector space V. Let $S(W) =$ the set of all linear combinations of elements of W. Then $S(W)$ is a vector space.

The set $S(W)$ in Theorem 1.1 is called the *subspace spanned by the set W*. Also, W is said to *span $S(W)$*, or to be a *spanning set* for $S(W)$.

Given a vector space, there may be an infinite number of spanning sets. For example, in E^3 the line L in Figure 1.1 is a subspace of E^3, and it is spanned by any point (vector) other than $(0, 0, 0)$ on the line.

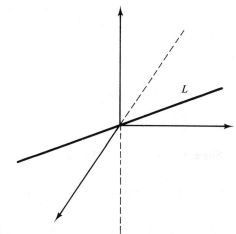

Figure 1.1

b. LINEAR DEPENDENCE AND INDEPENDENCE

Of particular importance is the concept of a linearly independent set of vectors.

Definition. Let V be a vector space in E^k. Let W be a subset of V with $W = \{v_1, v_2, \ldots, v_n\}$. W is said to be a *linearly independent* set of vectors if

$$\lambda_1 v_1 + \lambda_2 v_2 + \cdots + \lambda_n v_n = 0$$

implies that

$$\lambda_1 = \lambda_2 = \cdots = \lambda_n = 0$$

That is, the vectors are linearly independent if the *only* way a linear combination of them can equal the zero vector, or *null vector*, of V is for each coefficient of the linear combination to be zero.

Definition. A set of vectors that is not linearly independent is said to be *linearly dependent.*

Several remarks follow almost immediately from these definitions and should be verified by the student.

1. Any set of vectors from a vector space V that contains the null vector is linearly dependent.
2. Every subset of a linearly independent set is linearly independent.
3. A set consisting of a single nonnull vector is linearly independent.
4. If $S_1 \subseteq S_2$ and S_1 is linearly dependent, so is S_2.

The notion of linear independence is intimately related to the ability to express one element of a set of vectors as a linear combination of the others.

For example, consider a set of linearly dependent vectors $v_1, v_2, \ldots, v_n$. Then we have

$$\lambda_1 v_1 + \lambda_2 v_2 + \cdots + \lambda_n v_n = 0 \tag{1.1}$$

and $\lambda_r \neq 0$ for some integer r. Rewriting (1.1), we have then

$$\lambda_r v_r = -\lambda_1 v_1 - \lambda_2 v_2 - \cdots - \lambda_{r-1} v_{r-1} - \lambda_{r+1} v_{r+1} - \cdots - \lambda_n v_n \tag{1.2}$$

Since $\lambda_r \neq 0$, we may divide by it in (1.2), obtaining

$$v_r = -\frac{\lambda_1}{\lambda_r}(v_1) - \frac{\lambda_2}{\lambda_r}(v_2) - \cdots - \frac{\lambda_{r-1}}{\lambda_r}(v_{r-1}) - \frac{\lambda_{r+1}}{\lambda_r}(v_{r+1}) - \cdots - \frac{\lambda_n}{\lambda_r}(v_n)$$

Conversely, if v_r may be so expressed, for some r, then the set is linearly dependent; for if we have

$$v_r = c_1 v_1 + c_2 v_2 + \cdots + c_{r-1} v_{r-1} + c_{r+1} v_{r+1} + \cdots + c_n v_n$$

then

$$c_1 v_1 + c_2 v_2 + \cdots + c_{r-1} v_{r-1} - 1 v_r + c_{r+1} v_{r+1} + \cdots + c_n v_n = 0$$

and not all the coefficients are zero.

Thus, we have proved

Theorem 1.2

A set of vectors is linearly dependent if and only if one of them may be expressed as a linear combination of the others.

c. BASES AND DIMENSION

Let us now relate the concepts of spanning sets and linear independence.

Definition. If V is a vector space, then a *basis* for V is a linearly independent set of vectors from V that spans V.

As already stated, the vector space E^n (n finite) is the most important vector space for the purposes of this book. If we define the unit vector e_j, $j = 1, 2, \ldots, n$, to be a vector in E^n whose jth component is 1 and all other elements are zero, that is,

$$
\mathbf{e}_j = \begin{pmatrix} 0 \\ \cdot \\ \cdot \\ \cdot \\ 0 \\ 1 \\ 0 \\ \cdot \\ \cdot \\ \cdot \\ 0 \end{pmatrix} \longleftarrow j\text{th component}
$$

then the set of *n unit vectors* $\mathbf{e}_1, \mathbf{e}_2, \ldots, \mathbf{e}_n$ is a basis for E^n.

First, $\{\mathbf{e}_1, \mathbf{e}_2, \ldots, \mathbf{e}_n\}$, is linearly independent, for

$$
\lambda_1 \mathbf{e}_1 + \lambda_2 \mathbf{e}_2 + \cdots + \lambda_n \mathbf{e}_n = \begin{pmatrix} \lambda_1 \\ \lambda_2 \\ \cdot \\ \cdot \\ \cdot \\ \lambda_n \end{pmatrix},
$$

and if

$$
\begin{pmatrix} \lambda_1 \\ \lambda_2 \\ \cdot \\ \cdot \\ \cdot \\ \lambda_n \end{pmatrix} = \begin{pmatrix} 0 \\ 0 \\ \cdot \\ \cdot \\ \cdot \\ 0 \end{pmatrix} = \mathbf{0},
$$

then $\lambda_1 = \lambda_2 = \ldots = \lambda_n = 0$. Furthermore this set spans E^n, for if

$$\begin{pmatrix} \lambda_1 \\ \lambda_2 \\ \cdot \\ \cdot \\ \cdot \\ \lambda_n \end{pmatrix} = \boldsymbol{\alpha} \quad \text{is } any \text{ vector in } E^n,$$

then

$$\boldsymbol{\alpha} = \lambda_1 \mathbf{e}_1 + \lambda_2 \mathbf{e}_2 + \cdots + \lambda_n \mathbf{e}_n$$

Theorem 1.3 limits the size of a basis for a space spanned by a finite set of vectors.

Theorem 1.3

Let V be a vector space that is spanned by a *finite* set of vectors $\{\mathbf{v}_1, \mathbf{v}_2, \ldots, \mathbf{v}_n\}$. Then every linearly independent set of vectors from V contains no more than n elements.

Proof: Let $\{\mathbf{w}_1, \mathbf{w}_2, \ldots, \mathbf{w}_k\}$ with $k > n$ be a linearly independent set in V. Since $\mathbf{v}_1, \ldots, \mathbf{v}_n$ spans V, we have scalars such that

$$\mathbf{w}_j = c_{j1}\mathbf{v}_1 + c_{j2}\mathbf{v}_2 + \cdots + c_{jn}\mathbf{v}_n, \quad j = 1, \ldots, k$$

The completion of the proof is left for the reader.

As a corollary to Theorem 1.3, we have the following:

Corollary 1.1

If B_1 and B_2 are finite bases for a vector space V, then B_1 and B_2 have the same number of elements.

For suppose that B_1 has m vectors and B_2 has k vectors. By Theorem 1.3, we have both $m \leq k$ and $k \leq m$; thus $m = k$.

Aside from its spanning properties and linear independence, a basis has another important property.

Let $(\mathbf{b}_1, \mathbf{b}_2, \ldots, \mathbf{b}_n)$ be a basis for a vector space V. Let $\boldsymbol{\alpha} \in V$ and, suppose that $\boldsymbol{\alpha}$ may be expressed as two *different* linear combinations of the $\mathbf{b}_j$. That is,

$$\boldsymbol{\alpha} = c_1\mathbf{b}_1 + c_2\mathbf{b}_2 + \cdots + c_n\mathbf{b}_n$$
$$\boldsymbol{\alpha} = d_1\mathbf{b}_1 + d_2\mathbf{b}_2 + \cdots + d_n\mathbf{b}_n$$

Subtracting,

$$0 = (c_1 - d_1)\mathbf{b}_1 + (c_2 - d_2)\mathbf{b}_2 + \cdots + (c_n - d_n)\mathbf{b}_n$$

But since the $\mathbf{b}_j$ are linearly independent, $c_j = d_j$ for all j. Thus we have

Theorem 1.4

The representation of a vector in terms of a basis for that space is *unique*.

Definition. A vector space is called *finite dimensional* if it has a finite basis.

Definition. If a basis for a finite dimensional vector space V has m elements, then V is said to be m *dimensional*, or to have dimension m. One also writes dim $V = m$.

1.3. Matrices

i. *Definitions*

Definition. An $m \times n$ (read "m by n") matrix is a rectangular array of mn elements arranged in m rows, each having n elements; that is, there are m rows and n columns.

For example, an $m \times n$ matrix A over the real numbers will be expressed by

$$A = \begin{pmatrix} a_{11} & a_{12} & \cdots & a_{1n} \\ a_{21} & a_{22} & \cdots & a_{2n} \\ \cdot & \cdot & & \cdot \\ \cdot & \cdot & & \cdot \\ \cdot & \cdot & & \cdot \\ a_{m1} & a_{m2} & \cdots & a_{mn} \end{pmatrix}$$

and the real number a_{ij} is the element of A in the ith row and jth column.

Definition. If A is $m \times n$ and B is $r \times s$, then A is said to be equal to B, $A = B$, if $m = r$, $n = s$, and $a_{ij} = b_{ij}$ for $i = 1, \ldots, m$ and $j = 1, \ldots, n$.

Definition. The $m \times n$ *null matrix* is the $m \times n$ matrix that has every element zero.

Definition. If A is an $m \times n$ matrix, then the *transpose of A*, written here A^T, has as its ith *column* the ith *row* of A. Thus A^T is $n \times m$.

Example 1.3

If

$$A = \begin{pmatrix} 6 & 2 & 1 \\ 0 & 3 & 4 \\ -1 & -\frac{3}{2} & 8 \\ 7 & 7 & 0 \end{pmatrix}$$

then

$$A^T = \begin{pmatrix} 6 & 0 & -1 & 7 \\ 2 & 3 & -\frac{3}{2} & 7 \\ 1 & 4 & 8 & 0 \end{pmatrix}$$

ii. *Addition and Scalar Multiplication of Matrices*

As in the case of vectors, we define matrix addition and scalar multiplication.

Definition. Let A, B be $m \times n$ matrices with elements a_{ij}, b_{ij}, respectively. Then $A + B = C$ is an $m \times n$ matrix with elements $c_{ij} = a_{ij} + b_{ij}$.

Example 1.4

$$\begin{pmatrix} 1 & 0 & -1 \\ \frac{3}{2} & 4 & 10 \\ -\frac{1}{2} & 0 & 7 \end{pmatrix} + \begin{pmatrix} 2 & \frac{1}{2} & -\frac{1}{2} \\ \frac{1}{2} & -1 & -3 \\ \frac{1}{2} & 5 & -\frac{1}{2} \end{pmatrix} = \begin{pmatrix} 3 & \frac{1}{2} & -\frac{3}{2} \\ 2 & 3 & 7 \\ 0 & 5 & \frac{13}{2} \end{pmatrix}$$

Remark 1: $A + B = B + A$, whenever addition in the set is commutative.

Remark 2: If A is $m \times n$ and B is $r \times s$, then $A + B$ is defined when, *and only when, $m = r$ and $n = s$.* Thus

$$\begin{pmatrix} 1 & 2 \\ 3 & 4 \end{pmatrix} + \begin{pmatrix} 1 & 0 & -1 \\ 2 & 1 & 1 \\ 0 & 0 & 0 \end{pmatrix}$$

is not defined.

Definition. If A is an $m \times n$ matrix and c is a scalar, then $cA = M$ is an $m \times n$ matrix with elements $m_{ij} = ca_{ij}$.

Example 1.5

$$-2 \begin{pmatrix} 1 & 2 & 0 \\ 0 & -\frac{1}{2} & -1 \\ 1 & 3 & 0 \end{pmatrix} = \begin{pmatrix} -2 & -4 & 0 \\ 0 & 1 & 2 \\ -2 & -6 & 0 \end{pmatrix}$$

iii. *Row and Column Vectors*

It is frequently convenient to consider the rows and columns of an $m \times n$ matrix A as vectors in the spaces of dimension n and m, respectively. For example, if A is the 3×5 matrix

$$
\begin{pmatrix}
\frac{1}{2} & \pi & -4 & 28 & 9 \\
0 & 0 & 0 & 1 & 2 \\
-1 & -1 & 3 & 6 & \frac{\pi}{2}
\end{pmatrix}
$$

then

$$
\begin{pmatrix} \frac{1}{2} \\ 0 \\ -1 \end{pmatrix}, \quad
\begin{pmatrix} \pi \\ 0 \\ -1 \end{pmatrix}, \quad
\begin{pmatrix} -4 \\ 0 \\ 3 \end{pmatrix}, \quad
\begin{pmatrix} 28 \\ 1 \\ 6 \end{pmatrix}, \quad
\begin{pmatrix} 0 \\ 2 \\ \frac{\pi}{2} \end{pmatrix}
$$

are vectors in E^3 and

$$
\left(\frac{1}{2}, \pi, -4, 28, 9 \right), \quad (0, 0, 0, 1, 2), \quad \left(-1, -1, 3, 6, \frac{\pi}{2} \right)
$$

are vectors in E^5.

An n-tuple vector may itself be considered an $n \times 1$ matrix or a $1 \times n$ matrix, depending upon whether it is written as a column or a row.

We shall speak of the row vectors and column vectors of an array A, although, as above, we may write them as either rows or columns. The same *vector* may be denoted by $(x_1, x_2, \ldots, x_n)$ or by

$$
\begin{pmatrix} x_1 \\ x_2 \\ \cdot \\ \cdot \\ \cdot \\ x_n \end{pmatrix}
$$

As matrices, however, these are different objects, one being $n \times 1$, the other $1 \times n$. The difference can be important; for example, strictly speaking

$$
(x_1, x_2, \ldots, x_n) + \begin{pmatrix} y_1 \\ y_2 \\ \cdot \\ \cdot \\ \cdot \\ y_n \end{pmatrix}
$$

is not defined, whereas

$$(x_1, \ldots, x_n) + (y_1, \ldots, y_n) \quad \text{and} \quad \begin{pmatrix} x_1 \\ \cdot \\ \cdot \\ \cdot \\ x_n \end{pmatrix} + \begin{pmatrix} y_1 \\ \cdot \\ \cdot \\ \cdot \\ y_n \end{pmatrix}$$

both are.

The difference between vector addition of two n-tuples and matrix addition of them is that vector addition as shown (Section 1.2-i) has nothing to do with the representation of the components, whereas arrangement of the numbers is meaningful in a matrix.

iv. *Matrix Multiplication*

In particular instances it is possible to *define* a multiplication of matrices.†

Definition. Let A be an $m \times n$ matrix with elements a_{ij} and B be an $r \times s$ matrix with elements b_{ij}. The matrix product AB exists if and only if $n = r$ and is defined to be $AB = C$, which has elements

$$c_{ij} = \sum_{k=1}^{n} a_{ik} b_{kj}, \qquad i = 1, \ldots, m, \qquad j = 1, \ldots, s$$

In general, we observe that c_{ij} is *defined as the inner product of the ith row of A and the jth column of B.* We see immediately that AB is $m \times s$. This is easily remembered, for juxtaposing the dimensions $(m \times n)$ and $(r \times s)$, one notices that if the two middle integers are equal, AB is defined, and has dimensions represented by the two outside integers—in that order; that is, $m \times s$. Here "dimension" is used to indicate array size only.

If we consider an example of matrix multiplication, we see that this is an operation more easily accomplished than the defining equation for c_{ij} would indicate.

Example 1.6

$$\overset{3 \times 2}{A = \begin{pmatrix} 1 & 2 \\ -1 & 1 \\ 4 & 5 \end{pmatrix}}, \qquad \overset{2 \times 2}{B = \begin{pmatrix} 1 & 2 \\ -1 & 4 \end{pmatrix}}$$

---

†The definition, at first encounter, will probably seem rather strange and not intuitively satisfying. There is, however, strong motivation for this definition of matrix multiplication, although we shall not observe that here.

We see that $AB = C$ is defined and that the product will be 3×2. By definition,

$$c_{11} = \sum_{k=1}^{2} a_{1k}b_{k1}$$
$$= a_{11}b_{11} + a_{12}b_{21}$$
$$= 1(1) + 2(-1) = -1$$
$$AB = \begin{pmatrix} -1 & 10 \\ -2 & 2 \\ -1 & 28 \end{pmatrix}$$

Remark 1: If A is $m \times n$ and B is $n \times s$, then AB is defined; but is BA? If BA is to be defined also, we require $s = m$, that is, that A be $m \times n$ and B be $n \times m$. In general, then, if AB is defined, BA is not.

Remark 2: Suppose that AB and BA are both defined. Is $AB = BA$? First, AB and BA will be of dimensions $m \times m$ and $n \times n$, respectively, so there is no chance for equality unless $m = n$, that is, unless A, B are both $n \times n$. Furthermore, even if A and B are both $n \times n$,† it is not generally the case that $AB = BA$. For example,

$$\begin{pmatrix} 2 & 1 \\ -1 & 1 \end{pmatrix}\begin{pmatrix} 2 & 0 \\ 1 & -1 \end{pmatrix} = \begin{pmatrix} 5 & -1 \\ -1 & -1 \end{pmatrix}$$

while

$$\begin{pmatrix} 2 & 0 \\ 1 & -1 \end{pmatrix}\begin{pmatrix} 2 & 1 \\ -1 & 1 \end{pmatrix} = \begin{pmatrix} 4 & 2 \\ 3 & 0 \end{pmatrix}$$

Although not commutative, matrix multiplication is associative; that is $(AB)C = A(BC)$, assuming the dimensions are such that the multiplications are defined. This property may be proved by pure manipulation of symbols, using the definition of the matrix product, and we omit the proof.

v. *Inverse Matrices*

a. IDENTITY MATRICES

Definition. The $n \times n$ *identity matrix* I_n is the $n \times n$ matrix with elements $(i, i) = 1$ for $i = 1, \ldots, n$, and elements $(i, j) = 0$ for $j \neq i$.

---

†A matrix with dimensions $n \times n$ is said to be *square*.

For example,

$$I_6 = \begin{pmatrix} 1 & 0 & 0 & 0 & 0 & 0 \\ 0 & 1 & 0 & 0 & 0 & 0 \\ 0 & 0 & 1 & 0 & 0 & 0 \\ 0 & 0 & 0 & 1 & 0 & 0 \\ 0 & 0 & 0 & 0 & 1 & 0 \\ 0 & 0 & 0 & 0 & 0 & 1 \end{pmatrix}$$

Alternative descriptions are possible.

1. I_n has 1's on the *main diagonal* and zeros elsewhere.
2. The ith column of I_n is $\mathbf{e}_i$, the ith unit vector in E^n.

As is easily verified, if A is $m \times n$, then $AI_n = A$ and $I_mA = A$. This is the reason for the terminology *identity matrix*; the identity matrix is to matrix multiplication as the real number 1 is to the multiplication of real numbers.

If A is $n \times n$, then $I_nA = AI_n = A$. We drop reference to the dimensions of the identity matrix when there is no ambiguity, and write $I_n = I$.

b. Defining the Inverse

Definition. Let A be a square matrix. If there exists a square matrix B such that $BA = I$, then B is called *a left inverse of A*. A square matrix B such that $AB = I$ is called *a right inverse of A*.

Theorem 1.5

Suppose that A has a left inverse B and a right inverse B'. Then $B = B'$.

Proof: By assumption, $BA = I$ and $AB' = I$. Then

$$B = BI = B(AB') = (BA)B' = IB' = B'$$

When A has both a left and right inverse B, we say A is *invertible*, or that A has an inverse, and denote the inverse of A by A^{-1}.

In a similar fashion we show that inverses are *unique*. That is, if

$$AA^{-1} = A^{-1}A = I$$

and $$BA = AB = I,$$

then $$A^{-1} = IA^{-1} = (BA)A^{-1} = B(AA^{-1}) = BI = B$$

Not all square matrices have inverses, of course, the most obvious example being a square null matrix. Many nonnull matrices lack inverses, also, the reason for which we shall observe later. For another example, though, consider the matrix

$$\begin{pmatrix} 1 & 2 \\ 2 & 4 \end{pmatrix}$$

If there exists an inverse

$$\begin{pmatrix} x_1 & x_2 \\ x_3 & x_4 \end{pmatrix}$$

then we must have

$$\begin{pmatrix} 1 & 2 \\ 2 & 4 \end{pmatrix}\begin{pmatrix} x_1 & x_2 \\ x_3 & x_4 \end{pmatrix} = \begin{pmatrix} 1 & 0 \\ 0 & 1 \end{pmatrix}$$

Thus a system of equations must be satisfied; that is,

$$x_1 + 2x_3 = 1$$
$$x_2 + 2x_4 = 0$$
$$2x_1 + 4x_3 = 0$$
$$2x_2 + 4x_4 = 1$$

The system obviously has no solution, for we require simultaneously

$$x_1 + 2x_3 = 1$$

and
$$2x_1 + 4x_3 = 2(x_1 + 2x_3) = 2 = 0$$

c. AN APPLICATION OF THE INVERSE

Consider a system of n linear equations in n variables, $x_1, \ldots, x_n$.

$$a_{11}x_1 + a_{12}x_2 + \cdots + a_{1n}x_n = b_1$$
$$a_{21}x_1 + a_{22}x_2 + \cdots + a_{2n}x_n = b_2$$
$$\vdots$$
$$a_{n1}x_1 + a_{n2}x_2 + \cdots + a_{nn}x_n = b_n$$

(1.3)

In matrix notation, let

$$A = \begin{pmatrix} a_{11} & a_{12} & \cdots & a_{1n} \\ a_{21} & a_{22} & \cdots & a_{2n} \\ \vdots & & & \\ a_{n1} & a_{n2} & \cdots & a_{nn} \end{pmatrix}, \quad \mathbf{x} = \begin{pmatrix} x_1 \\ x_2 \\ \vdots \\ x_n \end{pmatrix}, \quad \mathbf{b} = \begin{pmatrix} b_1 \\ b_2 \\ \vdots \\ b_n \end{pmatrix}$$

The system (1.3) may then be expressed as

$$Ax = b \tag{1.4}$$

Suppose that A^{-1} exists. Multiply both sides of (1.4) by A^{-1} on the left, obtaining

$$A^{-1}Ax = A^{-1}b$$

or $$x = A^{-1}b$$

Thus knowledge of A^{-1} allows easy solution of the system.

d. A NOTATIONAL CONVENTION

An exponential notation is used for denoting certain matrix products. If A is $n \times n$, then $A^2 = AA$, $A^3 = AAA$, ..., $A^p = AA^{p-1}$. A^1 merely denotes A, and A^0 is usually taken to be I.

1.4. Computing Inverses and Solving Systems of Linear Equations

i. *Elementary Row Operations*

To speak of efficient methods for computing the inverse of a matrix, we need to first discuss certain matrix operations, the *elementary row operations*.

Let A be an $m \times n$ matrix. An elementary row operation on A is one of the following three operations.

1. Multiplication of any row of A by a nonzero scalar.
2. Multiplication of any row i of A by a nonzero scalar with the result added, component by component, to row $j \neq i$, row i unchanged.
3. Interchanging two rows.

(We could as easily work with column operations.)

Now consider the system $Ax = 0$. Perform any elementary row operation on A, obtaining A', and consider the system $A'x = 0$. Any solution x_0 to $Ax = 0$ also solves $A'x = 0$, and vice versa (verify), so the sets of solutions to the respective systems are identical; and the same holds of course if A' is obtained from A by a *succession* of elementary row operations, meaning we obtain B_1 from A, B_2 from B_1, ..., and eventually A' from B_k.

More generally, for the $m \times n$ matrix A, one is interested in solving the inhomogeneous system

$$Ax = b \tag{1.5}$$

where b is a given $m \times 1$ vector.

Suppose that an elementary row operation is performed on the $m \times (n + 1)$ array $(A, \mathbf{b})$, where $\mathbf{b}$ merely becomes a new last column of the new array. The operation yields an array $(A', \mathbf{b}')$ and, from the definition of elementary row operation, it should be verified that $\mathbf{x}_0$ satisfies

$$A\mathbf{x}_0 = \mathbf{b}$$

if and only if $\mathbf{x}_0$ satisfies

$$A'\mathbf{x}_0 = \mathbf{b}'$$

This result may be utilized to good advantage to obtain very simple methods for solving systems of linear equations.

ii. *Solving Systems of Linear Equations*

Consider, for example, the system

$$\begin{aligned}
x_1 - 2x_2 + x_3 + x_4 &= 3 \\
2x_1 \qquad - x_3 - 2x_4 &= 2 \\
x_2 + x_3 - x_4 &= 4 \\
x_1 + x_2 + x_3 - 4x_4 &= 0
\end{aligned}$$

From Section 1.3-v-c, we know that if the array of coefficients

$$\begin{pmatrix}
1 & -2 & 1 & 1 \\
2 & 0 & -1 & -2 \\
0 & 1 & 1 & -1 \\
1 & 1 & 1 & -4
\end{pmatrix}$$

has an inverse, the unique solution to the system may be obtained easily once the inverse is calculated. Here we proceed without requiring the calculation of the inverse.

a. METHOD

The goal is to obtain, by elementary row operations beginning with the array $(A, \mathbf{b})$, an array $(A', \mathbf{b}')$ of the form, where we include the variables for ease of interpretation and use primes to indicate arbitrary values,

$$\begin{aligned}
1x_1 + a'_{12}x_2 + a'_{13}x_3 + a'_{14}x_4 &= b'_1 \\
0 + \quad 1x_2 + a'_{23}x_3 + a'_{24}x_4 &= b'_2 \\
0 + \quad 0 + \quad 1x_3 + a'_{34}x_4 &= b'_3 \\
0 + \quad 0 + \quad 0 + \quad 1x_4 &= b'_4
\end{aligned} \qquad (1.6)$$

Looking first at the *last* equation of (1.6), we find $x_4 = b'_4$; this value is substituted into the third equation, and we find $x_3 = b'_3 - a'_{34}b'_4$. With values for both x_4, x_3, the second equation yields x_2, and, finally, the first equation can be solved for x_1, since x_2, x_3, x_4 are at that point all known.

b. EXAMPLE—A UNIQUE SOLUTION

With the form (1.6) in mind, let us operate on the given $(A, \mathbf{b})$,

$$\begin{pmatrix} 1 & -2 & 1 & 1 & 3 \\ \boxed{2} & 0 & -1 & -2 & 2 \\ 0 & 1 & 1 & -1 & 4 \\ 1 & 1 & 1 & -4 & 0 \end{pmatrix} \tag{1.7}$$

It is advisable to work from left to right and top to bottom, for then, as we shall see, a later operation *never* ruins a part of the desired form obtained at an earlier time. With regard to (1.7), we have already a 1 in the (1, 1) position, so we need operate so as to obtain a 0 where the 2 presently exists in the (2, 1) position. To get this, we can multiply row 1 by -2 and add, term by term, to row 2, obtaining a new row 2. After that the array appears as

$$\begin{array}{ccccc} 1 & -2 & 1 & 1 & 3 \\ 0 & 4 & -3 & -4 & -4 \\ \boxed{0} & 1 & 1 & -1 & 4 \\ 1 & 1 & 1 & -4 & 0 \end{array} \tag{1.8}$$

Still looking at column 1 of (1.8), we find the (3, 1) element already 0; so we observe the (4, 1) element, find a 1 there, and, consequently, multiply row 1 by -1, add term by term to row 4, and obtain a new row 4 with a 0 in location (4, 1):

$$\begin{array}{ccccc} 1 & -2 & 1 & 1 & 3 \\ 0 & \boxed{4} & -3 & -4 & -4 \\ 0 & 1 & 1 & -1 & 4 \\ 0 & 3 & 0 & -5 & -3 \end{array} \tag{1.9}$$

Having completed column 1, we proceed to column 2. The form (1.6) demands a 1 in position (2, 2) and 0's in (3, 2) and (4, 2). The 1 is obtained first; therefore, we multiply row 2 of (1.9) by $\frac{1}{4}$. [Had there been a 0 in position (2, 2) at this point, we would need to consider the interchange of rows; and to preserve the structure *already* obtained, we would consider only interchanges involving row 2 with row j, $j = 3, 4$. Interchanging rows 2 and 1 at this point would eliminate the desired column 1 form. The conclusion associated with both (3, 2) and (4, 2) *also* being 0 in that case is left for later.] Thus we obtain

$$
\begin{matrix}
1 & -2 & 1 & 1 & 3 \\
0 & 1 & -\frac{3}{4} & -1 & -1 \\
0 & 1 & 1 & -1 & 4 \\
0 & 3 & 0 & -5 & -3
\end{matrix}
\qquad (1.10)
$$

and next set about obtaining 0's in locations (3, 2) and (4, 2) in (1.10).
Continuing the process to the end, we obtain the array

$$
\begin{matrix}
1 & -2 & 1 & 1 & 3 \\
0 & 1 & -\frac{3}{4} & -1 & -1 \\
0 & 0 & 1 & 0 & \frac{20}{7} \\
0 & 0 & 0 & 1 & \frac{45}{14}
\end{matrix}
$$

which gives the solution

$$
\begin{aligned}
x_4 &= \tfrac{45}{14} \\
x_3 &= \tfrac{20}{7} \\
x_2 &= -1 + \tfrac{45}{14} + \tfrac{3}{4}(\tfrac{20}{7}) = \tfrac{61}{14} \\
x_1 &= 3 - \tfrac{45}{14} - \tfrac{20}{7} + 2(\tfrac{61}{14}) = \tfrac{79}{14}
\end{aligned}
$$

The foregoing method, called triangularization,† can become very unwieldy for hand computation, but is particularly well suited for machine execution.

c. EXAMPLE—AN INFINITY OF SOLUTIONS

It is instructive to consider next the example

$$
\begin{aligned}
x_1 + 2x_2 - x_3 &= 4 \\
-2x_1 - x_2 &= -4 \\
-x_1 + x_2 - x_3 &= 0
\end{aligned}
\qquad (1.11)
$$

Beginning with

$$
\begin{matrix}
1 & 2 & -1 & 4 \\
-2 & -1 & 0 & -4 \\
-1 & 1 & -1 & 0
\end{matrix}
$$

---

†An $n \times n$ matrix A that has either the form $a_{ij} = 0$ for all $i > j$ or the form $a_{ij} = 0$ for all $i < j$ is called a triangular matrix. The first form was obtained when applying the method to the square array.

and proceeding as before, one obtains

$$
\begin{array}{cccc}
1 & 2 & -1 & 4 \\
0 & 1 & -\tfrac{2}{3} & \tfrac{4}{3} \\
0 & 0 & 0 & 0
\end{array}
\qquad (1.12)
$$

at which point the process must end—the desired form (1.6) cannot be obtained. But we still know that any solution of (1.12) is a solution to the original system (1.11), and vice versa. In (1.12) let x_3 be *any* real number. Having fixed $x_3 = x_3^0$, we find

$$x_2^0 = \tfrac{4}{3} + \tfrac{2}{3}x_3^0$$

Then $\qquad\qquad x_1^0 = 4 + x_3^0 - 2x_2^0$

So once x_3 is fixed, x_1 and x_2 are uniquely determined, and the resulting triple is a solution of (1.11). But x_3 may assume *any* real value, so the system (1.11) has an *infinite* number of solutions.

d. EXAMPLE—NO SOLUTIONS

As a next example consider

$$
\begin{aligned}
x_1 + 2x_2 - x_3 &= 4 \\
-2x_1 - x_2 &= -4 \\
-x_1 + x_2 - x_3 &= 1
\end{aligned}
\qquad (1.13)
$$

Comparing (1.13) to (1.11), we see our method had best show that (1.13) has *no* solution, since the third equation of (1.11) is just the sum of the first two, and in (1.13), since the constant only has been changed, it should be impossible that (1.13) be solvable.

Triangularization yields the array

$$
\begin{array}{cccc}
1 & 2 & -1 & 4 \\
0 & 1 & -\tfrac{2}{3} & \tfrac{4}{3} \\
0 & 0 & 0 & 1
\end{array}
$$

and the corresponding third equation cannot be satisfied.

Remark 1: It was the case that whenever we set out to obtain zeros in positions $(j+1, j), (j+2, j), \ldots, (n, j)$ after obtaining the 1 in (j, j), we used multiples of the row j just obtained. This is not essential. For example, given

$$
\begin{array}{ccc}
1 & 2 & 3 \\
2 & 3 & 5 \\
4 & 8 & 0
\end{array}
$$

and desiring a 0 in the (2, 1) position, we can multiply row 3 by $-\frac{1}{2}$ and add term by term to row 2, obtaining

$$
\begin{array}{rrr}
1 & 2 & 3 \\
0 & -1 & 5 \\
4 & 8 & 0
\end{array}
$$

The advantages to the method first shown, however, are

1. The proper *multiple* is just the negative of the element to be zeroed, owing to the presence of the 1.
2. Using always multiples of the same row provides a more systematic procedure—more desirable for manual computation and especially so for computer solution.

Remark 2: For manual computation, the arithmetic can become quite laborious, and a single arithmetical error can, in addition to providing an erroneous solution, cause a great deal of effort to be wasted. A simple error-detecting procedure is easily incorporated, and although it will not detect all possible combinations of errors, it is well worth the small increase in computation.

One incorporates a *new* column in the matrix, and row j of the new column contains the sum of all other elements of row j. Elements of the new column are subjected to all row operations performed, so that at *all* times the new elements remain the sums of all other elements in the respective rows.

Remark 3: Rather than obtaining the triangular structure described, one can pursue what is called *complete elimination* and, by means of row operations, transform

$$
\begin{array}{ccccc}
a_{11} & a_{12} & \cdots & a_{1n} & b_1 \\
a_{21} & a_{22} & \cdots & a_{2n} & b_2 \\
\cdot & \cdot & & \cdot & \cdot \\
\cdot & \cdot & & \cdot & \cdot \\
\cdot & \cdot & & \cdot & \cdot \\
a_{n1} & a_{n2} & \cdots & a_{nn} & b_n
\end{array}
$$

into

$$
\begin{array}{cccccc}
1 & 0 & 0 & \cdots & 0 & b_1' \\
0 & 1 & 0 & \cdots & 0 & b_2' \\
0 & 0 & 1 & \cdots & 0 & b_3' \\
\cdot & \cdot & 0 & \cdot & & \cdot \\
\cdot & \cdot & & \cdot & \cdot & \cdot \\
\cdot & \cdot & \cdot & & 0 & \cdot \\
0 & 0 & 0 & \cdots & 1 & b_n'
\end{array}
$$

Equation (1.14) may be written

$$A_1^1 \mathbf{x}_1 = \mathbf{b}^1 - A_2^1 \mathbf{x}_2$$

and since A_1^1, having rank k, has an inverse (by Theorem 1.6), it may be rewritten as

$$\mathbf{x}_1 = (A_1^1)^{-1} \mathbf{b}^1 - (A_1^1)^{-1} A_2^1 \mathbf{x}_2 \tag{1.15}$$

In (1.15), if we fix $\mathbf{x}_2 = \mathbf{x}_2^0$, there is determined a *unique* vector

$$\mathbf{x}_1 = \mathbf{x}_1^0 = (A_1^1)^{-1} \mathbf{b}^1 - (A_1^1)^{-1} A_2^1 \mathbf{x}_2^0 \tag{1.16}$$

and $(\mathbf{x}_1^0, \mathbf{x}_2^0)$ is a solution to (1.14).

Consider now the relative magnitudes of k, n.

Case a. $k = n$

Equation (1.15) becomes

$$\mathbf{x}_1 = (A_1^1)^{-1} \mathbf{b}^1$$

since A_2 is then nonexistent, and the system $A\mathbf{x} = \mathbf{b}$ has a unique solution.

Case b. $k < n$

The system $A\mathbf{x} = \mathbf{b}$ has an infinite number of solutions, for $\mathbf{x}_2$ may be specified in an infinite number of ways, *for each of which* the resulting vector $\mathbf{x}_1$, along with $\mathbf{x}_2$, comprises a solution to $A\mathbf{x} = \mathbf{b}$.

It must be observed that our results here are in terms of $k = r(A)$ and n, rather than m and n; that is, we might have $m > n$ and yet have an infinity of solutions.

A situation of particular interest for case b is that in which $\mathbf{x}_2$ is made the *null* vector of $n - k$ components so that the resulting solution has the form

$$\mathbf{x} = (x_1, x_2, \ldots, x_k, 0, 0, \ldots, 0)$$

and *at most* k of the variables are nonzero.† Such a solution is called a *basic solution* of the system $A\mathbf{x} = \mathbf{b}$ for the simple reason that the columns of A associated with variables not explicitly set equal to zero, being a linearly independent set, constitute a *basis* for E^k. The variables $x_1, \ldots, x_k$ are called *basic variables*, whereas those set to zero, $x_{k+1}, \ldots, x_n$, are called *nonbasic variables*.

We shall have need for these concepts in the chapter on linear programming.

---

†$n - k$ of the variables are made to equal zero, the other k being then uniquely determined. Some of these k variables, however, may also be found to be zero.

iv. *Computing the Inverse*

Elementary row operations, as we have observed, may be used to solve systems of linear equations. They also provide a computational procedure for computing inverses of square matrices or for determining when inverses do not exist.

We state the result in the form of a theorem.

Theorem 1.7

Write the $n \times n$ matrix A to the left of the $n \times n$ identity matrix I_n, obtaining an $n \times 2n$ matrix (A, I_n). Apply elementary row operations on the $n \times 2n$ matrix so as to transform (A, I_n) into (I_n, B), where B is the resulting $n \times n$ array on the right. If this can be accomplished, then $B = A^{-1}$. If the first n columns cannot be transformed into I_n, then A^{-1} does not exist.

We illustrate the theorem with two examples.

Example 1.7

Let

$$A = \begin{pmatrix} 1 & 2 & -1 \\ 0 & -1 & 0 \\ -1 & 3 & 2 \end{pmatrix}$$

The initial 3×6 array is then

$$\begin{array}{rrr|rrr} 1 & 2 & -1 & 1 & 0 & 0 \\ 0 & -1 & 0 & 0 & 1 & 0 \\ -1 & 3 & 2 & 0 & 0 & 1 \end{array}$$

We obtain successively

$$\begin{array}{rrr|rrr} 1 & 2 & -1 & 1 & 0 & 0 \\ 0 & -1 & 0 & 0 & 1 & 0 \\ 0 & 5 & 1 & 1 & 0 & 1 \end{array}$$

$$\begin{array}{rrr|rrr} 1 & 0 & -1 & 1 & 2 & 0 \\ 0 & 1 & 0 & 0 & -1 & 0 \\ 0 & 0 & 1 & 1 & 5 & 1 \end{array}$$

$$\begin{array}{rrr|rrr} 1 & 0 & 0 & 2 & 7 & 1 \\ 0 & 1 & 0 & 0 & -1 & 0 \\ 0 & 0 & 1 & 1 & 5 & 1 \end{array}$$

Thus
$$\begin{pmatrix} 2 & 7 & 1 \\ 0 & -1 & 0 \\ 1 & 5 & 1 \end{pmatrix} = B$$

should comprise A^{-1}, and

$$AB = \begin{pmatrix} 1 & 2 & -1 \\ 0 & -1 & 0 \\ -1 & 3 & 2 \end{pmatrix} \begin{pmatrix} 2 & 7 & 1 \\ 0 & -1 & 0 \\ 1 & 5 & 1 \end{pmatrix} = \begin{pmatrix} 1 & 0 & 0 \\ 0 & 1 & 0 \\ 0 & 0 & 1 \end{pmatrix}$$

so $B = A^{-1}$.

Example 1.8

Letting

$$A = \begin{pmatrix} 1 & 2 & -1 \\ 0 & -1 & 0 \\ 2 & 3 & -2 \end{pmatrix}$$

we begin with

$$\begin{array}{rrr|rrr} 1 & 2 & -1 & 1 & 0 & 0 \\ 0 & -1 & 0 & 0 & 1 & 0 \\ 2 & 3 & -2 & 0 & 0 & 1 \end{array}$$

and obtain

$$\begin{array}{rrr|rrr} 1 & 2 & -1 & 1 & 0 & 0 \\ 0 & -1 & 0 & 0 & 1 & 0 \\ 0 & -1 & 0 & -2 & 0 & 1 \end{array}$$

$$\begin{array}{rrr|rrr} 1 & 0 & -1 & 1 & 2 & 0 \\ 0 & 1 & 0 & 0 & -1 & 0 \\ 0 & 0 & 0 & -2 & -1 & 1 \end{array}$$

and the left portion cannot be transformed into I_3. Thus the given A has no inverse (A's third row is twice the first row added to the second row).

1.5. Determinants

i. *Definition*

Associated with every square matrix A is a unique element called the determinant of A and denoted here by det A. If A has integer elements, det A is an integer; if A has complex elements, det A is a complex number; and so on.

ii. Computing the Determinant

For a 2×2 matrix A, the reader will recall the determinant to be simply the product of the main diagonal elements minus the product of the other two elements. That is, the determinant of

$$\begin{pmatrix} a_{11} & a_{12} \\ a_{21} & a_{22} \end{pmatrix}$$

is just $a_{11}a_{22} - a_{21}a_{12}$.

For a 3×3 matrix A, a simple familiar computational procedure is to juxtapose the first two columns of A with A itself, to consider six lines as being drawn through the resulting array as follows

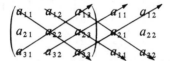

and to compute the product of the three elements lying on each line. Then the sum of the upward lines is subtracted from the sum of the downward lines.

Of course, for large n, systematic and efficient methods for determinant computation are needed. We proceed to give one that allows the expression of the determinant in terms of determinants of smaller matrices.

Define A_{ij} to be the $(n - 1) \times (n - 1)$ matrix obtained by *deleting the ith row and jth column of A.*

Considering j fixed, it may be shown that

$$\det A = \sum_{i=1}^{n} (-1)^{i+j}(\det A_{ij})a_{ij} \tag{1.17}$$

or holding i fixed, it may be shown that

$$\det A = \sum_{j=1}^{n} (-1)^{i+j}(\det A_{ij})a_{ij} \tag{1.18}$$

Thus, for example, the determinant of

$$A = \begin{pmatrix} 0 & 1 & -1 & 2 \\ 2 & 0 & -2 & 3 \\ 0 & 2 & 4 & -3 \\ 2 & 3 & 0 & -3 \end{pmatrix} \tag{1.19}$$

may be represented four ways according to (1.17), and an identical number of representations is afforded by (1.18). Of course, if a row or column with a large number of zeros exists, the expansion (1.17) or (1.18) in terms of that row or column will produce fewer nonzero terms and, consequently, less computational effort.

At any rate, we include several expressions for the determinant of (1.19).

$$\det A = \quad 0 \cdot \det \begin{pmatrix} 0 & -2 & 3 \\ 2 & 4 & -3 \\ 3 & 0 & -3 \end{pmatrix} \quad - 2 \det \begin{pmatrix} 1 & -1 & 2 \\ 2 & 4 & -3 \\ 3 & 0 & -3 \end{pmatrix}$$

$$+ 0 \cdot \det \begin{pmatrix} 1 & -1 & 2 \\ 0 & -2 & 3 \\ 3 & 0 & -3 \end{pmatrix} \quad - 2 \det \begin{pmatrix} 1 & -1 & 2 \\ 0 & -2 & 3 \\ 2 & 4 & -3 \end{pmatrix}$$

$$\det A = \quad -\det \begin{pmatrix} 2 & 0 & 3 \\ 0 & 2 & -3 \\ 2 & 3 & -3 \end{pmatrix} \quad + 2 \det \begin{pmatrix} 0 & 1 & 2 \\ 0 & 2 & -3 \\ 2 & 3 & -3 \end{pmatrix}$$

$$+ 4 \det \begin{pmatrix} 0 & 1 & 2 \\ 2 & 0 & 3 \\ 2 & 3 & -3 \end{pmatrix}$$

$$\det A = \quad -2 \det \begin{pmatrix} 2 & 0 & -2 \\ 0 & 2 & 4 \\ 2 & 3 & 0 \end{pmatrix} \quad + 3 \det \begin{pmatrix} 0 & 1 & -1 \\ 0 & 2 & 4 \\ 2 & 3 & 0 \end{pmatrix}$$

$$+ 3 \det \begin{pmatrix} 0 & 1 & -1 \\ 2 & 0 & -2 \\ 2 & 3 & 0 \end{pmatrix} \quad - 3 \det \begin{pmatrix} 0 & 1 & -1 \\ 2 & 0 & -2 \\ 0 & 2 & 4 \end{pmatrix}$$

(Verify that these expressions provide the same result.) The term $(-1)^{i+j} \det A_{ij}$ in (1.17) and (1.18), which is called the *i, j cofactor* of A, merits further examination. First, however, we cite a result that will be of later use.

Theorem 1.8

Let A, B be $n \times n$ matrices. Then $\det(AB) = (\det A)(\det B)$.

We illustrate Theorem 1.8. Let

$$A = \begin{pmatrix} 1 & 2 \\ -1 & 0 \end{pmatrix}$$

$$B = \begin{pmatrix} 1 & 3 \\ 2 & -1 \end{pmatrix}$$

Then
$$\det A = \det\begin{pmatrix} 1 & 2 \\ -1 & 0 \end{pmatrix} = 2$$

$$\det B = \det\begin{pmatrix} 1 & 3 \\ 2 & -1 \end{pmatrix} = -7$$

$$\det(AB) = \det\left[\begin{pmatrix} 1 & 2 \\ -1 & 0 \end{pmatrix}\begin{pmatrix} 1 & 3 \\ 2 & -1 \end{pmatrix}\right] = \det\begin{pmatrix} 5 & 1 \\ -1 & -3 \end{pmatrix} = -14$$

Theorem 1.8 also implies that $\det(AB) = \det(BA)$. With the same 2×2 matrices, we find

$$\det(BA) = \det\left[\begin{pmatrix} 1 & 3 \\ 2 & -1 \end{pmatrix}\begin{pmatrix} 1 & 2 \\ -1 & 0 \end{pmatrix}\right] = \det\begin{pmatrix} -2 & 2 \\ 3 & 4 \end{pmatrix} = -14$$

a. ADJOINTS AND INVERSES

The $n \times n$ matrix whose (i, j) element is $(-1)^{i+j} \det A_{ji}$ is called the *adjoint of A* and denoted adj A. For example, if

$$A = \begin{pmatrix} 1 & 0 & 2 \\ -1 & 2 & 1 \\ 3 & 0 & 2 \end{pmatrix}$$

then
$$\text{adj } A = \begin{pmatrix} 4 & 0 & -4 \\ 5 & -4 & -3 \\ -6 & 0 & 2 \end{pmatrix}$$

An important result involving the adjoint is given in Theorem 1.9.

Theorem 1.9

If A is an $n \times n$ matrix, then

$$(\text{adj } A)A = A(\text{adj } A) = (\det A)I_n \tag{1.20}$$

Aside from stating that every such square matrix commutes with its adjoint, Theorem 1.9 has other interesting implications. First, suppose that det A has a multiplicative inverse. Denote it $(\det A)^{-1}$, so that $(\det A)(\det A)^{-1} = 1$. Equation (1.20) then says that

$$[(\det A)^{-1}(\text{adj } A)]A = I_n$$

and
$$A[(\det A)^{-1}(\text{adj } A)] = I_n$$

Thus the matrix

$$(\det A)^{-1}(\operatorname{adj} A) = A^{-1} \tag{1.21}$$

On the other hand, suppose that A is invertible; that is,

$$A^{-1}A = I$$

Now $\qquad\qquad \det(A^{-1}A) = \det I = 1 \tag{1.22}$

But by Theorem 1.8, we have, with (1.22),

$$(\det A^{-1})(\det A) = 1$$

Thus we have obtained Theorem 1.10,

Theorem 1.10

If A is an $n \times n$ matrix, then A has an inverse if and only if $\det A \neq 0$.

Definition. A square matrix is said to be *singular* if $\det A = 0$. Otherwise, it is said to be *nonsingular*.

Now let us utilize (1.21) to compute an inverse matrix.

Example 1.9

Let $\qquad\qquad A = \begin{pmatrix} 0 & -2 & 1 \\ 3 & 0 & 2 \\ 2 & -1 & 1 \end{pmatrix}$

Then $\qquad\qquad \operatorname{adj} A = \begin{pmatrix} 2 & 1 & -4 \\ 1 & -2 & 3 \\ -3 & -4 & 6 \end{pmatrix}$

and $\det A = -5$, giving

$$A^{-1} = \begin{pmatrix} -\frac{2}{5} & -\frac{1}{5} & \frac{4}{5} \\ -\frac{1}{5} & \frac{2}{5} & -\frac{3}{5} \\ \frac{3}{5} & \frac{4}{5} & -\frac{6}{5} \end{pmatrix}$$

$$\frac{1}{\det A} A(\operatorname{adj} A) = -\frac{1}{5}\begin{pmatrix} -5 & 0 & 0 \\ 0 & -5 & 0 \\ 0 & 0 & -5 \end{pmatrix} = \begin{pmatrix} 1 & 0 & 0 \\ 0 & 1 & 0 \\ 0 & 0 & 1 \end{pmatrix}$$

The disadvantage to the foregoing computation of the inverse, of course, is that the calculation of det A for $n \times n$ A requires an effort which increases far more rapidly with n than does the row-operation technique.

Combining Theorem 1.10 with the previous result that A has an inverse if and only if $r(A) = n$, we see that A is nonsingular if and only if A has rank n, that is, if and only if A's rows (columns) are linearly independent.

1.6. Eigenvalues of Matrices

Let A be an arbitrary square matrix of dimension n, and let I be the $n \times n$ identity matrix. Let c be a real number, and let $\mathbf{x} \neq \mathbf{0}$ belong to E^n. If

$$(A - cI)\mathbf{x} = \mathbf{0} \qquad (1.23)$$

then c is called an *eigenvalue* or *characteristic value* of A, and $\mathbf{x}$ is called an *eigenvector* or *characteristic vector* of A associated with c.

In (1.23) suppose that the matrix $(A - cI)$ has an inverse. Then multiplying on the left of both sides of (1.23) by $(A - cI)^{-1}$, we find

$$\mathbf{x} = \mathbf{0}$$

But $\mathbf{x}$ was assumed to be nonzero, and thus $(A - cI)$ does not possess an inverse if c is an eigenvalue. On the other hand, suppose that for an arbitrary $n \times n$ matrix A and an n-tuple $\mathbf{x} \neq \mathbf{0}$, we find

$$(A - cI)\mathbf{x} = \mathbf{0}$$

But then the columns of $(A - cI)$ are linearly dependent, and $(A - cI)$ is singular.

Thus the real number c is an eigenvalue or characteristic value of A if and only if

$$\det(A - cI) = 0 \qquad (1.24)$$

If we expand the left side of (1.24) according to the defining expression for the determinant, we obtain an nth degree polynomial in c, and this polynomial is called the *characteristic polynomial* of A, (1.24) being the characteristic *equation* of A. Thus the $n \times n$ matrix A has exactly n eigenvalues, although these need not be unique or even real.

To find all eigenvalues of a square matrix A, one can solve the characteristic equation (1.24), which is, in general, a nontrivial task.

Example 1.10

Let
$$A = \begin{pmatrix} 2 & 0 & -1 \\ 3 & 1 & 2 \\ 0 & -2 & 4 \end{pmatrix}$$

$$(A - cI) = \begin{pmatrix} 2-c & 0 & -1 \\ 3 & 1-c & 2 \\ 0 & -2 & 4-c \end{pmatrix}$$

$$\det(A - cI) = (2-c)(1-c)(4-c) + 6 + 4(2-c)$$
$$= -c^3 + 7c^2 - 18c + 22$$

whose roots are the three eigenvalues of the given A.

Example 1.11

It is easily seen that a real matrix might have only complex characteristic values. For

$$A = \begin{pmatrix} 0 & 1 \\ -1 & 0 \end{pmatrix}$$

$$\det(A - cI) = \det\begin{pmatrix} -c & 1 \\ -1 & -c \end{pmatrix} = c^2 + 1$$

which has roots $c = \pm\sqrt{-1}$.

Example 1.12

Eigenvalues need not be unique. To see this let

$$A = \begin{pmatrix} 1 & 0 & 0 & 0 \\ 0 & 2 & 0 & 0 \\ 0 & 0 & 1 & 0 \\ 0 & 0 & 0 & 2 \end{pmatrix}$$

whose characteristic equation is

$$(1-c)^2(2-c)^2 = 0$$

and which therefore has four eigenvalues, $c = 1, 1, 2, 2$.

REFERENCES

[1] CARNAHAN, B., et al., *Applied Numerical Methods*, John Wiley & Sons, Inc., New York, 1969.

[2] HADLEY, G., *Linear Algebra*, Addison-Wesley Publishing Company, Inc., Reading, Mass., 1961.

[3] HAMMING, R., *Numerical Methods for Scientists and Engineers*, McGraw-Hill Book Company, New York, 1962.

[4] HOFFMAN, K., and R. KUNZE, *Linear Algebra*, Prentice-Hall, Inc., Englewood Cliffs, N.J., 1961.

PROBLEMS

1. Show that in the plane E^2 the inner product of any two vectors at a 90° angle is zero.

2. Show that a vector $\mathbf{y}$ and its negative $-\mathbf{y}$ cannot both belong to a linearly independent set.

3. Prove that any three vectors in E^2 are linearly dependent.

4. Which of the following sets of vectors are linearly independent? Linearly dependent?
   (a) $(1, -2, \frac{1}{2})$, $(\frac{1}{2}, 1, 3)$, $(1, 0, \frac{13}{4})$ in E^3.
   (b) $(3, 2, 4)$, $(-3, 0, -2)$, $(0, 0, 1)$ in E^3.
   (c) $(1, 0, \frac{1}{2}, -\frac{1}{2})$, $(0, 3, -1, 1)$, $(2, \frac{1}{2}, \frac{1}{2}, -\frac{1}{2})$, $(-1, -1, 0, 1)$ in E^4.
   (d) $2x^4 - x^3 + \frac{1}{2}x^2 + x - \frac{3}{4}$, $-x^4 + 2x^2 - 3x + \frac{2}{3}$, $\frac{2}{3}x^4 - 2x^3 - 4x + 1$, $-\frac{1}{2}x^4 - x^3 - x^2 + \frac{3}{4}x - 2$ in the vector space of fourth-degree polynomial functions with rational coefficients.

5. Prove that two vectors $\mathbf{v}_1$, $\mathbf{v}_2$, with $\mathbf{v}_1 \neq \mathbf{0} \neq \mathbf{v}_2$, are linearly dependent if and only if $\mathbf{v}_1 = \lambda\mathbf{v}_2$ for some scalar λ.

6. Prove that any subset of a linearly independent set is, again, linearly independent.

7. Let $B = (\mathbf{b}_1, \mathbf{b}_2, \ldots, \mathbf{b}_m)$ be a basis for a vector space V. Let $\boldsymbol{\alpha} \in V$ be expressible (uniquely) in terms of B by

$$\boldsymbol{\alpha} = c_1\mathbf{b}_1 + c_2\mathbf{b}_2 + \cdots + c_m\mathbf{b}_m$$

for some scalars $c_1, \ldots, c_m$. Then show that

$$B_1 = (\mathbf{b}_1, \ldots, \mathbf{b}_{k-1}, \boldsymbol{\alpha}, \mathbf{b}_{k+1}, \ldots, \mathbf{b}_m)$$

is a basis for the vector space V if and only if $c_k \neq 0$.

8. Let $B = (\mathbf{b}_1, \mathbf{b}_2, \ldots, \mathbf{b}_m)$ be a linearly independent set. Then for any scalar $c \neq 0$, show that both sets

$$B_1 = (\mathbf{b}_1, \mathbf{b}_2, \ldots, \mathbf{b}_{i-1}, c\mathbf{b}_i, \mathbf{b}_{i+1}, \ldots, \mathbf{b}_m)$$

$$B_2 = (\mathbf{b}_1, \mathbf{b}_2, \ldots, \mathbf{b}_i, \mathbf{b}_{i+1}, \ldots, \mathbf{b}_{k-1}, c\mathbf{b}_i + \mathbf{b}_k, \mathbf{b}_{k+1}, \ldots, \mathbf{b}_m)$$

are linearly independent, where $i \neq k$ in B_2.

9. If $(\mathbf{b}_1, \ldots, \mathbf{b}_k)$ is a linearly independent set, and $\boldsymbol{\alpha}$ is a vector from the same space, prove that $(\mathbf{b}_1, \ldots, \mathbf{b}_k, \boldsymbol{\alpha})$ is linearly independent if and only if $\boldsymbol{\alpha}$ cannot be expressed as a linear combination of $\mathbf{b}_1, \mathbf{b}_2, \ldots, \mathbf{b}_k$.

10. Suppose that $\mathbf{B} = (\mathbf{b}_1, \mathbf{b}_2, \ldots, \mathbf{b}_n)$ is a basis for V. For $\boldsymbol{\alpha} \in V$, the unique scalars $c_1, c_2, \ldots, c_n$ satisfying

$$\boldsymbol{\alpha} = c_1 \mathbf{b}_1 + c_2 \mathbf{b}_2 + \cdots + c_n \mathbf{b}_n$$

are called the coordinates of $\boldsymbol{\alpha}$ relative to B.

Suppose that $V = E^n$, that there is a second basis $B' = (\mathbf{b}'_1, \mathbf{b}'_2, \ldots, \mathbf{b}'_n)$, and that the coordinates of $\boldsymbol{\alpha}$ relative to B are $(c_1, c_2, \ldots, c_n)$. Show how you would find the coordinates of $\boldsymbol{\alpha}$ relative to B', that is, $(c'_1, c'_2, \ldots, c'_n)$.

11. For the vector of real numbers $(c_1, \ldots, c_n) = \mathbf{c}$, define a subset of E^n, X, by

$$X = \{\mathbf{x} = (x_1, \ldots, x_n) \in E^n \,|\, c_1 x_1 + \cdots + c_n x_n = 0\}$$

(a) Show that X is a subspace of E^n.
(b) What is the dimension of X?
(c) What is the geometric relationship between $\mathbf{c}$ and X? (Consider first a particular $\mathbf{c}$ in two or three dimensions.)

12. For E^3 find a basis $\{\mathbf{b}_1, \mathbf{b}_2, \mathbf{b}_3\}$ such that $\mathbf{b}_i > 0$, $i = 1, 2, 3$;

$$\mathbf{b}_i \mathbf{b}_j = \begin{cases} 1 & \text{for } i = j \\ 0 & \text{for } i \neq j \end{cases}$$

13. Find a 2×2 matrix A of real numbers such that $a_{ij} \neq 0$ for all i, j and such that $A^2 = 0$, the 2×2 null matrix.

14. What is a necessary and sufficient condition that $A^T = A$ for an $n \times n$ matrix?

15. Compute the number of additions and the number of multiplications required to find AB, where A is $m \times n$ and B is $n \times r$.

16. Find two real 2×2 matrices A, B with $A \neq I \neq B$, $A \neq B^{-1}$, and neither A nor B null, such that $AB = BA$.

17. Let n be a positive integer, and let P be a set of real matrices defined as follows: P consists of all $n \times n$ real matrices that have exactly one 1 in each row and exactly one 1 in each column, and every other element zero.
(a) How may members has P?
(b) Show that if $P_1, P_2 \in P$, then $P_1 P_2 \in P$.
(c) If $P_i \in P$, then for some integer $k > 1$, $P_i^k = P_i$.

18. Illustrate the associativity of matrix multiplication with

$$A = \begin{pmatrix} 1 & -1 & 2 & 0 \\ 3 & 4 & 7 & -5 \\ 2 & 1 & 0 & -1 \end{pmatrix}$$

$$B = \begin{pmatrix} 0 & -1 & 3 \\ 4 & 5 & -3 \end{pmatrix}$$

$$C = \begin{pmatrix} 3 & 2 \\ -2 & 4 \\ 1 & 0 \\ -1 & 3 \end{pmatrix}$$

19. Show that if A is a square matrix then AA^T is a *symmetric matrix*, that is, one in which $a_{ij} = a_{ji}$ for all i, j.

20. If

$$B = \begin{pmatrix} b_{11} & b_{12} & \cdots & b_{1m} \\ b_{21} & b_{22} & \cdots & b_{2m} \\ \cdot & \cdot & & \cdot \\ \cdot & \cdot & & \cdot \\ \cdot & \cdot & & \cdot \\ b_{m1} & b_{m2} & \cdots & b_{mm} \end{pmatrix}$$

is a matrix with inverse B^{-1}, what is a simple expression for the inverse of

$$B_1 = \begin{pmatrix} b_{11} & b_{12} & \cdots & b_{1m} & 0 \\ b_{21} & b_{22} & \cdots & b_{2m} & 0 \\ \cdot & \cdot & & \cdot & \cdot \\ \cdot & \cdot & & \cdot & \cdot \\ \cdot & \cdot & & \cdot & \cdot \\ b_{m1} & b_{m2} & \cdots & b_{mm} & 0 \\ 0 & 0 & \cdots & 0 & 1 \end{pmatrix}$$

21. Suppose that A is an invertible matrix and that A^{-1} is known. What is a simple method for determining A_1^{-1}, where A_1 is simply A with every element in rows $i_1, i_2, \ldots, i_k$ multiplied by -1?

22. If A has inverse A^{-1}, what is the inverse of $A_1 = cA$ for a scalar $c \neq 0$?

23. Let A, B be invertible $n \times n$ matrices. Show that AB and BA are invertible.

24. Show whether or not there are any real square matrices such that $A = A^{-1}$, but $A \neq I$.

25. Find the following matrix products by inspection:

(a)
$$\begin{pmatrix} -3 & 0 & 1 & 2 \\ 1 & -1 & 2 & 3 \\ -2 & 0 & 1 & 0 \end{pmatrix} \begin{pmatrix} -2 \\ 1 \\ -1 \\ -1 \end{pmatrix}$$

(b)
$$\begin{pmatrix} 4 & 1 & -3 & \frac{1}{2} & 1 \\ 2 & 4 & 1 & 0 & -1 \end{pmatrix} \begin{pmatrix} 0 & -2 & -1 \\ 1 & 2 & -1 \\ -1 & 0 & 0 \\ -2 & 3 & 3 \\ 2 & -1 & -4 \end{pmatrix}$$

(c)
$$\begin{pmatrix} 1 & 0 & 3 & 2 \\ -1 & 1 & 2 & -1 \\ 2 & 2 & 0 & 0 \\ -3 & 2 & 1 & 0 \end{pmatrix} \begin{pmatrix} 1 & 0 & 3 & -1 \\ -1 & 2 & -1 & -2 \\ 1 & 1 & 2 & 3 \\ 0 & -1 & 1 & 0 \end{pmatrix}$$

26. Show that if A has $0 \leq a_{ij} < 1$ for all i, j, and

$$\sum_{j=1}^{n} a_{ij} < 1, \quad \text{all } i,$$

then $a_{ij}^{(k)} \to 0$ for all i, j as $k \to \infty$, where $a_{ij}^{(k)}$ is the (i, j)th element of A^k.

27. For certain real numbers a, we know that the expression

$$\frac{1}{1-a} = 1 + a + a^2 + a^3 + \cdots + a^n + \cdots$$

is valid. (When?)

A comparable result is available for matrices. We have, for example,

$$(I - A)(I + A) = I - A^2$$
$$(I - A)(I + A + A^2) = I - A^3$$
$$(I - A)(I + A + A^2 + A^3 + \cdots + A^q) = I - A^{q+1}$$

If A^{q+1} were the null matrix, then

$$I + A + A^2 + \cdots + A^q$$

would be $(I - A)^{-1}$.

Show that if Problem 26 is applicable, then $(I - A)^{-1}$ is

$$I + \sum_{j=1}^{\infty} A^j$$

28. Prove that the set of solutions to $A\mathbf{x} = \mathbf{b}$ is identical to those for $A_1\mathbf{x} = \mathbf{b}$, where A_1 is obtained by performing a single row operation on A.

29. Describe precisely the set of solutions to each of the following systems of equations.

(a)
$$\begin{aligned} 2x_1 - x_2 + 2x_3 \quad\;\;\; &= 2 \\ -x_1 + 2x_2 - x_3 + 3x_4 &= -1 \\ 4x_2 - 6x_3 - x_4 &= 0 \\ x_1 \quad\quad + x_3 - 2x_4 &= 4 \end{aligned}$$

(b) $\begin{aligned} -x_1 - 2x_2 + 2x_3 - 3x_4 &= -4 \\ \tfrac{1}{2}x_1 + \tfrac{1}{2}x_2 - 2x_3 - x_4 &= -3 \\ 2x_1 + \tfrac{1}{2}x_2 \qquad\quad + x_4 &= 3 \\ x_1 - x_2 + 3x_3 + \tfrac{1}{2}x_4 &= 4 \end{aligned}$

(c) $\begin{aligned} -x_1 + 2x_2 + \tfrac{1}{2}x_3 \qquad\quad - 2x_5 &= 2 \\ 2x_1 - x_2 + 2x_3 - 2x_4 + x_5 &= 4 \\ x_2 \qquad\quad + x_4 - x_5 &= -3 \\ x_1 + 2x_2 - 2x_3 \qquad\quad + 4x_5 &= 0 \\ \tfrac{1}{2}x_1 \qquad\quad + 3x_3 - 4x_4 &= 1 \end{aligned}$

(d) $\begin{aligned} x_1 + 2x_2 - x_3 - \tfrac{1}{2}x_4 &= 1 \\ -2x_1 + 4x_2 + 2x_3 - 6x_4 &= 0 \\ -\tfrac{1}{6}x_1 - \tfrac{2}{3}x_2 + \tfrac{17}{3}x_3 - 10x_4 &= -4 \\ -\tfrac{1}{2}x_1 + \tfrac{2}{3}x_2 - x_3 - 3x_4 &= -2 \end{aligned}$

(e) $\begin{aligned} x_1 + x_2 - 2x_3 - 3x_4 &= 8 \\ 4x_1 + x_2 - x_3 - 7x_4 &= 10 \\ -2x_1 + x_2 - 3x_3 + x_4 &= 6 \\ 8x_1 - x_2 - 5x_3 - 9x_4 &= -2 \end{aligned}$

(f) $\begin{aligned} x_1 + \tfrac{1}{2}x_2 - 3x_3 &= 2 \\ 2x_1 - x_2 - \tfrac{5}{2}x_3 &= \tfrac{3}{2} \\ 2x_1 - 3x_2 + x_3 &= -1 \\ 3x_1 - 7x_2 &= -4 \\ -x_1 - x_2 - 2x_3 &= -2 \end{aligned}$

(g) $\begin{aligned} x_2 + 2x_3 - 3x_4 &= 4 \\ -x_1 - x_2 + 4x_3 - 7x_4 &= -2 \\ x_1 + 2x_2 - x_3 + 4x_4 &= 6 \\ -2x_1 - 3x_2 + 4x_3 - 11x_4 &= -8 \\ - x_2 - 4x_3 + 3x_4 &= -4 \end{aligned}$

30. Let A be an $n \times n$ real matrix. Define

$$V_1 = \{A\mathbf{x} \,|\, \text{all } \mathbf{x} \in E^n\}$$

and $\qquad\qquad\qquad V_2 = \{\mathbf{x} \in E^n \,|\, A\mathbf{x} = \mathbf{0}\}$

Prove that dim V_1 + dim $V_2 = n$.

31. Suppose that one is in the process of triangularizing the square matrix A. For some $j < n$, it is found that the (j, j), $(j + 1, j)$, $(j + 2, j)$, ..., (n, j) elements of the current array are *all* zero. Prove that $r(A) < n$.

32. Suppose that for an $n \times n$ matrix A [with $r(A) = n = r(A, \mathbf{b})$] we solve $A\mathbf{x} = \mathbf{b}$ twice, first by triangularization, then by complete elimination. For each method calculate
(a) The number of additions and subtractions required.
(b) The number of multiplications required.
    Using, then, representative times for computer execution of these operations, compare the two methods. (Here, of course, we have neglected questions of accuracy.)

33. Let A, B be two $n \times n$ matrices. Let $r(A) = k$ and $r(B) = m$. What can be said about $r(A + B)$? About $r(AB)$?

34. Systems of linear equations can be solved by means of "column operations" quite as easily as by row operations. Define the corresponding elementary column operations and resolve Problem 29(a), (b), (d), and (g) using them.

35. Denote the columns of an $m \times n$ matrix A by $\mathbf{a}_1, \mathbf{a}_2, \ldots, \mathbf{a}_n$ and let $x_1, x_2, \ldots, x_n$ be a solution to $A\mathbf{x} = \mathbf{b}$; that is,

$$\sum_{i=1}^{n} x_i \mathbf{a}_i = \mathbf{b}$$

Show that it is possible to proceed from this solution to one in which the columns associated with positive variables comprise a linearly independent set.

Of course, the set may initially be linearly independent; otherwise, show that it is possible to drive at least one variable at a time to zero until the set of columns remaining is linearly independent.

36. Suppose that we begin with an $m \times n$ matrix A and perform a succession of elementary row operations, eventually obtaining the $m \times n$ matrix B. Show that there exists an $m \times m$ matrix C, where C has an inverse, such that

$$B = CA$$

37. Let A be an $m \times n$ matrix. Prove that

$$A\mathbf{v} = \mathbf{b}$$

has a solution if and only if every solution $\mathbf{x}$ to $A^T\mathbf{x} = \mathbf{0}$ satisfies $\mathbf{b}^T\mathbf{x} = 0$.

38. With A an $m \times n$ matrix, discuss the maximum number of basic solutions that the system $A\mathbf{x} = \mathbf{b}$ might have.

39. Using elementary row operations, compute the inverse for each of the following matrices or determine that none exists.

(a) $\begin{pmatrix} \frac{3}{2} & 2 & -1 \\ 1 & -3 & 1 \\ \frac{7}{2} & \frac{5}{2} & -\frac{3}{2} \end{pmatrix}$

(b) $\begin{pmatrix} -2 & 1 & 3 & 0 \\ -1 & \frac{1}{2} & 0 & -2 \\ 2 & 3 & 3 & -2 \\ 1 & -5 & 1 & -3 \end{pmatrix}$

(c) $\begin{pmatrix} -1 & -1 & 2 & 3 \\ 2 & 0 & 4 & 0 \\ -7 & 0 & -7 & 1 \\ -2 & 1 & -1 & -2 \end{pmatrix}$

(d) $\begin{pmatrix} -1 & 2 & 0 & 1 & -1 \\ 3 & -1 & 0 & 1 & -2 \\ 1 & 1 & 0 & 4 & -2 \\ -3 & -2 & 3 & -1 & 0 \\ 0 & 0 & -3 & 4 & 2 \end{pmatrix}$

40. Compute the determinant of each of the following matrices:

(a) $\begin{pmatrix} 2 & 3 & 3 & -1 & 3 \\ -1 & 4 & 1 & 0 & -2 \\ 4 & 0 & -1 & 0 & 4 \\ -3 & 0 & 2 & 0 & 6 \\ 2 & 1 & 3 & 2 & 1 \end{pmatrix}$

(b) $\begin{pmatrix} 2 & -1 & 1 & 0 \\ 6 & 6 & 0 & 0 \\ 3 & 2 & 0 & -1 \\ 2 & 1 & 1 & 2 \end{pmatrix}$

(c) $\begin{pmatrix} 1 & 2 & 4 & 1 \\ 1 & 6 & 3 & -2 \\ -4 & 2 & -3 & -1 \\ 4 & 3 & -1 & 2 \end{pmatrix}$

41. Let A and B be $m \times m$ matrices with $r(A) = m > k = r(B)$. What can be said about $r(AB)$?

42. With A an $m \times m$ matrix, use the definition of the determinant and find the determinant of each of the following matrices in terms of det A:
(a) $A_1 = cA$, $c \neq 0$.
(b) $A_2 = A$ with two rows interchanged.
(c) $A_3 = A$ with two columns interchanged.

43. Let A' be a square matrix obtained from A by a rearrangement of columns. How can det A' be described in terms of the rearrangement? Answer the question in terms of a particular A of small dimension.

44. Let $Ax = b$ be a system in which all scalars are integers. A is $m \times n$ with $r(A) = m < n$ and has the property that every set of m columns from A has determinant 1, -1, or zero. What property has every basic solution to the system?

45. In Problem 42(a) and (b) the value of the determinant under two of the three kinds of elementary row operations is obtained in terms of that of the initial matrix. Derive the corresponding expression for the other variety of row operation.

46. A Hadamard matrix of order n is an $n \times n$ matrix H of the elements ± 1 that satisfies

$$HH^T = nI$$

(a) What is det H?

(b) How many Hadamard matrices of order 2 are there? What are they?

(c) Construct, for small n, an $n \times n$ array whose elements are no larger in absolute value than 1 and whose determinant has absolute value larger than the number obtained in part (a). What might one conclude?

(d) Do H, H^T commute?

47. An $n \times n$ matrix A is said to be a triangular matrix if either

$$a_{ij} = 0 \quad \text{for all } i > j$$

or
$$a_{ij} = 0 \quad \text{for all } i < j$$

Write an expression for the determinant of a triangular matrix.

48. Compute the inverse of each of the following matrices, or determine that none exists, by first finding the adjoint matrix.

(a) $\begin{pmatrix} 2 & -1 \\ \frac{3}{2} & -3 \end{pmatrix}$

(b) $\begin{pmatrix} 0 & 2 & -1 \\ \frac{1}{2} & 3 & 2 \\ -\frac{1}{2} & -1 & -3 \end{pmatrix}$

(c) $\begin{pmatrix} -\frac{1}{2} & \frac{1}{4} & 0 \\ 5 & -1 & 4 \\ -2 & 5 & -10 \end{pmatrix}$

(d) $\begin{pmatrix} 1+i & -2i & 3+2i \\ 2i & 1-2i & 4-5i \\ 1-i & 2+6i & 3+i \end{pmatrix}$

(e) $\begin{pmatrix} -1 & 2 & -1 & 0 \\ 0 & 3 & 4 & -2 \\ -7 & 5 & 1 & -1 \\ 6 & 3 & -2 & 0 \end{pmatrix}$

49. Find all the eigenvalues for each of the following matrices, and locate a characteristic vector for each.

(a) $\begin{pmatrix} 1 & 2 \\ -1 & -1 \end{pmatrix}$

(b) $\begin{pmatrix} 2 & 2 \\ 2 & 2 \end{pmatrix}$

(c) $\begin{pmatrix} 0 & 1 & 1 \\ 1 & 1 & 0 \\ 1 & 0 & 1 \end{pmatrix}$

(d) $\begin{pmatrix} 0 & 1 & 2 \\ 1 & 0 & -1 \\ 2 & -1 & 0 \end{pmatrix}$

50. An interesting and important theorem is the Cayley–Hamilton theorem, which says that if $p(c)$ is the characteristic polynomial A, then $p(A) = 0_{n \times n}$, where products of A and scalar multiples of A are performed according to the powers and coefficients of p. $0_{n \times n}$, of course, is the $n \times n$ null matrix. See if the result holds for each of the following matrices:

(a) $\begin{pmatrix} 1 & -1 \\ -1 & 1 \end{pmatrix}$

(b) $\begin{pmatrix} -2 & 1 & 0 \\ -2 & 0 & 1 \\ 0 & 1 & -1 \end{pmatrix}$

(c) $\begin{pmatrix} 3 & -1 & 0 \\ 2 & 0 & -2 \\ 3 & -1 & 1 \end{pmatrix}$

51. Test the following theorems:
  (a) If A, B are $n \times n$ matrices, then AB and BA have identical eigenvalues.
  (b) If A is a symmetric matrix, all its eigenvalues are real.
  (c) For a symmetric matrix A with nonnegative elements the largest eigenvalue is nonnegative and is less than or equal to the largest row sum in A.

CLASSICAL MATHEMATICS
OF OPTIMIZATION

2.1. Introduction

Much of the body of knowledge and many areas of study comprising operations research deal with optimization, a generic term that connotes maximization or minimization. Typically, one thinks of examples from business or industry—a manufacturing concern would like to schedule production so as to minimize costs; a business would like to choose from a set of investment alternatives so as to maximize expected return.

Examples abound in virtually every area of study. Scientists may seek trajectories to Mars so that fuel requirements for rockets can be minimized; a student may allocate his available study time among his courses so as to maximize his expected grade average; the military may select attack strategies so as to maximize the probability of striking a target.

The motivation for these optimizations may assume a variety of forms—profit, survival, convenience; but implicit in all our examples is the assumption that there are certain variables in the problem under the control of the problem solver and that it is possible to measure the results of choosing specific values for these decision variables. In qualitative terms, the optimiza-

2

tion problem is one of deciding which values of these decision variables will produce the best or optimal result.

Mathematics has for centuries been concerned with optimization problems. The reader will recall from his study of calculus the finding of maximal and minimal function values and the variable values that produce them. It is this sort of optimization to which the present chapter is devoted. A great many of the results contained herein are of eighteenth century vintage, and hence we employ "classical" in our title. The title is somewhat inaccurate, since we have included some more recent material, but its relationship to the classical analysis is sufficiently intimate as to permit its inclusion here.

2.2. Functions of a Single Variable

Toward introducing some terminology, let us agree that we have f, a real-valued function of a single real variable defined everywhere on an open interval I.

43

Definition. We say f has a *global maximum* at $x_0 \in I$ if $f(x_0) \geq f(x)$ for all $x \in I$.

Definition. Similarly, f is said to have a *global minimum* at x_0 if $f(x_0) \leq f(x)$ for all $x \in I$.

Definition. If one of the above inequalities holds for some point x_0 and all of I, then we say f has a *global extremum* at x_0 (*absolute extremum* is often used).

Definition. If for $x_0 \in I$ there exists a δ neighborhood of x_0 ($\delta > 0$) such that $f(x_0) \geq f(x)$ for all x satisfying $|x - x_0| \leq \delta$, then f is said to have a *relative maximum* at x_0.

Definition. f is said to have a *relative minimum* at $x_0 \in I$ if $f(x_0) \leq f(x)$ for all x satisfying $|x - x_0| \leq \delta$ for some positive δ.

Definition. If x_0 satisfies one of the two preceding definitions, it is said to be a *relative extremum*. (Can x_0 be both a relative max *and* a relative min?) (*Local extremum* is often used.)

The definitions for relative extrema need a slight modification in case I is *closed*, $I = [a, b]$, or in case I is half-closed, $I = [a, b)$, or $I = (a, b]$, for an end point does not have a neighborhood lying entirely within I. For the end points, then, we consider a δ neighborhood to be the appropriate half of the usual one—as in Figure 2.1. When I is an open interval, $I = (a, b)$, there are no such problems.

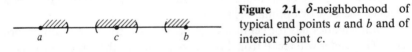

Figure 2.1. δ-neighborhood of typical end points a and b and of interior point c.

We observe that a point x_0 that is a relative or global minimum for $f(x)$ is also a relative or global maximum, respectively, for $-f(x)$. Consequently, one can choose to look always at maximization or always at minimization problems if he chooses, multiplying the function by -1 if necessary to obtain the desired form.

i. Necessary Conditions

In optimization problems we are interested in global extrema, of course, for a relative extremum might be far from optimal. Unfortunately, analysis is not equipped for locating global extrema directly, and the quest for global extrema must be carried out in terms of relative extrema.

Furthermore, unless f is reasonably well behaved, the methods of analysis will fail. Therefore, let us suppose that on the interval I the function f is *differentiable*, meaning that it has a derivative everywhere on the interior of I.

(In general, we cannot speak of derivatives at end points, since f may not be defined outside I.)

However, let x_0 be an interior point of I and suppose that x_0 is a relative maximum. Since f is differentiable, $f'(x_0) = \alpha$, for some real number α. Since f' exists at x_0, so do the right-and left-hand derivatives, which are, respectively,

$$\lim_{h \to 0+} \left[\frac{f(x_0 + h) - f(x_0)}{h} \right] \tag{2.1}$$

(where $h \to 0+$ means h decreases to zero; that is, passes through positive numbers to zero), and

$$\lim_{h \to 0-} \left[\frac{f(x_0 + h) - f(x_0)}{h} \right] = \lim_{h \to 0+} \left[\frac{f(x_0 - h) - f(x_0)}{-h} \right] \tag{2.2}$$

If x_0 is a relative max, then the bracketed quotient in expression (2.1) is ≤ 0 inside some neighborhood of x_0 and the bracketed quotients in equation (2.2) are ≥ 0 inside some neighborhood of x_0, but in the limit $f'(x_0)$ exists, so the two expressions must be identical. Therefore,

$$\alpha = 0 = f'(x_0)$$

The same conclusion is obtained under the assumption that x_0 is a relative minimum, and we have the following familiar theorem

Theorem 2.1

If f is differentiable at x_0, then in order that x_0 be a relative extremum it is necessary that $f'(x_0) = 0$.

It is not necessary, of course, that f be differentiable everywhere on I, since we are interested only in a neighborhood of x_0.

ii. Sufficiency Conditions

The reader is also reminded of the *nonsufficiency* of Theorem 2.1, for an *inflection point* will satisfy Theorem 2.1, as in Figure 2.2, without being

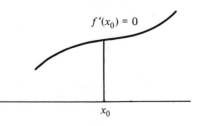

Figure 2.2. Typical inflection point of a curve.

a relative extremum. In general, a point where f' vanishes is called a *critical point*.

Also, Theorem 2.1 is not helpful at end points, because we cannot speak of right- or left-handed derivatives in a half-neighborhood where f is not defined. But we ask the reader to assume that right (left) hand derivatives exist about a left (right) end point and to obtain necessary and sufficient conditions that such an end point be a relative extremum.

If f is differentiable everywhere upon I, the following straightforward tactic will yield the global maximum (minimum) if there is one. Find all points where $f'(x) = 0$, and evaluate f at each. Then evaluate f at end points if there are any. The point corresponding to the largest (smallest) function value obtained is the desired point.

We say "if there is one," owing to the fact that there may not be. For example, let $I = (a, b)$ and define f on I by $f(x) = 2x + 1$, $a < x < b$. Then, since the interval is open, f has no smallest value and no largest value on I. In such a case, one might be interested in the infimum (written inf) or *greatest lower bound* of f and a point that provides it. Here $\inf f(x) = f(a)$. Also, we could discuss the supremum (sup) or *least upper bound*, since there is no maximum. We have $\sup f(x) = f(b)$.

It is possible, of course, to determine in other ways whether a point x_0 at which $f'(x_0) = 0$ is a true relative extremum. If, for example, f has a derivative at every point of some interval $(x_0 - h, x_0 + h)$, the signs of the derivative in the two half-intervals can be revealing.

All the possibilities are included in Table 2.1, as well as some of the conclusions. The reader is asked to complete the table.

Table 2.1

Interval $(x_0 - h, x_0)$	Interval $(x_0, x_0 + h)$	Conclusion
$f' = 0$	$f' = 0$	x_0 both a relative max and a
$f' = 0$	$f' > 0$	relative min
$f' = 0$	$f' < 0$	
$f' > 0$	$f' = 0$	
$f' > 0$	$f' > 0$	x_0 not a relative extremum
$f' > 0$	$f' < 0$	x_0 a relative max
$f' < 0$	$f' = 0$	
$f' < 0$	$f' > 0$	
$f' < 0$	$f' < 0$	

Exercise: Construct an example where f is differentiable, $f'(x_0) = 0$, $f' = 0$ in $(x_0 - h, x_0)$, and $f' > 0$ in $(x_0, x_0 + h)$.

The foregoing procedure for determining the nature of a critical point is not practicable, since locating a half-interval in which the derivative retains the same sign is not always simple.

a. Using Higher Derivatives

Second derivatives may be used to identify the nature of critical points, if any. For example, if f has a second derivative at x_0 and if $f'(x_0) = 0$, then if $f''(x_0) < 0$, the first derivative is decreasing at x_0; but being zero at x_0, it must therefore be passing from positive values through zero to negative values in some neighborhood of x_0. Thus it must be that x_0 is a point of relative *maximum*.

Similarly, $f''(x_0) > 0 \Rightarrow x_0$ is a relative *minimum*. Unfortunately, if $f''(x_0) = 0$, x_0's character remains in doubt.

More generally, suppose that $f'(x_0) = f''(x_0) = \cdots = f^{(n-1)}(x_0) = 0$, and $f^{(n)}(x_0) \neq 0$; that is, $f^{(n)}$ is the first nonvanishing derivative at x_0. We want to assume that $f^{(n)}$ is continuous in a neighborhood of x_0 [consequently, then, so are $f', f'', \ldots, f^{(n-1)}$].

Letting x be an arbitrary point within the assumed neighborhood of continuity of $f^{(n)}$, about x_0, Taylor's formula says

$$f(x) - f(x_0) = (x - x_0)f'(x_0) + \frac{1}{2!}(x - x_0)^2 f''(x_0)$$

$$+ \cdots + \frac{1}{(n-1)!}(x - x_0)^{n-1} f^{(n-1)}(x_0) \qquad (2.3)$$

$$+ \frac{f^{(n)}(x')}{n!}(x - x_0)^n$$

[is equation (2.3) valid if $f^{(n)}$ is not continuous?], where x' is an unknown point, but lies between x_0 and x; that is, $x_0 < x' < x$ or $x < x' < x_0$, depending upon the relative locations of x_0, x.

But, by assumption, equation (2.3) reduces to

$$f(x) - f(x_0) = \frac{1}{n!} f^{(n)}(x')(x - x_0)^n \qquad (2.4)$$

Since $f^{(n)}$ is continuous and $f^{(n)}(x_0) \neq 0$, then throughout *some neighborhood* of x_0, $f^{(n)}$ retains the same sign it possesses at x_0—another fundamental theorem of calculus.

If then one chooses x sufficiently near x_0, $f^{(n)}(x')$ will have the same sign as $f^{(n)}(x_0)$, since x' lies between x, x_0; all we are doing is ensuring that x' will fall within the "constant sign" neighborhood of x_0, whose existence is guaranteed by the continuity of $f^{(n)}$.

If n is odd, then $(x - x_0)^n$ *changes* sign as x approaches x_0 and passes beyond it. But for some range of such x values $f^{(n)}$ does *not* change sign, as we have observed, and, consequently, from equation (2.4), $f(x) - f(x_0)$ changes sign in a neighborhood of x_0, regardless of how small a neighborhood is chosen. Therefore, x_0 is not a relative extremum.

If n is even, $(x - x_0)^n$ does not change sign, and since $f^{(n)}$ does not, then there exists a neighborhood of x_0 in which $f(x) - f(x_0)$ has the same sign [of course, $f(x) - f(x_0) = 0$ for $x = x_0$]. Therefore, if $f^{(n)}(x_0) > 0$, $f(x) - f(x_0) > 0$, and x_0 is a relative min, and if $f^{(n)}(x_0) < 0$, x_0 is a relative max. (What if f has derivatives of all orders and all vanish at x_0?) Summarizing, we have

Theorem 2.2

If the first nonvanishing derivative of f at x_0 has order n and f has continuous derivatives of order at least n, then

(i) if n is odd, x_0 is not a relative extremum,

(ii) if n is even, x_0 is a relative max or relative min as $f^{(n)}(x_0) < 0$ or $f^{(n)}(x_0) > 0$, respectively.

iii. *Sufficiency of the Necessary Conditions*

For some classes of functions the necessary condition of Theorem 2.1 is also sufficient, so that as soon as x_0 is found with $f'(x_0) = 0$, one may conclude that x_0 is a global maximum or minimum; there is even no necessity for testing x_0 to discover whether or not it is a relative extremum.

Two such classes of functions, which are of interest not only because of their desirable properties, but also because they arise frequently in real problems, are defined next.

a. CONVEX AND CONCAVE FUNCTIONS

Definition. Let f be defined everywhere on an interval I. The function f is said to be *convex* on I if for every $x_1, x_2, \in I$ with $x_1 < x_2$

$$f(\lambda x_2 + (1 - \lambda)x_1) \leq \lambda f(x_2) + (1 - \lambda)f(x_1) \quad \text{for all } \lambda, 0 \leq \lambda \leq 1$$
$$(2.5)$$

To understand the definition, let $\mathbf{x}_1, \mathbf{x}_2$ be any two vectors in the plane, as in Figure 2.3.

It is clear that the expression

$$\lambda(\mathbf{x}_2 - \mathbf{x}_1)$$

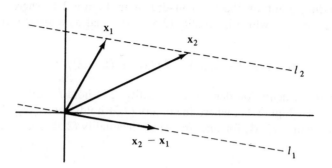

Figure 2.3. Parallel lines l_1 and l_2 generated by $\mathbf{x}_1$, $\mathbf{x}_2$ and by $\mathbf{x}_2 - \mathbf{x}_1$.

generates the line l_1 through $\mathbf{x}_2 - \mathbf{x}_1$, as λ is allowed to assume all possible real values; the lines l_1 and l_2 being parallel, then,

$$\lambda(\mathbf{x}_2 - \mathbf{x}_1) + \mathbf{x}_1$$

generates the line l_2, which passes through $\mathbf{x}_1$, $\mathbf{x}_2$. And, clearly, if we restrict λ to $0 \le \lambda \le 1$, then

$$\lambda(\mathbf{x}_2 - \mathbf{x}_1) + \mathbf{x}_1 \tag{2.6}$$

generates the line segment that joins $\mathbf{x}_1$, $\mathbf{x}_2$. Finally,

$$\lambda(\mathbf{x}_2 - \mathbf{x}_1) + \mathbf{x}_1 = \lambda\mathbf{x}_2 + (1 - \lambda)\mathbf{x}_1$$

The expression (2.6) may be obtained by purely algebraic means, with no reliance on the geometric interpretation of vector addition in the plane.

Consider the line represented in Figure 2.4. The inequality (2.5) then simply states that points $(x, f(x))$ on the curve lie always on or below the

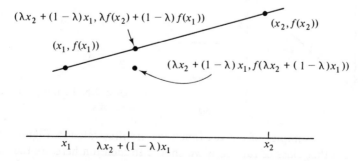

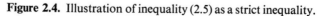

Figure 2.4. Illustration of inequality (2.5) as a strict inequality.

corresponding point on the segment drawn in Figure 2.4, where we have illustrated a case in which inequality (2.5) is satisfied as a *strict* inequality; that is,

$$f(\lambda x_2 + (1 - \lambda)x_1) < \lambda f(x_2) + (1 - \lambda)f(x_1)$$

Notice that the definition demands inequality (2.5) hold *for all* x_1, x_2, with $x_1 < x_2$, so in Figure 2.5, where inequality (2.5) holds for x_1, x_2, there are many such pairs, x_1', x_2', for one, for which inequality (2.5) fails miserably.

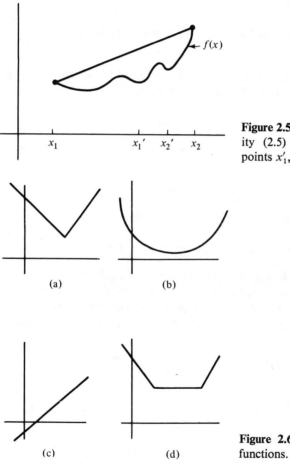

Figure 2.5. A case where inequality (2.5) fails for the pair of points x_1', x_2'.

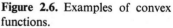

Figure 2.6. Examples of convex functions.

Several examples of convex functions are illustrated in Figure 2.6.

The other class of functions we choose to mention here are the *concave* functions.

Definition. The function f is said to be *concave* on I if for every $x_1, x_2 \in I$ with $x_1 < x_2$

$$f(\lambda x_2 + (1 - \lambda)x_1) \geq \lambda f(x_2) + (1 - \lambda)f(x_1) \qquad (2.7)$$

for all λ, $0 \leq \lambda \leq 1$.

The only difference is that the sense of the defining inequality has been reversed. Several examples of concave functions are illustrated in Figure 2.7.

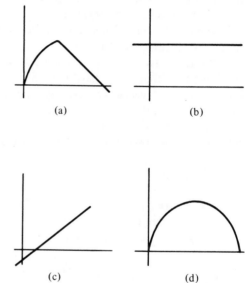

(a)

(b)

(c)

(d)

Figure 2.7. Examples of concave functions.

In the event either inequality (2.5) or (2.7) is satisfied as a strict inequality for all $x_1, x_2 \in I$ whenever $0 < \lambda < 1$, then f is called *strictly convex* (Figures 2.6a and b) or *strictly concave* (Figures 2.7a and d). Inequalities (2.5) and (2.7) are always satisfied as equalities for $\lambda = 0$, $\lambda = 1$.

Exercise 2.1

Prove that f is concave on I if and only if $-f$ is convex on I.

Exercise 2.2

Show that linear functions are both convex and concave over the entire real line.

Exercise 2.3

If $\{f_1, f_2, \ldots, f_k\}$ are convex (concave) functions on I, then $f = \sum_{i=1}^{k} f_i$ is a convex (concave) function on I.

Exercise 2.4

Remembering that f is to be defined everywhere on I, can you sketch a convex or concave function on I that has a discontinuity on the interior of I? What might one suspect?

b. GLOBAL EXTREMA FOR CONVEX AND CONCAVE FUNCTIONS

Now, if a convex or concave function has a point x_0 on its interval of definition where the derivative vanishes, then x_0 must be a relative extremum, for otherwise there would be a neighborhood of x_0 in which f assumed values both greater and less than $f(x_0)$ in such fashion as to contradict the convexity or concavity of f. This statement requires proof, but a geometric illustration should convince the reader.

Of course, as attested by Figures 2.6 and 2.7, there are convex and concave functions that never have their derivatives zero—indeed, that do not have derivatives everywhere. At any rate, the following stronger statement then implies that the vanishing of the derivative is a *sufficient* condition for a global extremum of a convex or concave function and is also applicable independently of the existence of derivatives.

Theorem 2.3

If x_0 is a relative minimum of a convex function f on an interval I, then x_0 is a global minimum of f on I.

Proof: Let $x^* \in I$ be such that

$$f(x^*) < f(x_0)$$

Since x_0 is a relative min, there is a neighborhood N of x_0 in which $f(x) \geq f(x_0)$ for all x within that neighborhood. Next consider the segment joining $(x^*, f(x^*))$, with $(x_0, f(x_0))$, a typical point of which has representation

$$(\lambda x_0 + (1 - \lambda)x^*, \lambda f(x_0) + (1 - \lambda)f(x^*))$$

Clearly, unless $\lambda = 1$,

$$\lambda f(x_0) + (1 - \lambda)f(x^*) < f(x_0)$$

Consider those $\lambda \neq 1$ which correspond to points x within N. For all these we have

$$\lambda f(x_0) + (1 - \lambda)f(x^*) < f(x_0) \leq f(x) = f(\lambda x_0 + (1 - \lambda)x^*)$$

which contradicts f's convexity over I, which proves the theorem, since there must be no such x^* as assumed. Exercise 2.1 then provides the analogous result for concave functions; that is, if f is concave over I and x_0 is a relative max of f, then x_0 is a global max of f.

2.3. Functions of Several Variables—No Constraints

We want next to consider functions of n real variables, $f(x_1, x_2, \ldots, x_n)$, so that the points at which f is defined comprise some subset of E^n, the n-dimensional Euclidean vector space.

i. *Some Geometry and Definitions in E^n*

A few remarks and definitions regarding Euclidean geometry in n dimensions will be appropriate from time to time.

Definition. If x_1, x_2 are points (vectors) in E^n, then the line in E^n passing through them consists of the set $L = \{P \mid P = \lambda x_2 + (1 - \lambda)x_1,$ all real $\lambda\}$, and the line segment joining them is $l = \{P \mid P = \lambda x_2 + (1 - \lambda)x_1, 0 \leq \lambda \leq 1\}$.

All we have done, clearly, to define line and line segment in E^n is to copy our two-dimensional definitions. It is simple to verify that the objects defined in one, two, and three dimensions look like lines are supposed to, and possess all the necessary properties. Beyond three dimensions geometric intuition is insufficient—statements that are obviously true in three or fewer dimensions are sometimes most difficult to establish in spaces of higher dimension. However, the objects we have defined behave in higher dimensions as lines and line segments should.

Definition. A set $S \subset E^n$ is said to be *connected* if for every pair of points $x_0, x_{n+1} \in S$ there exists a sequence of points $x_1, x_2, \ldots, x_n$ and a sequence of line segments $l_1, \ldots, l_{n+1}$ with l_j connecting x_{j-1} and x_j, $1 \leq j \leq n + 1$, such that each of $l_1, \ldots, l_{n+1}$ lies entirely within S.

In Figure 2.8, an example of a connected set is S_1, whereas S_2 is not connected.
    Sets of interest to us will, in fact, be *simply connected*, which means that closed surfaces lying entirely within the connected set S contain *only* points of S. Thus sets like S_1 in Figure 2.8 are eliminated.
    Suppose then that our sets are simply connected; we shall term such a set a *region*, be it open or closed, although a region is sometimes defined as an open simply connected set.

Performing the indicated differentiations, we find

$$F''(t) = \left(\theta_1 \frac{\partial^2 f}{\partial x_1^2} + \theta_2 \frac{\partial^2 f}{\partial x_1 \partial x_2} + \cdots + \theta_n \frac{\partial^2 f}{\partial x_1 \partial x_n}\right)\theta_1$$

$$+ \left(\theta_1 \frac{\partial^2 f}{\partial x_2 \partial x_1} + \theta_2 \frac{\partial^2 f}{\partial x_2^2} + \cdots + \theta_n \frac{\partial^2 f}{\partial x_2 \partial x_n}\right)\theta_2 + \cdots$$

$$+ \left(\theta_1 \frac{\partial^2 f}{\partial x_n \partial x_1} + \theta_2 \frac{\partial^2 f}{\partial x_n \partial x_2} + \cdots + \theta_n \frac{\partial^2 f}{\partial x_n^2}\right)\theta_n$$

In more compact notation

$$F''(t) = \sum_{i=1}^{n} \sum_{j=1}^{n} \theta_i \theta_j \frac{\partial^2 f}{\partial x_i \partial x_j}$$

We also notice that $F'''(t)$ has the form

$$F'''(t) = \sum_{i=1}^{n} \sum_{j=1}^{n} \sum_{k=1}^{n} \theta_i \theta_j \theta_k \frac{\partial^3 f}{\partial x_i \partial x_j \partial x_k}$$

Now we apply Taylor's formula to the function F; in particular,

$$F(1) - F(0) = (1-0)F'(0) + \frac{(1-0)^2 F''(0)}{2!} + \frac{(1-0)^3 F'''(\xi)}{3!} \quad (2.8)$$

where $0 < \xi < 1$.
We know that

$$F'(0) = \sum_{i=1}^{n} \theta_i \frac{\partial f(y_1, \ldots, y_n)}{\partial x_i} \quad (2.9)$$

$$F''(0) = \sum_{i=1}^{n} \sum_{j=1}^{n} \theta_i \theta_j \frac{\partial^2 f(y_1, \ldots, y_n)}{\partial x_i \partial x_j} \quad (2.10)$$

and $F'''(\xi)$ may be written

$$F'''(\xi) = \sum_{i=1}^{n} \sum_{j=1}^{n} \sum_{k=1}^{n} \theta_i \theta_j \theta_k \frac{\partial^3 f(y_1 + \theta_1 \xi, \ldots, y_n + \theta_n \xi)}{\partial x_i \partial x_j \partial x_k} \quad (2.11)$$

    Thus the Taylor representation for F is perfectly valid for any choice of $\theta_1, \ldots, \theta_n$ for which $(\partial^3 f/\partial x_i \partial x_j \partial x_k)(y_1 + \theta_1 \xi, \ldots, y_n + \theta_n \xi)$ exists for all $\xi, 0 < \xi < 1$; for then F'' is continuous and F''' exists in a neighborhood of $(y_1, \ldots, y_n)$, by virtue of the continuity of $\partial^2 f/\partial x_i \partial x_j$ and the existence of $\partial^3 f/\partial x_i \partial x_j \partial x_k$ in a neighborhood of $(y_1, \ldots, y_n)$. These properties were assumed, recall, for *some* neighborhood of $(y_1, \ldots, y_n)$.
    Substituting expressions (2.9), (2.10), and (2.11) into the Taylor representation (2.8), we get

$$F(1) - F(0) = \sum_{i=1}^{n} \theta_i \frac{\partial f(y_1, \ldots, y_n)}{\partial x_i}$$

$$+ \frac{1}{2!} \sum_{i=1}^{n} \sum_{j=1}^{n} \theta_i \theta_j \frac{\partial^2 f(y_1, \ldots, y_n)}{\partial x_i \, \partial x_j} \tag{2.12}$$

$$+ \frac{1}{3!} \sum_{i=1}^{n} \sum_{j=1}^{n} \sum_{k=1}^{n} \frac{\partial^3 f(y_1 + \theta_1 \xi, \ldots, y_n + \theta_n \xi) \theta_i \theta_j \theta_k}{\partial x_i \, \partial x_j \, \partial x_k}$$

for some ξ, $0 < \xi < 1$.
But

$$F(1) - F(0) = f(y_1 + \theta_1, y_2 + \theta_2, \ldots, y_n + \theta_n) - f(y_1, \ldots, y_n) \tag{2.13}$$

Now by *varying* $\theta_1, \ldots, \theta_n$, simultaneously, the expression $(y_1 + \theta_1, y_2 + \theta_2, \ldots, y_n + \theta_n)$ generates *all* points in a neighborhood of $(y_1, \ldots, y_n)$. It is precisely the difference on the right side of equation (2.13), throughout a neighborhood of $(y_1, \ldots, y_n)$, in which we must be interested if we are to determine whether $(y_1, \ldots, y_n)$ is a relative extremum or not, and if it is, whether it is a maximum or a minimum. More accurately, the *sign* of the difference is the pertinent item; for if $f(y_1 + \theta_1, \ldots, y_n + \theta_n) - f(y_1, \ldots, y_n)$ *changes sign* in a neighborhood of $(y_1, \ldots, y_n)$, then $(y_1, \ldots, y_n)$ is not a relative extremum; if the difference is always nonnegative, we have a relative minimum; if always nonpositive, a relative maximum.

But since the right side of equation (2.13) equals the right side of equation (2.12), we ask whether or not anything can be said about the signs of the right sides of equations (2.12) or (2.13), as $\theta_1, \ldots, \theta_n$ are allowed to simultaneously vary, so that $(y_1 + \theta_1, \ldots, y_n + \theta_n)$ generates a neighborhood of $(y_1, \ldots, y_n)$.

It turns out that the right side of equation (2.12) can give such information. Since the first partials all vanish at $(y_1, \ldots, y_n)$, equation (2.12) may be rewritten

$$f(y_1 + \theta_1, \ldots, y_n + \theta_n) - f(y_1, \ldots, y_n)$$

$$= \frac{1}{2!} \sum_{i=1}^{n} \sum_{j=1}^{n} \theta_i \theta_j \frac{\partial^2 f(y_1, \ldots, y_n)}{\partial x_i \, \partial x_j} \tag{2.14}$$

$$+ \frac{1}{3!} \sum_{i=1}^{n} \sum_{j=1}^{n} \sum_{k=1}^{n} \theta_i \theta_j \theta_k \frac{\partial^3 f(y_1 + \theta_1 \xi, \ldots, y_n + \theta_n \xi)}{\partial x_i \, \partial x_j \, \partial x_k}$$

For notational convenience, let us denote the first term on the right side of equation (2.14) by Q, the second by R. If we agree to let all of $\theta_1, \ldots, \theta_n$ be quite small, then Q dominates $Q + R$, for each term of Q possesses a product of two θ's and each term of R, a product of three θ's, and, importantly, all the second and third partials are finite quantities. Thus, for sufficiently small $\theta_1, \ldots, \theta_n$, a sum of finite terms each involving a product of three small quantities $\theta_i\theta_j\theta_k$ will be dominated by a sum of finite terms each involving a product of two small quantities $\theta_i\theta_j$.

Thus the sign of the difference $f(y_1 + \theta_1, \ldots, y_n + \theta_n) - f(y_1, \ldots, y_n)$ will be exactly that of Q, except in the unfortunate and troublesome case where for some

$$\boldsymbol{\theta} = (\theta_1, \ldots, \theta_n) \neq (0, 0, \ldots, 0)$$

Q vanishes also. Then the sign of the difference will be that of the expression R, which is actually unknown, owing to the fact that ξ is known only within an interval.

Thus, from the representation (2.8), and consequently equation (2.14), a concrete conclusion can be drawn if and only if we can discover that

1. Q is always positive for all $\boldsymbol{\theta} \neq \boldsymbol{0}$; then **y** is a relative minimum.
2. Q is always negative for all $\boldsymbol{\theta} \neq \boldsymbol{0}$; then **y** is a relative maximum.
3. Q changes sign for some choices of $\boldsymbol{\theta}$; then **y** is not an extremum.

a. QUADRATIC FORMS

It is the case, however, that Q possesses a structure for which conclusions like 1, 2, and 3 may be readily drawn. Let us digress to examine more closely the expression Q. Q has the general form

$$\sum_{i=1}^{n} \sum_{j=1}^{n} a_{ij}\theta_i\theta_j,$$

which is known as a *homogeneous quadratic form* in the variables $\theta_1, \ldots, \theta_n$.

The prefix *homogeneous* arises from the fact that the general quadratic form in $\theta_1, \ldots, \theta_n$ is

$$\sum_{i=1}^{n} \sum_{j=1}^{n} a_{ij}\theta_i\theta_j + \sum_{i=1}^{n} b_i\theta_i + d$$

That is, in addition to the quadratic term, there is a linear term and a constant term. When

$$b_1 = b_2 = \cdots = b_n = d = 0$$

the quadratic form is called homogeneous. In this chapter our quadratic forms will always be homogeneous, but we shall refer to them as *quadratic forms*, not *homogenous* quadratic forms.

The reader is certainly familiar with quadratic forms, for consider the quadratic equation

$$x_1^2 + 2x_1x_2 + 3x_2^2 = 0$$

the left side of which is a quadratic form in x_1, x_2 with $a_{11} = 1, a_{12} = 2, a_{21} = 0$, and $a_{22} = 3$. Observe that

$$\sum_{i=1}^{n}\sum_{j=1}^{n} a_{ij}\theta_i\theta_j$$

$$= (\theta_1, \theta_2 \ldots, \theta_n) \begin{pmatrix} a_{11} & a_{12} & \cdots & a_{1n} \\ a_{21} & \cdot & \cdot & a_{2n} \\ \vdots & & & \vdots \\ a_{n1} & \cdot & \cdot & a_{nn} \end{pmatrix} \begin{pmatrix} \theta_1 \\ \theta_2 \\ \vdots \\ \theta_n \end{pmatrix} = \boldsymbol{\theta}^T A \boldsymbol{\theta}$$

The following definitions, then, are pertinent.

Definition. The quadratic form $\boldsymbol{\theta}^T A \boldsymbol{\theta}$ is said to be *positive (negative) definite* if for *all* $\boldsymbol{\theta} \neq \mathbf{0}$, $\boldsymbol{\theta}^T A \boldsymbol{\theta} > 0 (\boldsymbol{\theta}^T A \boldsymbol{\theta} < 0)$. Of course, $\boldsymbol{\theta}^T A \boldsymbol{\theta} = 0$ for $\boldsymbol{\theta} = \mathbf{0}$.

Definition. $\boldsymbol{\theta}^T A \boldsymbol{\theta}$ is said to be *positive semidefinite (negative semidefinite)* if for all $\boldsymbol{\theta} \neq \mathbf{0}$, $\boldsymbol{\theta}^T A \boldsymbol{\theta} \geqslant 0 (\boldsymbol{\theta}^T A \boldsymbol{\theta} \leqslant 0)$, but $\boldsymbol{\theta}^T A \boldsymbol{\theta} = 0$ for some value of $\boldsymbol{\theta} \neq \mathbf{0}$.

Definition. $\boldsymbol{\theta}^T A \boldsymbol{\theta}$ is said to be *indefinite* if it changes sign, regardless of how small

$$|\boldsymbol{\theta}| = \left| \sum_{i=1}^{n} \theta_i^2 \right|^{1/2}$$

is constrained to be.

We also point out that it is common to speak of the matrix A as positive definite, indefinite, and so on, rather than the corresponding quadratic form, the meanings being the same.

For our purposes, then, knowing A to be definite or indefinite will permit the drawing of the appropriate conclusions 1, 2, and 3, and it is the *semidefinite* cases about which we are uncertain. The structure of the matrix A, is for our problem,

$$A = \begin{pmatrix} \dfrac{\partial^2 f(\mathbf{y})}{\partial x_1^2} & \dfrac{\partial^2 f(\mathbf{y})}{\partial x_1 \partial x_2} & \cdots & \dfrac{\partial^2 f(\mathbf{y})}{\partial x_1 \partial x_n} \\ \dfrac{\partial^2 f(\mathbf{y})}{\partial x_2 \partial x_1} & \dfrac{\partial^2 f(\mathbf{y})}{\partial x_2^2} & \cdots & \dfrac{\partial^2 f(\mathbf{y})}{\partial x_2 \partial x_n} \\ \vdots & \vdots & & \vdots \\ \dfrac{\partial^2 f(\mathbf{y})}{\partial x_n \partial x_1} & \dfrac{\partial^2 f(\mathbf{y})}{\partial x_n \partial x_2} & \cdots & \dfrac{\partial^2 f(\mathbf{y})}{\partial x_n^2} \end{pmatrix}$$

This matrix of second partials has a special name; it is called the Hessian (after the German mathematician Hesse) of f at y; let us denote it by $H_f(y)$.

It is important to notice that, owing to the continuity of the second partials, $A = H_f(y)$ is a *symmetric* matrix; that is, $a_{ij} = a_{ji}$ or

$$\frac{\partial^2 f}{\partial x_i \, \partial x_j} = \frac{\partial^2 f}{\partial x_j \, \partial x_i}, \quad \text{for all } 1 \le i \le n, \, 1 \le j \le n$$

b. DETERMINING DEFINITENESS

There are available certain tests for determining the definiteness of $H_f(y)$, and we give these without proof of their validity.

Test 1

Let A be an $n \times n$ matrix of real numbers. Let $A_1, A_2, \ldots, A_n$ be the *principal submatrices* of A; that is,

$$A = \begin{pmatrix} A_1 & A_2 & A_3 & & A_i & & A_n \\ a_{11} & a_{12} & a_{13} & \cdots & a_{1i} & \cdots & a_{1n} \\ a_{21} & a_{22} & a_{23} & \cdots & a_{2i} & \cdots & a_{2n} \\ a_{31} & a_{32} & a_{33} & \cdots & a_{3i} & \cdots & a_{3n} \\ \cdot & & & & \cdot & & \cdot \\ \cdot & & & & \cdot & & \cdot \\ \cdot & & & & \cdot & & \cdot \\ a_{i1} & a_{i2} & \cdot & \cdot & a_{ii} & \cdots & a_{in} \\ \cdot & & & & \cdot & & \cdot \\ \cdot & & & & \cdot & & \cdot \\ \cdot & & & & \cdot & & \cdot \\ a_{n1} & a_{n2} & \cdot & \cdot & a_{ni} & \cdots & a_{nn} \end{pmatrix}$$

and

$$A_j = \begin{pmatrix} a_{11} & a_{12} & \cdots & a_{1j} \\ a_{21} & \cdot & \cdot & a_{2j} \\ \cdot & & & \\ \cdot & & & \\ \cdot & & & \\ a_{j1} & a_{j2} & \cdots & a_{jj} \end{pmatrix}$$

(a) Then $\det(A_1) = a_{11}$, $\det(A_2), \ldots, \det(A_n)$ are all positive if and only if A is positive definite.

Now the definitions of positive and negative definite imply that A is negative definite if and only if $-A$ is positive definite. Also, $\det(-A_j) = (-1)^j \det(A_j)$, so we have

(b) A is negative definite if and only if

$$\det(A_1) < 0, \ \det(A_2) > 0, \ \det(A_3) < 0, \ldots, (-1)^n \det(A_n) > 0$$

In other words, the determinants of the principal submatrices of A must alternate in sign, negative, positive, negative, and so on.

It is correct to deduce that implementing this test is not a trivial computation; calculating the determinants of fairly large arrays can be most time consuming. Furthermore, (a) and (b) say nothing about semidefiniteness and indefiniteness.

Test 2

A second test involves the *characteristic polynomial* of a square matrix A (see, e.g., Chapter 1).

Letting c be a variable, consider the expression $\det(A - cI)$, where I is the $n \times n$ identity matrix. Then

$$\det(A - cI) = \det \begin{pmatrix} a_{11} - c & a_{12} & \cdots & a_{1n} \\ a_{21} & a_{22} - c & \cdots & a_{2n} \\ \vdots & & & \\ a_{n1} & a_{n2} & \cdots & a_{nn} - c \end{pmatrix}.$$

$$= n\text{th-degree polynomial in } c, \text{ say } p(c)$$

The polynomial $p(c)$ is called the *characteristic polynomial of A*, and the roots of $p(c) = 0$ are called the *characteristic values* or *eigenvalues* of A.

The following conclusions may then be obtained

(a) A is positive (negative) definite iff the eigenvalues of A are all positive (negative).

(b) A is positive (negative) semidefinite iff the eigenvalues of A are all $\geqslant 0$ ($\leqslant 0$) and at least one *is* zero.

(c) A is indefinite iff A has both positive and negative eigenvalues.

c. COMMENTS ON THE TESTS

Some comments are appropriate. First, in general an nth-degree polynomial may have complex roots, in which case our conclusions would be meaningless. But for *symmetric A*, the eigenvalues are always real, so there is no such difficulty. Second, the computational difficulties of Test 2 are not significantly less than those of Test 1; finding the roots of an nth-degree polynomial is not frequently a simple problem. But, of course, we are not interested in the values of the eigenvalues, only their signs; and there are some classical results that give information about the signs of the roots of an nth-degree polynomial.

Two such results, both of which the reader has probably seen previously, are

1. Descartes' rule of sign: The nth-degree polynomial

$$p(x) = a_n x^n + a_{n-1} x^{n-1} + \cdots + a_1 x + a_0$$

has a number of positive roots less than or equal to the number of variations in sign in the terms of $p(x)$. The number of negative roots of $p(x)$ is less than or equal to the number of variations of sign in the terms of $p(-x)$.

2. Descartes' rule of sign—extended: $p(x)$ has a number of positive roots equal to the number σ of sign variations in the terms of $p(x)$ or equal to σ minus an even integer. Similarly, the number of negative roots is either σ', the number of sign variations in the terms of $p(-x)$, or σ' minus an even integer.

These results can be more useful in giving information about the roots or our characteristic polynomials, since they will have no complex roots. Furthermore the presence and the number of zero roots is available upon inspection.

For example, suppose that the characteristic equation for a Hessian A is

$$p(c) = c^4 - 4c^3 + 3c^2 + 4c - 4 = 0$$

Since the Hessian is a symmetric matrix, all four roots of $p(c)$ are real. $p(0) = -4$; hence $c = 0$ is not a root. The sequence of signs in the successive terms of $p(c)$ is $+ - + + -$, with a total of three sign variations. Thus the number of positive roots is either 3 or $3 - 2 = 1$. Hence $p(c)$ must have at least one negative root. Now the sequence of signs in the successive terms of $p(-c)$ is $+ + + - -$, with a total of one sign variation. Thus the number of negative roots must be at most one. Hence $p(c)$ has exactly one negative root and, consequently, exactly three positive roots. [The reader may verify that the four roots of $p(c)$ are 2, 2, 1, -1.]

Using the results obtained in applying Test 2 for characterizing the Hessian, we note that the Hessian A for this problem is indefinite, since it has both positive and negative characteristic values.

We do wish to point out, however, that eigenvalues may be approximated by several methods of numerical analysis *without recourse to the characteristic equation, $p(c) = 0$.* These methods require only the matrix A, and some are computationally quite simple. See References [1] and [2].

Now let us apply our tests to some sample functions.

Example 2.1

Consider

$$f(x, y, z) = -x^2 - y^2 - z^2$$

The only critical point is $x = 0$, $y = 0$, $z = 0$, and we find

$$\frac{\partial^2 f}{\partial x^2} = \frac{\partial^2 f}{\partial y^2} = \frac{\partial^2 f}{\partial z^2} = -2 \quad \text{at } (0, 0, 0) = \mathbf{y}_0$$

$H_f(\mathbf{y}_0)$ then is

$$\begin{pmatrix} -2 & 0 & 0 \\ 0 & -2 & 0 \\ 0 & 0 & -2 \end{pmatrix}$$

$$\det(A_1) = -2 < 0, \quad \det(A_2) = 4 > 0, \quad \det(A_3) = -8 < 0,$$

so $H_f(\mathbf{y}_0)$ is negative definite and $(0, 0, 0)$ is a relative maximum (in fact, a global maximum) of f.

Applying Test 2 to the same function, we set

$$\det \begin{pmatrix} -2 - c & 0 & 0 \\ 0 & -2 - c & 0 \\ 0 & 0 & -2 - c \end{pmatrix} = 0$$

that is $(-2 - c)^3 = 0$, and the roots are -2, -2, and -2. Thus $H_f(\mathbf{y}_0)$ is negative definite and $\mathbf{y}_0 = (0, 0, 0)$ is a relative maximum.

Example 2.2

Let us consider a less obvious example with, again, f being a function of three variables.
Let

$$f(x_1, x_2, x_3) = x_1^2 - 2x_1x_2 + x_1x_3 - x_2^2 + 4x_2x_3 + 3x_3 - \frac{23}{8}x_3^2$$

In this case $f(x_1, x_2, x_3)$ is itself a quadratic form (inhomogeneous), and the first task is to solve the necessary conditions

$$\frac{\partial f}{\partial x_1} = \frac{\partial f}{\partial x_2} = \frac{\partial f}{\partial x_3} = 0$$

a task that it must be understood is, in general, nontrivial, for there are no algorithms for explicitly solving a system of n equations in n variables.
At any rate, in this case, the necessary conditions are

$$2x_1 - 2x_2 + \quad x_3 = 0$$
$$-2x_1 - 2x_2 + 4x_3 = 0$$
$$x_1 + 4x_2 - \frac{23}{4}x_3 = -3$$

There are *no* solutions to this system, since the equations are inconsistent; f has no critical points.
Let us therefore redefine f as

$$f(x_1, x_2, x_3) = x_1^2 - 2x_1x_2 - 4x_1x_3 - x_2^2 + 4x_2x_3 + 3x_3$$

The necessary conditions, then, are

$$2x_1 - 2x_2 - 4x_3 = 0$$
$$-2x_1 - 2x_2 + 4x_3 = 0$$
$$-4x_1 + 4x_2 + 3 \quad = 0$$

which has the unique solution

$$x_1 = \tfrac{3}{4}, \qquad x_2 = 0, \qquad x_3 = \tfrac{3}{8}$$

The second partial derivatives are

$$\frac{\partial^2 f}{\partial x_1^2} = 2, \qquad \frac{\partial^2 f}{\partial x_2^2} = -2, \qquad \frac{\partial^2 f}{\partial x_3^2} = 0$$

$$\frac{\partial^2 f}{\partial x_1 \partial x_2} = \frac{\partial^2 f}{\partial x_2 \partial x_1} = -2, \qquad \frac{\partial^2 f}{\partial x_1 \partial x_3} = \frac{\partial^2 f}{\partial x_3 \partial x_1} = -4$$

$$\frac{\partial^2 f}{\partial x_2 \partial x_3} = \frac{\partial^2 f}{\partial x_3 \partial x_2} = 4$$

The matrix of interest is

$$\begin{pmatrix} 2 & -2 & -4 \\ -2 & -2 & 4 \\ -4 & 4 & 0 \end{pmatrix}$$

The reader is asked to verify that the point $(\tfrac{3}{4}, 0, \tfrac{3}{8})$ is not a relative extremum for the function f.

d. OBSERVATIONS

The reader should notice that for $n = 1$ our sufficient conditions become exactly as those obtained for functions of a single variable. Furthermore, the semidefinite case for several variables is entirely analogous to the vanishing of the second derivative in the case of a single variable, only much more difficult to resolve. When the second derivative vanished at an inflection point for a function of a single variable, we were able to resolve the uncertainty by examining the first nonvanishing derivative at the inflection point, the results of which are our Theorem 2.2.

For functions of several variables, the vanishing of the quadratic form implies that the remainder term of equation (2.14) must be examined. If it too vanished for some θ, then a fourth-order term would need to be treated. But higher order terms in the series representation for functions of several variables are exceedingly more complicated than their one-dimensional counterparts, as one observes. Nevertheless, sufficient conditions may be obtained to resolve the semidefinite case, although these are inappropriate to an introduction of this sort and will be omitted.

It must be pointed out, however, that since our ultimate goal is locating global extrema, locating relative extrema is but the first step—if f is bounded from above (below) the global max (min) will be the best of the relative maxima (minima) found. Here again we would have a problem in dealing with the function on the boundary of its region of definition.

2.4. Constrained Optimization—Equality Constraints

When we assumed our functions to be defined on specific intervals or regions, we essentially were *constraining* the values that the variables could assume.

At present we wish to consider optimization problems in which the admissible variable values are limited by explicit mathematical expressions and limited *only in this way*.

The simplest kind of interesting problem we can construct is the problem of, say, maximizing $f(x, y)$ over all real values (x, y) that satisfy the equation

$$g(x, y) = 0,$$

which will be called a constraint. We shall write

$$\max f(x, y)$$

subject to

$$g(x, y) = 0$$

where "max" abbreviates "maximize" and not "maximum." Also, of course, we are again interested in global extrema, where global now implies best from among all points satisfying the constraint. Also, we shall hereafter refer to the function being optimized as the *objective function* and the set of points satisfying the constraints as the *constraint set*, the *feasible set*, or the *set of feasible solutions*.

i. *Method of Substitution*

An obvious method for solving this particular type of optimization problem would be to first solve the constraint equation $g(x, y) = 0$ for either of the variables x and y, thus obtaining an explicit expression of the form $y = h(x)$ or $x = k(y)$. Substituting one of the resulting expressions in the objective function would yield the unconstrained optimization problem in one variable:

$$\max f(x, h(x))$$

or
$$\max f(k(y), y)$$

The previously discussed methods of solution could then be applied. We illustrate the procedure by an example.

Example 2.3

Consider the problem

$$\min f(x, y) = (x - 2)^2 + (y - 3)^2 + 2$$

subject to
$$y = x^2 + 2$$

The reader should note that the function $g(x, y)$ is

$$g(x, y) = y - x^2 - 2$$

We can picture this constrained optimization problem in the geometry of three dimensions. Referring to Figure 2.9, the function $z = f(x, y)$ represents a surface of revolution, more specifically a paraboloid. The constraint equation $y = x^2 + 2$ represents another surface of revolution, a parabola sheet,

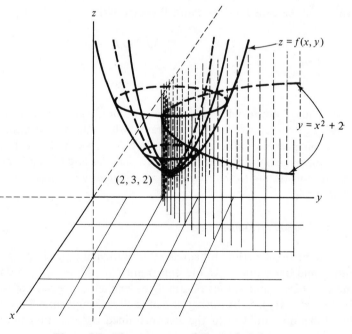

Figure 2.9. The geometry of Example 2.3.

with axis parallel to the z axis. The problem then addresses itself to determining that point on the curve of intersection of the two surfaces having the lowest altitude.

Owing to the nature of f, the minimizing point will be that point on the parabola nearest (2, 3). With a more precise diagram one could closely approximate the optimal (x, y) by inspection.

Also, let us pause to show that a two-dimensional representation of the problem is possible and is quite helpful for dealing with three-dimensional problems.

The *contour map*, as Figure 2.10 is called, and in which we have included the constraint $g(x, y) = 0$, essentially allows one to look down upon the function of Figure 2.9 from above.

This two-dimensional representation is obtained by drawing the locus of points for which $f(x, y) = c$, a constant, for various fixed values of c.

With the present f, for example, let

$$f(x, y) = 2$$
$$(x - 2)^2 + (y - 3)^2 + 2 = 2$$

It is clear that the set of points satisfying that equation is the single point $x = 2, y = 3$.

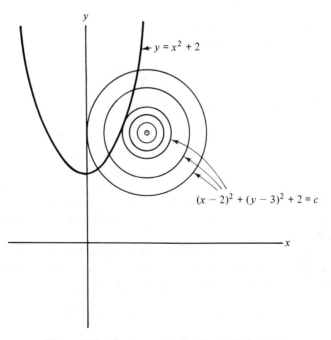

Figure 2.10. Contour map for Example 2.3.

If we write

$$(x - 2)^2 + (y - 3)^2 + 2 = 6$$

or

$$(x - 2)^2 + (y - 3)^2 \quad\;\; = 4$$

the set of points defined is the circle of radius 2 centered at (2, 3), and so on.

The set of points satisfying $f(x, y) = c$ is called a *level contour;* for the given f, each is simply a circle in the x, y plane.

Clearly, an optimal solution to our problem is easily approximated by even a crude contour map. It is also apparent that the level contour $f(x, y) = c^*$, where c^* is the minimum objective function value subject to the constraint, will be the unique circle that is *tangent to the parabola*

$$y = x^2 + 2$$

At any rate, returning to the mathematical statement of the problem, we may trade our constrained problem in two variables for an unconstrained problem in a single variable, solve the constraining equation for y, and substitute that expression $y = x^2 + 2$ into the expression for f, obtaining

$$\min (x - 2)^2 + (x^2 + 2 - 3)^2 + 2$$

or

$$\min (x - 2)^2 + (x^2 - 1)^2 + 2$$

In making this substitution, *we have essentially built the constraint into the objective function* to obtain

$$f(x, y) = f(x, x^2 + 2) = w(x)$$

Differentiating w and setting the derivative to zero, we find

$$
\begin{aligned}
w'(x) &= 2(x - 2) + 2(x^2 - 1) \cdot 2x \\
&= 2x - 4 + 4x^3 - 4x = 0 \\
&= 4x^3 - 2x - 4 = 0
\end{aligned}
$$

This equation has two complex roots and one real root $x_r \cong \frac{7}{6}$ that is of interest to us. We find $w''(x_r) > 0$, which implies that x_r is indeed a relative minimum.

This procedure is quite effective in this case; but certainly one can construct many equations $g(x, y) = 0$ that are not explicitly solvable for one of the variables in terms of the other: for example,

$$e^{\sin xy^2} + y^2 x^x + \log xy^2 = 0$$

appears to offer a bit of a challenge.

At any rate, we pursue an alternative approach, which we shall see *essentially accomplishes the same goal without the need for explicitly solving the constraining equation.*

For the moment let us assume the same problem form—one constraint, two variables.

ii. *Method of Lagrange Multipliers*

Suppose that (x_0, y_0) is a relative extremum for $f(x, y)$ subject to $g(x, y) = 0$. Thus $g(x_0, y_0) = 0$, and over all points satisfying the constraint in a neighborhood of (x_0, y_0), f has its extreme value at (x_0, y_0).

Assume that g has continuous first partials in some neighborhood of (x_0, y_0), and that one of

$$\frac{\partial g(x_0, y_0)}{\partial x}, \qquad \frac{\partial g(x_0, y_0)}{\partial y}$$

is nonzero; for example, assume that

$$\frac{\partial g(x_0, y_0)}{\partial y} \neq 0$$

At this point we invoke a very powerful theorem of analysis, whose exact statement we defer temporarily. At any rate,

$$\frac{\partial g(x_0, y_0)}{\partial y} \neq 0$$

implies that there exists a differentiable function u such that $y = u(x)$ in some neighborhood of x_0; that is, for x in a neighborhood of x_0, $y = u(x)$ and $g(x, u(x)) = 0$. What we are saying is that in a neighborhood of x_0 there is a differentiable function u which allows, essentially, the solving of $g(x, y) = 0$ for y in terms of x. The theorem does not state what u is, or how to find it, only that it exists. The function u will not be utilized explicitly—its existence is all that is needed.

Writing $g(x, y) = g(x, u(x)) = G(x)$, we notice that $G(x) \equiv 0$ in a neighborhood of x_0 at least, so

$$\frac{dG}{dx} = \frac{\partial g}{\partial x} \cdot \frac{dx}{dx} + \frac{\partial g}{\partial y}\frac{du}{dx} = 0 \quad \text{at } (x_0, y_0)$$

$$= \frac{\partial g}{\partial x} + \frac{\partial g}{\partial y}\frac{du}{dx} = 0 \quad \text{at } (x_0, y_0) \tag{2.15}$$

Also, since (x_0, y_0) is a constrained extremum for $f(x, y)$, if we write $Q(x) = f(x, u(x))$, we must have

$$\frac{dQ}{dx} = 0 = \frac{\partial f}{\partial x} + \frac{\partial f}{\partial y}\frac{du}{dx} \quad \text{at } (x_0, y_0) \tag{2.16}$$

Since, by assumption, $(\partial g/\partial y)(x_0, y_0) \neq 0$, the equation

$$\frac{\partial f(x_0, y_0)}{\partial y} + \lambda\frac{\partial g(x_0, y_0)}{\partial y} = 0 \tag{2.17}$$

is solvable for λ,

$$\lambda = -\frac{\dfrac{\partial f(x_0, y_0)}{\partial y}}{\dfrac{\partial g(x_0, y_0)}{\partial y}}$$

Multiplying equation (2.15) by λ and adding it to equation (2.16) gives, using equation (2.17)

$$\frac{\partial f(x_0, y_0)}{\partial x} + \lambda\frac{\partial g(x_0, y_0)}{\partial x} = 0 \tag{2.18}$$

which, together with equation (2.17) and the fact that $g(x_0, y_0) = 0$, gives a set of three conditions that are necessary for (x_0, y_0) to be a relative extremum of $f(x, y)$ subject to the constraint $g(x, y) = 0$.

a. NECESSARY CONDITIONS AND THE LAGRANGIAN

Theorem 2.5

The following three conditions are *necessary* for the point (x_0, y_0) to solve the problem

$$\max f(x, y)$$

subject to

$$g(x, y) = 0$$
$$g(x_0, y_0) = 0$$
$$\frac{\partial f(x_0, y_0)}{\partial y} + \lambda\frac{\partial g(x_0, y_0)}{\partial y} = 0 \tag{2.19}$$
$$\frac{\partial f(x_0, y_0)}{\partial x} + \lambda\frac{\partial g(x_0, y_0)}{\partial x} = 0$$

Thus we are interested in all solutions to the system of three *equations* in the three variables x, y, and λ,

$$g(x, y) = 0$$

$$\frac{\partial f}{\partial y} + \lambda \frac{\partial g}{\partial y} = 0 \tag{2.20}$$

$$\frac{\partial f}{\partial x} + \lambda \frac{\partial g}{\partial x} = 0$$

The system (2.20) may be conveniently generated by constructing the *Lagrangian function* of three variables x, y, and λ,

$$F(x, y, \lambda) = f(x, y) + \lambda g(x, y) \tag{2.21}$$

and writing the necessary conditions for a relative extremum of the *unconstrained* function F:

$$\frac{\partial F}{\partial x} = \frac{\partial F}{\partial y} = \frac{\partial F}{\partial \lambda} = 0$$

It should be observed (and shown) that the assumption that $(\partial g/\partial x)(x_0, y_0) \neq 0$ leads also to conditions (2.19).

Furthermore equation (2.17) could as easily have been written

$$\frac{\partial f(x_0, y_0)}{\partial y} - \lambda \frac{\partial g(x_0, y_0)}{\partial y} = 0$$

in which case the system of interest would become

$$g(x, y) = 0$$

$$\frac{\partial f}{\partial y} - \lambda \frac{\partial g}{\partial y} = 0 \tag{2.22}$$

$$\frac{\partial f}{\partial x} - \lambda \frac{\partial g}{\partial x} = 0$$

which may be generated by the Lagrangian function

$$f(x, y) - \lambda g(x, y) \tag{2.23}$$

The forms (2.22) and (2.23) are to be found frequently, but we shall use here equations (2.20) and (2.21), noticing that the only difference is the sign of the Lagrange multiplier λ in a solution (x, y, λ) to equations (2.20) or (2.22).

It is emphasized that given (x_0, y_0) the value of λ is determined by equation (2.17); on the other hand, the system (2.20) is a set of necessary conditions, which will be used to *find* such points (x_0, y_0) as well as values for λ. Such values for λ will satisfy equation (2.17), of course, but will not be known in advance. The variable λ is just another variable in the system of equations (2.20). We reiterate that the conditions obtained are necessary, not sufficient.

Example 2.4

Let us approach our previous example

$$\min (x - 2)^2 + (y - 3)^2 + 2$$

subject to

$$y - x^2 - 2 = 0$$

by the Lagrange multiplier method. The Lagrangian becomes

$$F(x, y, \lambda) = (x - 2)^2 + (y - 3)^2 + 2 + \lambda(y - x^2 - 2)$$

and the necessary conditions are

$$\frac{\partial F}{\partial x} = 2(x - 2) - 2\lambda x = 0$$

$$\frac{\partial F}{\partial y} = 2(y - 3) + \lambda = 0$$

$$\frac{\partial F}{\partial \lambda} = y - x^2 - 2 = 0$$

Let us solve both the first and second equations for λ and equate the results. We obtain

$$6 - 2y = \frac{x - 2}{x} \tag{2.24}$$

Solving the third equation for y and substituting into equation (2.24), we have

$$6 - 2(x^2 + 2) = \frac{x - 2}{x}$$

which, when solved for x, yields the *same equation as when the problem was solved by direct substitution;* that is,

$$2x^3 - x - 2 = 0$$

Obtaining $x \cong \frac{7}{6}$, one finds unique values of $y \cong \frac{10}{3}$ and $\lambda \cong -\frac{2}{3}$ from the system of equations.

Thus the same answer is obtained when Lagrange's procedure is utilized, and moreover, we have obtained the same equation to solve for x as when the constraint was solved for y and substituted into $f(x, y)$. That fact is no proof, of course, but does lend credence to the claim that the Lagrange multiplier technique is, in fact, equivalent to the substitution approach.

Let us now cite the theorem that has made the derivation of our Lagrange multiplier method possible and that allows the extension of the method to the case of n variables and m constraints.

b. IMPLICIT FUNCTION THEOREM

Let $\mathbf{x}_0 = (x_1^0, x_2^0, \ldots, x_n^0)$ be a point in n-dimensional Euclidean space such that

1. The function $g_i(x_1, x_2, \ldots, x_n)$ has continuous first partials in a neighborhood of $\mathbf{x}_0$, $i = 1, \ldots, m$.

2. $g_i(\mathbf{x}_0) = 0$, $i = 1, \ldots, m$.

3. For some set of m distinct integers $\{i_1, i_2, \ldots, i_m\} \subset \{1, 2, \ldots, n\}$, the matrix

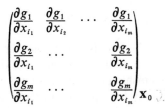

is nonsingular; that is, when each term of the array is evaluated at $\mathbf{x}_0$, the determinant is nonzero. This matrix is called the Jacobian, after the German mathematician Jacobi, of the system

$$g_i(\mathbf{x}) = 0, \qquad i = 1, \ldots, m$$

with respect to $x_{i_1}, x_{i_2}, \ldots, x_{i_m}$. For a given set of integers $i_1, i_2, \ldots, i_m$, the singularity or nonsingularity of the resulting matrix is *not* a function of the *order* of these integers, since permuting the order of the integers results only in permuting columns of the matrix. Neglecting such permutations, one can write $\binom{n}{m}$ Jacobians from the given system. More about the relative magnitudes of m, n later, but for the present consider $n > m$.

The implicit function theorem then states in part that there is an ϵ neighborhood of the point

$$\mathbf{x}_0' = (x_{j_1}^0, x_{j_2}^0, \ldots, x_{j_{n-m}}^0)\dagger \in E^{n-m}$$

$\dagger \{j_1, j_2, \ldots, j_{n-m}\} \cup \{i_1, i_2, \ldots, i_m\} = \{1, 2, \ldots, n\}$

upon which there exist differentiable functions $u_1, u_2, \ldots, u_m$, with

$$x_{i_p} = u_p(x_{j_1}, x_{j_2}, \ldots, x_{j_{n-m}}) \tag{2.25}$$

and such that for any

$$\mathbf{x}' = (x'_{j_1}, x'_{j_2}, \ldots, x'_{j_{n-m}})$$

in the ϵ neighborhood of $\mathbf{x}'_0$, the values

$$x'_{j_1}, x'_{j_2}, \ldots, x'_{j_{n-m}}$$

along with the values

$$x'_{i_p} = u_p(x'_{j_1}, x'_{j_2}, \ldots, x'_{j_{n-m}})$$

constitute a point

$$(x'_1, x'_2, \ldots, x'_n)$$

with

$$g_i(x'_1, x'_2, \ldots, x'_n) = 0, \quad i = 1, \ldots, m$$

The implicit function theorem is not simple to prove, and has only been stated here. We reiterate that no information about the functions $u_1, u_2, \ldots, u_m$ is given. But the nonvanishing of a Jacobian at a point $\mathbf{x}_0$ *guarantees* that some m of the variables may be expressed in terms of the remaining $n - m$ variables of the system everywhere within a neighborhood of the point $\mathbf{x}_0$.

In terms of our example with two variables, we have $n = 2$, $m = 1$, so the two possible Jacobians (1×1 matrices, here) are

$$\frac{\partial g(\mathbf{x}_0)}{\partial x_1}, \qquad \frac{\partial g(\mathbf{x}_0)}{\partial x_2}$$

Thus demanding that one of these be nonzero guarantees the existence of the function u in the derivation of the necessary conditions (2.19) for the case of two variables and one constraint.

c. Necessary Conditions for the General Problem

Without deriving the necessary conditions for the problem

$$\begin{array}{c} \max \\ \min \end{array} f(x_1, x_2, \ldots, x_n)$$

subject to

$$g_i(x_1, \ldots, x_n) = 0, \quad i = 1, \ldots, m$$

we state them. They are conveniently generated by writing the Lagrangian

$$f(x_1, \ldots, x_n) + \sum_{i=1}^{m} \lambda_i g_i(x_1, x_2, \ldots, x_n) = F(x_1, \ldots, x_n, \lambda_1, \ldots, \lambda_m)$$

and writing the necessary conditions for the unconstrained function F.

Theorem 2.6

A set of conditions necessary for $(x_1^0, \ldots, x_n^0) = \mathbf{x}_0$ to solve the problem

$$\begin{matrix} \max \\ \min \end{matrix} f(x_1, \ldots, x_n)$$

subject to

$$g_i(x_1, \ldots, x_n) = 0, \quad i = 1, \ldots, m$$

are as follows:

$$\frac{\partial F}{\partial x_j} = 0 = \frac{\partial f}{\partial x_j} + \sum_{i=1}^{m} \lambda_i \frac{\partial g_i}{\partial x_j}, \quad j = 1, \ldots, n$$

$$\frac{\partial F}{\partial \lambda_k} = 0 = g_k(x_1, \ldots, x_n), \quad k = 1, \ldots, m$$
(2.26)

d. FAILURE OF THE METHOD

The reader was asked to show for the case of one constraint and two variables that the form of the necessary conditions is invariant to which partial derivative is nonvanishing. The same is true for m constraints and n variables—the necessary conditions (2.26) are obtained regardless of which Jacobian(s) is (are) nonsingular.

However, if *all* the Jacobians are singular, the derivation of the necessary conditions is not valid—the procedure *may* fail.

In the trivial problem

$$\min x^2 + y^2 - 2xy$$

subject to

$$x^2 - y^2 = 0 = g(x, y)$$

an optimal solution is $x = 0$, $y = 0$ at which point

$$\frac{\partial g}{\partial x} = \frac{\partial g}{\partial y} = 0$$

However, $x = 0$, $y = 0$ is a solution to the resulting necessary conditions, so singularity of all Jacobians does not imply certain failure. On the other hand, consider the problem

$$\min 2(x + 1)^2 + 3y^2 + 2$$

subject to

$$(x - 1)^3 - y^2 = 0 \tag{2.27}$$

whose optimal solution $x = 1$, $y = 0$ may be found easily via a sketch. That solution, however, does *not* satisfy the necessary conditions, as is easily verified.

e. A Modification

What is needed then is a way to generate those points at which *all* Jacobians are singular—the Lagrange procedure given will turn up all those points at which at least one Jacobian is nonsingular. The global extremum, if there is one, must lie within one of these two sets.

We know that (see Chapter 1) if A is a square matrix and is nonsingular (has an inverse) then the system $A\mathbf{x} = \mathbf{0}$ has only the trivial solution. Thus if $A\mathbf{x} = \mathbf{0}$ has a nontrivial solution, A is singular.

Thus if we can find a solution

$$(\lambda_1^0, \lambda_2^0, \ldots, \lambda_m^0, x_1^0, x_2^0, \ldots, x_n^0) = (\boldsymbol{\lambda}^0, \mathbf{x}^0)$$

in which $\boldsymbol{\lambda}^0 \neq \mathbf{0}$ for the system

$$(\lambda_1, \lambda_2, \ldots, \lambda_m) \begin{vmatrix} \dfrac{\partial g_1}{\partial x_{i_1}} & \dfrac{\partial g_1}{\partial x_{i_2}} & \cdots & \dfrac{\partial g_1}{\partial x_{i_m}} \\ \dfrac{\partial g_2}{\partial x_{i_1}} & \cdots & & \dfrac{\partial g_2}{\partial x_{i_m}} \\ \dfrac{\partial g_m}{\partial x_{i_1}} & \cdots & & \dfrac{\partial g_m}{\partial x_{i_m}} \end{vmatrix} = \underbrace{(0, \ldots, 0)}_{m}$$

we have a point $\mathbf{x}^0$ at which that particular Jacobian is singular.

But we are interested in points where *all* the Jacobians are singular; that is, the integers $\{i_1, \ldots, i_m\}$ may be *any* integers from $\{1, 2, \ldots, n\}$. Thus we want solutions $(\boldsymbol{\lambda}^0, \mathbf{x}^0)$ with $\boldsymbol{\lambda}^0 \neq \mathbf{0}$ for the system

$$(\lambda_1, \ldots, \lambda_m) \begin{vmatrix} \dfrac{\partial g_1}{\partial x_1} & \dfrac{\partial g_1}{\partial x_2} & \cdots & \dfrac{\partial g_1}{\partial x_n} \\ \dfrac{\partial g_2}{\partial x_1} & \dfrac{\partial g_2}{\partial x_2} & \cdots & \dfrac{\partial g_2}{\partial x_n} \\ \vdots & & & \vdots \\ \dfrac{\partial g_m}{\partial x_1} & \dfrac{\partial g_m}{\partial x_2} & \cdots & \dfrac{\partial g_m}{\partial x_n} \end{vmatrix} = \underbrace{(0, 0, \ldots, 0)}_{n} \tag{2.28}$$

These and only these solutions provide points $\mathbf{x}^0$ where all Jacobians are singular.

Performing the indicated multiplications in equation (2.28), we see that the *system is precisely the set of necessary conditions that would be generated by the Lagrange procedure if the function f were identically zero.*

While $f \equiv 0$ does not make much sense in an optimization problem, it does suggest a simple way to think of generating *both systems* necessary for finding *all* the points that must be examined. Writing the Lagrangian

$$F(x_1, x_2, \ldots, x_n, \lambda_0, \lambda_1, \ldots, \lambda_m)$$
$$= \lambda_0 f(x_1, \ldots, x_n) + \sum_{i=1}^{m} \lambda_i g_i(x_1, \ldots, x_n)$$

let $\lambda_0 = 1$, and generate the necessary conditions for F. This gives the usual necessary conditions (2.26); then let $\lambda_0 = 0$, and write the necessary conditions for F. This gives the system (2.28).

Example 2.5

Let us apply this procedure to the problem [see (2.27)]

$$\min 2(x + 1)^2 + 3y^2 + 2$$

subject to

$$(x - 1)^3 - y^2 = 0$$

The Lagrangian function is written as

$$F(x, y, \lambda_0, \lambda_1) = \lambda_0(2(x + 1)^2 + 3y^2 + 2) + \lambda_1((x - 1)^3 - y^2)$$

For $\lambda_0 = 1$, we obtain the necessary conditions

$$4(x + 1) + 3\lambda_1(x - 1)^2 = 0$$
$$6y - 2\lambda_1 y = 0$$
$$(x - 1)^3 - y^2 = 0$$

The optimal solution $x = 1, y = 0$ previously obtained does not satisfy these equations. On the other hand, setting $\lambda_0 = 0$, we obtain the system

$$3\lambda_1(x - 1)^2 = 0$$
$$-2\lambda_1 y = 0$$
$$(x - 1)^3 - y^2 = 0$$

and $x = 1, y = 0$ does indeed solve the system.

f. RELATIVE MAGNITUDES OF m, n

When we consider the system of equations

$$g_i(x_1, \ldots, x_n) = 0, \quad i = 1, \ldots, m$$

what shall we say about the relative magnitudes of m, n?

For illustrative purposes, suppose that $n = 3$, and that

$$g_i(x_1, x_2, x_3) = a_{i1}x_1 + a_{i2}x_2 + a_{i3}x_3, \quad \text{all } i$$

so that the set of points satisfying $g_i(x_1, x_2, x_3) = 0$ is a plane in E^3. If $m = 1$, the constraint set is all the points on such a plane. If $m = 2$, the constraint set consists of the *intersection* of two such planes; the set may be null (*if the two equations are inconsistent*, which probably indicates the problem has not been properly formulated); it may be the same plane again (if one equation is simply a multiple of the other, which would indicate *redundant* constraints); most likely, the set will be a *line* in E^3. Assuming this "most likely" situation for two planes, suppose that we add a third constraint, whereupon the constraint set may again be a line (if the former line lay in the new plane, which again implies redundant constraints); it may be null; or, it may be a *single point.*

To summarize, if $m = 3$ and the constraints are independent, the constraint set is either null or consists of a single point. If nonempty, the constraint set will not offer much in the way of an optimization problem, since there is but a single point over which to optimize. If we continue to add linear constraints beyond three, the best we can do is maintain a single point, as we continue to form intersections. These constraint sets, also, are uninteresting. To create an interesting optimization problem, that is, to maintain a constraint set with an infinite number of elements, we shall need to have either one or two of our linear equations. (The reader will recall Sections 1.4-ii and 1.4-iii.) Similar reasoning applies to surfaces of higher dimension, and usually one has $m < n$ in problems of real interest.

Example 2.6

As another example, consider

$$\min x^3 + y^3 - 9xy$$

subject to

$$y - x = 0$$

The Lagrangian becomes

$$x^3 + y^3 - 9xy + \lambda(y - x),$$

and the necessary conditions

$$3x^2 - 9y - \lambda = 0$$
$$-9x + 3y^2 + \lambda = 0$$
$$-x + y = 0$$

There are two solutions to this system:

$$x = 0, \quad y = 0, \quad \lambda = 0$$

and $$x = 3, \quad y = 3, \quad \lambda = 0$$

It is clear by substitution and comparison of values that the solution to our problem is $x = 3$, $y = 3$. Since the solution—if there is a global extremum—must lie among the solutions of the necessary conditions, evaluating all possible solutions is effective. (Why may $\lambda_0 = 0$ be ignored here?)

g. Sufficiency Test

A more sophisticated technique—not always more desirable—is available in the form of a sufficient condition for local constrained extrema, which we shall state here in its general form.

For the problem

$$\begin{matrix} \max \\ \min \end{matrix} f(\mathbf{x})$$

subject to

$$g_i(\mathbf{x}) = 0, \quad i = 1, \ldots, m$$

let $(\mathbf{x}_0, \boldsymbol{\lambda}_0)$ be a point that satisfies the necessary conditions, and compute the determinant of an array analogous to that encountered in the unconstrained case. We solve the $(n - m)$-degree polynomial equation in c,

$$\det \begin{pmatrix}
\begin{array}{ccc|c}
F_{11}(\mathbf{x}_0, \boldsymbol{\lambda}_0) - c & F_{12}(\mathbf{x}_0, \boldsymbol{\lambda}_0) & \cdots & F_{1n}(\mathbf{x}_0, \boldsymbol{\lambda}_0) & \\
F_{21}(\mathbf{x}_0, \boldsymbol{\lambda}_0) & F_{22}(\mathbf{x}_0, \boldsymbol{\lambda}_0) - c & \cdots & F_{2n}(\mathbf{x}_0, \boldsymbol{\lambda}_0) & \\
& & & & G(\mathbf{x}_0) \\
\vdots & & & & \\
F_{n1}(\mathbf{x}_0, \boldsymbol{\lambda}_0) & \cdots & & F_{nn}(\mathbf{x}_0, \boldsymbol{\lambda}_0) - c & \\
\hline
& G^T(\mathbf{x}_0) & & & 0\, m \times m
\end{array}
\end{pmatrix} = 0†$$

(2.29)

---

†Partials of F with respect to $x_1, \ldots, x_n$ only.

where

$$G(\mathbf{x}_0) = \begin{pmatrix} \dfrac{\partial g_1(\mathbf{x}_0)}{\partial x_1} & \dfrac{\partial g_2(\mathbf{x}_0)}{\partial x_1} & \cdots & \dfrac{\partial g_m(\mathbf{x}_0)}{\partial x_1} \\ \cdot & \cdot & & \cdot \\ \cdot & \cdot & & \cdot \\ \cdot & \cdot & & \cdot \\ \dfrac{\partial g_1(\mathbf{x}_0)}{\partial x_n} & \dfrac{\partial g_2(\mathbf{x}_0)}{\partial x_n} & \cdots & \dfrac{\partial g_m(\mathbf{x}_0)}{\partial x_n} \end{pmatrix}$$

The conclusions, then, regarding the nature of $\mathbf{x}_0$ depend upon the signs of the $n - m$ roots c of equation (2.29), exactly as the signs of the eigenvalues determined the nature of a critical point in the unconstrained case.

Example 2.7

Let us treat the two solutions to Example 2.6 in this way. The array of interest becomes

$$\begin{pmatrix} 6x - c & -9 & \vdots & -1 \\ -9 & 6y - c & \vdots & 1 \\ \hdashline -1 & 1 & \vdots & 0 \end{pmatrix}$$

For $x = 0$, $y = 0$, $\lambda = 0$, the equation to be solved is

$$2c + 18 = 0$$

whence $c = -9$, and $x = 0$, $y = 0$ is a relative maximum.

For $x = 3$, $y = 3$, $\lambda = 0$, the equation is

$$2c - 18 = 0$$

giving $c = 9$, and, as we know, $x = 3$, $y = 3$ is a relative minimum for the constrained problem. The weakness of our sufficiency test is apparent; we still encounter inconclusive cases. Furthermore, one now needs to compute the determinant of $(n + m) \times (n + m)$ matrices, and, finally, we still have a test for relative extrema only; so we would be better advised to merely evaluate f at the points we obtain from solving the necessary conditions.

The example problem is further interesting owing to the fact that, as it turned out, the problem could have been solved by *neglecting* the constraint. (Verify.) The constraint did not actually do any constraining. Simultaneously, we have, corresponding to the optimal x, y values, the value $\lambda = 0$; and when $\lambda = 0$, the Lagrangian function is exactly the given objective function f, evidenced by equation (2.21).

We next explore in greater detail the significance of the Lagrange multipliers, including the events of the foregoing paragraph.

h. Interpreting the Lagrange Multipliers—A Proof

The Lagrange multipliers in our constrained optimization problems may appear to be superfluous entities, since obtaining values of the decision variables $x_1, x_2, \ldots, x_n$ that optimize some function is the actual aim of the optimization problem. It is the case, however, that the multipliers possess properties that are at once both theoretically interesting and of practical use.

Previously, any constants appearing in our constraints were incorporated into the functions g_i. That is, given a constraint

$$h_i(x_1, \ldots, x_n) = b_i^0$$

a constant, we defined

$$g_i = h_i - b_i^0$$

and treated the constraint

$$g_i(x_1, \ldots, x_n) = 0$$

For the present, we would like to utilize all these given constants (which may be zero, anyway) and consider the problem to be given as

$$\max f(x_1, \ldots, x_n)$$

subject to   (2.30)

$$h_i(x_1, \ldots, x_n) = b_i^0, \quad i = 1, \ldots, m$$

The Lagrangian function is then

$$F(x_1, \ldots, x_n, \lambda_1, \ldots, \lambda_m) = f(x_1, \ldots, x_n) + \sum_{i=1}^{m} \lambda_i [h_i(x_1, \ldots, x_n) - b_i^0]$$

Suppose that $x_1^*, \ldots, x_n^*$ is found to be optimal for the problem and that the corresponding values of the multipliers are $\lambda_1^*, \ldots, \lambda_m^*$. Then we have

$$F(x_1^*, \ldots, x_n^*, \lambda_1^*, \ldots, \lambda_m^*) = f(x_1^*, \ldots, x_n^*) + \sum_{i=1}^{m} \lambda_i^* [h_i^*(x_1^*, \ldots, x_n^*) - b_i^0] \dagger$$

In general, of course, the values $x_1^*, \ldots, x_n^*, \lambda_1^*, \ldots, \lambda_m^*$ depend upon the values $b_1^0, b_2^0, \ldots, b_m^0$. We would like next to allow b_i^0 to vary continuously

---

†Notice that for any feasible $(x_1, \ldots, x_n)$, including $(x_1^*, \ldots, x_n^*)$, the value of F is identical to the value of f, *since the constraints are satisfied.*

over some neighborhood that contains b_i^0, $i = 1, \ldots, m$. The "variable" constant is denoted b_i.

Rather than mere dependency, we shall make *the much stronger assumption that each x_j^* and λ_i^* is a differentiable function of the variables $b_1, \ldots, b_m$.* To explain more fully, as the b_i vary, we create new optimization problems, which have optimal solutions and associated multipliers

$$x_j^*(b_1, b_2, \ldots, b_m), \quad j = 1, \ldots, n$$
$$\lambda_i^*(b_1, b_2, \ldots, b_m), \quad i = 1, \ldots, m$$

We are further assuming that *there exist differentiable functions α_j, β_i* such that

$$x_j^* = \alpha_j(b_1, \ldots, b_m), \quad j = 1, \ldots, n$$
$$\lambda_i^* = \beta_i(b_1, \ldots, b_m), \quad i = 1, \ldots, m$$

hold for $(b_1, \ldots, b_m)$ in some neighborhood of $(b_1^0, \ldots, b_m^0)$.

Application of the implicit function theorem would give a condition sufficient for the existence of the α_j, β_i, but we omit this. Although our assumption is *not* always true, it is frequently satisfied and will allow the deduction of a worthwhile result.

If we write

$$z^* = f(x_1^*, x_2^*, \ldots, x_n^*)$$
$$= f(\alpha_1(b_1, \ldots, b_m), \alpha_2(b_1, \ldots, b_m), \ldots, \alpha_n(b_1, \ldots, b_m))$$

and use the facts that f is a differentiable function and $\alpha_1, \ldots, \alpha_n$ are differentiable functions, then the value, z^*, of the objective function for an optimal solution is a differentiable function of $b_1, \ldots, b_m$. Thus it is true that

$$\frac{\partial z^*}{\partial b_i} = \sum_{j=1}^{n} \frac{\partial f}{\partial x_j^*} \cdot \frac{\partial x_j^*}{\partial b_i} \tag{2.31}$$

at least in some neighborhood of $(b_1^0, \ldots, b_m^0)$. We also have $h_i(x_1^*, \ldots, x_n^*) = b_i$, so

$$\frac{\partial h_r}{\partial b_i} = \sum_{j=1}^{n} \frac{\partial h_r}{\partial x_j^*} \frac{\partial x_j^*}{\partial b_i} = \begin{cases} 1 & \text{if } i = r \\ 0 & \text{if } i \neq r \end{cases} \tag{2.32}$$

Multiplying in equation (2.32) by λ_r^* and summing over all $r = 1, \ldots, m$, we obtain

$$\sum_{r=1}^{m} \lambda_r^* \left(\sum_{j=1}^{n} \frac{\partial h_r}{\partial x_j^*} \cdot \frac{\partial x_j^*}{\partial b_i} \right) = \lambda_i^* \tag{2.33}$$

Adding equations (2.33) and (2.31),

$$\frac{\partial z^*}{\partial b_i} = \sum_{j=1}^{n} \frac{\partial f}{\partial x_j^*} \cdot \frac{\partial x_j^*}{\partial b_i} + \sum_{r=1}^{m} \lambda_r^* \left(\sum_{j=1}^{n} \frac{\partial h_r}{\partial x_j^*} \cdot \frac{\partial x_j^*}{\partial b_i} \right) - \lambda_i^*$$

$$= \sum_{j=1}^{n} \left[\frac{\partial f}{\partial x_j^*} + \sum_{r=1}^{m} \lambda_r^* \frac{\partial h_r}{\partial x_j^*} \right] \frac{\partial x_j^*}{\partial b_i} - \lambda_i^* \qquad (2.34)$$

But the bracketed term in equation (2.34) is zero, since $x_1^*, \ldots, x_n^*$, $\lambda_1^*, \ldots, \lambda_m^*$ satisfy the necessary conditions (the partial derivatives of constant terms in the constraints are zero); so we have

$$\frac{\partial z^*}{\partial b_i} = -\lambda_i^*$$

which is the result desired.

i. ECONOMIC INTERPRETATION OF THE LAGRANGE MULTIPLIERS

Thus we see that the rate of change of optimal objective function value with respect to the ith constraint constant is exactly the negative of the optimal ith multiplier.

The result allows one to predict the effect on optimal objective function values to problems which are perturbed versions of the original, that is, ones in which one of the constraint constants is varied slightly, and such knowledge can be of use.

For example, suppose that a minimization problem with a constraint

$$h_i(x_1, x_2, \ldots, x_n) = b_i$$

has been solved and z^* obtained as the minimum objective function value. Suppose also that $\lambda_i^* < 0$.

Consider the same problem with the ith constraint replaced by

$$h_i(x_1, \ldots, x_n) = b_i' > b_i$$

and where b_i' is but slightly larger than b_i. We know in advance, since $-\lambda_i^* > 0$, that the optimal objective function value for the new problem will be *greater* than that for the original.

Suppose that b_i represents a quantity of some expendable resource, all of which must be used in some resource allocation problem that is solved from time to time to produce an optimal operating policy of some sort. The effects of having available a little more or less of the resource would be known in advance and would be of use in planning for the future.

We emphasize that strong assumptions were needed to establish

$$\frac{\partial z^*}{\partial b_i} = -\lambda_i^*$$

and, furthermore, that the validity of the result is known only in some neighborhood of b_i; thus if the constant is varied a substantial amount, the conclusion drawn regarding z^* for the new problem may be entirely false. Even if true for a small neighborhood only, the result is of considerable theoretical importance, as we shall see.

Example 2.8

Consider the simple example

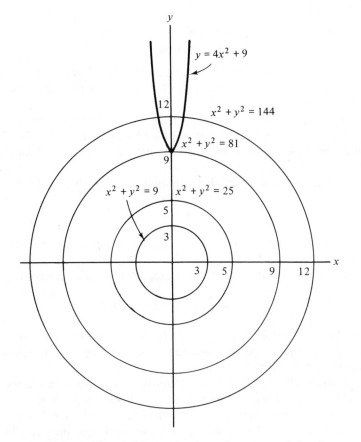

Figure 2.11. The problem: min $x^2 + y^2$, s.t. $y - 4x^2 = 9$.

$$\min x^2 + y^2$$

subject to

$$y - 4x^2 = 9$$

An optimal solution is easily found to be

$$x^* = 0, \qquad y^* = 9, \qquad \lambda^* = -18$$

so that by slightly increasing the constraint constant $b = 9$ and solving the new problem we know the resulting minimum will be *larger* than that of the given problem.

The validity of the prediction is easily verified geometrically (see Figure 2.11). We resort to a contour map, which allows representation of $z = x^2 + y^2$ in the plane for various values of z, resulting in a family of concentric circles with center at the origin and radius $\sqrt{z}$. The solution to the problem is then clearly that point on the parabola $y = 4x^2 + 9$ which lies on the smallest-valued level contour. Hence $x = 0$, $y = 9$ is the optimal solution. Increasing the constraint constant, then, amounts to translating the parabola farther up the y axis, and hence the minimum increases.

Example 2.9

It is easy to construct a geometric example where large changes in the constraint constant do not produce the result guaranteed for small changes.

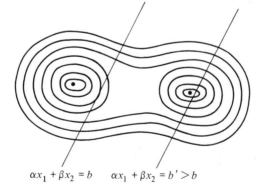

$$\alpha x_1 + \beta x_2 = b \qquad \alpha x_1 + \beta x_2 = b' > b$$

Figure 2.12. Contour map for a function with two relative minima subject to a linear constraint.

In Figure 2.12 we graph a function that we wish to minimize subject to a linear constraint and that has two relative minima with the same function value, with the direction of increasing b as indicated.

One need not always consider the effects of slight *increases* in b_i. He may be interested in the effects of decreasing b_i, which are also clear from the properties of λ_i^*. Certainly, for example, if z^* tends to increase at b_i, then decreasing b_i will produce a decrease in z^*.

j. ZERO MULTIPLIERS AND SADDLE POINTS

Of special concern is the case where $\lambda_i^* = 0$, from which we deduce that at b_i the rate of change of the optimal objective function value as a function of the constraint constant is zero.

It is possible to obtain an interesting implication of $\lambda_i^* = 0$ if a rather strong assumption regarding the nature of the Lagrangian is valid.

For a general Lagrangian function $F(\mathbf{x}, \boldsymbol{\lambda})$, F is said to have a *saddle point* at $(\mathbf{x}^*, \boldsymbol{\lambda}^*)$ if there is a neighborhood of $(\mathbf{x}^*, \boldsymbol{\lambda}^*)$ such that

$$F(\mathbf{x}, \boldsymbol{\lambda}^*) \leq F(\mathbf{x}^*, \boldsymbol{\lambda}^*) \leq F(\mathbf{x}^*, \boldsymbol{\lambda}) \qquad (2.35)$$

for all $(\mathbf{x}, \boldsymbol{\lambda})$ in the neighborhood.

If inequality (2.35) holds for *all* points $(\mathbf{x}, \boldsymbol{\lambda})$, then $(\mathbf{x}^*, \boldsymbol{\lambda}^*)$ is called a *global* saddle point of F.

The terminology saddle point is derived from the appearance of such a function. For $\mathbf{x}$, $\boldsymbol{\lambda}$ each of dimension 1, a portion of the corresponding surface will appear similar to Figure 2.13.

It is sometimes the case (see the problems) that the point $(\mathbf{x}^*, \boldsymbol{\lambda}^*)$, which provides the global max of our equality constrained optimization problem (2.30), is a global saddle point for the Lagrangian $F(\mathbf{x}, \boldsymbol{\lambda})$.

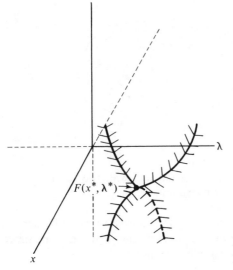

Figure 2.13. Example of a saddle-point.

Suppose that this is true, and return to the situation where $\lambda_i^* = 0$. What we would like to show is that the *i*th constraint *could have been neglected*, and that an optimal solution to the problem created would be an optimal solution to the original problem.

That is, if $\mathbf{x}^*$ is optimal for

$$\max f(\mathbf{x})$$

subject to   (2.36)

$$h_i(\mathbf{x}) = b_i, \quad i = 1, \ldots, m$$

and the corresponding $\boldsymbol{\lambda}^*$ has $\lambda_k^* = 0, k \in K \subseteq \{1, \ldots, m\}$, then the problem

$$\max f(\mathbf{x})$$

subject to   (2.37)

$$h_i(\mathbf{x}) = b_i, \quad i \notin K$$

has $\mathbf{x}^*$ as an optimal solution.

Let us suppose not. Suppose that $\mathbf{x}'$ is optimal for (2.37). It must be that $\mathbf{x}'$ violates at least one of the constraints $h_i(\mathbf{x}) = b_i$ for $i \in K$, for, otherwise, we would not find a different optimum. Furthermore, it must be that $f(\mathbf{x}') > f(\mathbf{x}^*)$; otherwise, $\mathbf{x}^*$ would be optimal for problem (2.37). Now

$$F(\mathbf{x}', \boldsymbol{\lambda}^*) = f(\mathbf{x}') + \sum_{i=1}^{m} \lambda_i^*[h_i(\mathbf{x}') - b_i] = f(\mathbf{x}') \qquad (2.38)$$

since any constraints k violated by $\mathbf{x}'$ have also $\lambda_k^* = 0$.

We also know that $F(\mathbf{x}^*, \boldsymbol{\lambda}^*) = f(\mathbf{x}^*)$ so equation (2.38) and $f(\mathbf{x}') > f(\mathbf{x}^*)$ give $F(\mathbf{x}', \boldsymbol{\lambda}^*) > F(\mathbf{x}^*, \boldsymbol{\lambda}^*)$.

But $(\mathbf{x}^*, \boldsymbol{\lambda}^*)$ is a global saddle point for F; so by inequality (2.35) it must be that $F(\mathbf{x}', \boldsymbol{\lambda}^*) \leq F(\mathbf{x}^*, \boldsymbol{\lambda}^*)$, and the contradiction implies that $\mathbf{x}^*$ is optimal for problem (2.37). Thus the constraints corresponding to $\lambda_k^* = 0$ could have been omitted, provided the assumptions regarding the saddle point were valid.

We emphasize that this interpretation of a zero multiplier required a strong assumption. Continue to imagine that assumption valid. Physically, what does it mean that a constraint could have been neglected? The meaning is that the other constraints *implied* the one in question. If there are no *other* constraints besides the ones with zero multipliers, it means that the unconstrained extremum happened to satisfy all the constraints.

We *may* then realize a zero multiplier when there are redundant constraints, but not necessarily.

Example 2.10

Consider the problem

$$\min x_1^2 + x_2^2 + x_3^2$$

subject to

$$2x_1 - x_2 - 2x_3 = 4$$
$$4x_1 - 2x_2 - 4x_3 = 8$$

which has one constraint a multiple of the other, and whose necessary conditions, being redundant, constrain only the sum $\lambda_1 + 2\lambda_2$. All points $\mathbf{x}$ that solve the necessary conditions are expressed in terms of $\lambda_1 + 2\lambda_2$. Hence, depending upon what value we assign one of the multipliers, there may be no zero multipliers.

Example 2.11

Another easily understood problem is

$$\min x_1^2 + x_2^2 + x_3^2$$

subject to

$$x_1 + x_2 + x_3 = 0$$
$$2x_1 - x_2 - 2x_3 = 0$$

The necessary conditions are satisfied *only* when all variables are zero. The optimal solution for the unconstrained problem

$$\min x_1^2 + x_2^2 + x_3^2$$

is $x_1 = 0 = x_2 = x_3$. Clearly, it satisfies both constraints. Thus we conclude that both constraints were unnecessary.

2.5. Constrained Optimization—Inequality Constraints

The subject of the present section is the problem

$$\max f(\mathbf{x})$$

subject to   (2.39)

$$g_i(\mathbf{x}) \leq 0, \quad i = 1, \ldots, m$$
$$\mathbf{x} \geq \mathbf{0}$$

which is called the general *mathematical programming problem* or, equivalently, the *general mathematical program*.

It is in terms of tangible problems that we justify the inclusion of nonnegativity constraints on the variables. In real world problems where the decision variables are tangible quantities, $\mathbf{x} \geq \mathbf{0}$ is often the only possibility. Since those constraints are of the same form as the others in (2.39), though, their presence or absence is not significant. It is in these terms also that we justify the importance of inequality constrained problems. Constraints placed upon optimization problems are obviously more likely to be of the form "the amount of money spent on activity i must be no more than c dollars," or "machine A is available up to k hours per week," or "during the next production period we must produce at least d units to satisfy anticipated demand," or "the reliability of the component must be at least p" rather than their respective equation counterparts, "exactly c dollars must be spent on activity i," or "machine A must be used exactly k hours each week," and so on.

Problems in which we find

$$g_i(\mathbf{x}) \geq 0$$

may be cast in the desired form by multiplying the constraint by -1.

An equality constraint of the form

$$g_i(\mathbf{x}) = 0$$

may easily be converted to a set of inequaltiy constraints by noting that

$$g_i(\mathbf{x}) = 0 \quad \text{if and only if} \quad g_i(\mathbf{x}) \geq 0 \quad \text{and} \quad g_i(\mathbf{x}) \leq 0$$

and then treating the first of the two in the above fashion.

Thus the formulation (2.39) is as general as any other.

Example 2.12

Assume that nonnegativity constraints are the only constraints and consider the problem

$$\max f(\mathbf{x})$$

subject to

$$x_1, x_2 \geq 0$$

Employing a contour diagram, suppose that we have f as in Figure 2.14, and that f has continuous first partial derivatives.

The way the diagram has been drawn, had we neglected the nonnegativity constraints we would find the relative maxima $\mathbf{p}_0, \mathbf{p}_1, \mathbf{p}_2, \mathbf{p}_3$, only one of which satisfies $x_1, x_2 \geq 0$.

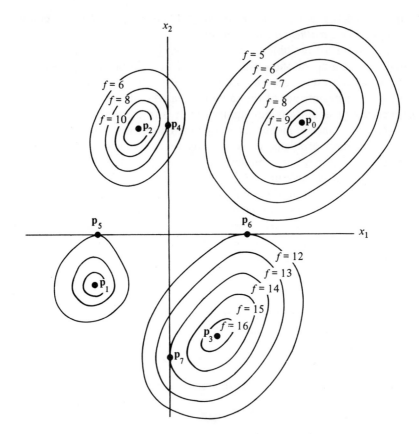

Figure 2.14. Contour diagram for Example 2.12.

If we decided that $\mathbf{p}_0$ were the optimal solution to the problem, however, we would be quite mistaken. The reason is the same as when we found points on a closed interval at which the derivative of a function of one variable vanished—if the global optimum occurs at an end point of the interval, it may very well be that the derivative does not vanish there. For functions of several variables with nonnegativity constraints, it may be that a global extremum is a point where not all the first partials vanish, if that extremum is a boundary point.

For functions of one variable it was necessary to evaluate the function at the end points of the interval. Again, for functions of several variables the procedure for checking the boundary is entirely analogous, though more involved.

To observe the behavior of f on the boundary in our present example, we need to examine $f(0, x_2)$ and $f(x_1, 0)$. From the drawing again, $f(0, x_2)$ produces the relative maxima $\mathbf{p}_7, \mathbf{p}_4$, while $f(x_1, 0)$ yields $\mathbf{p}_5, \mathbf{p}_6$. These four points are discovered by differentiating the functions of one variable. Finally, one computes $f(0, 0)$, since $(0, 0)$ may be a global extremum with neither first partial zero. That feasible point which provides the best function value from all those maxima produced is, then, the optimal solution to the problem.

i. *Method of Solving a Problem Subject Only to Nonnegativity Constraints*

Consider first the problem

$$\max f(x_1, x_2, x_3)$$

subject to

$$x_1, x_2, x_3 \geq 0$$

To solve this problem, we may use the following step-by-step procedure.

Step 1. Find maxima for $f(x_1, x_2, x_3)$.
Step 2. Find maxima for $f(x_1, 0, 0), f(0, x_2, 0), f(0, 0, x_3)$.
Step 3. Find maxima for $f(x_1, x_2, 0), f(x_1, 0, x_3), f(0, x_2, x_3)$.
Step 4. Evaluate $f(0, 0, 0)$.

Step 3 involves examining those portions of the boundary where exactly one variable is zero, while step 2 involves those portions where exactly two variables are zero.

Clearly, the work involved in examining the behavior of the function on the boundary increases *combinatorially* with the number of variables. In general, for the n variable maximization problem

$$\max f(x_1, x_2, \ldots, x_n)$$

subject to

$$x_1, x_2, \ldots, x_n \geq 0$$

the procedure would involve the following steps:

Step 1. Find maxima of $f(x_1, x_2, \ldots, x_n)$.
Step 2. Find the maxima of each of the n functions $f(x_1, 0, \ldots, 0)$, $f(0, x_2, \ldots, 0), \ldots, f(0, 0, \ldots, x_n)$.
Step 3. Find the maxima of each of the $\binom{n}{2}$ functions $f(x_1, x_2, 0, \ldots, 0)$, $f(x_1, 0, x_3, 0, \ldots, 0), \ldots, f(0, 0, \ldots, 0, x_{n-1}, x_n)$

$$\cdots$$

Step $k + 1$. Find the maxima of each of the $\binom{n}{k}$ functions $f(x_1, x_2, \ldots, x_k, 0, 0, \ldots, 0)$, and so on.

$$\cdots$$

Step $n + 1$. Compute $f(0, 0, \ldots, 0)$.

When we consider that solving the necessary conditions for each of steps 1 through n involves in general the solution of a system of several nonlinear equations in a like number of variables, we see that the procedure may require an extraordinary amount of work.

a. Extension of the Procedure When Equality Constraints Are Added

Next, we would like to investigate a comparable procedure for locating global optima of nonnegativity constrained problems, which have also the usual equality constraints. The procedure would be essentially the same, but when the equality constraints are also present, the combination of variables which are set equal to zero in the objective function must also be made zero in the constraints. Thus each of the family of smaller problems is also an equality-constrained problem, albeit of smaller dimension than the original. But, for the same reasons as given in the discussion of the relative magnitudes of m and n, we would not set to zero any combination of more than $n - m$ variables.

Example 2.13

Given the problem

$$\max f(x_1, x_2, x_3)$$

subject to

$$g(x_1, x_2, x_3) = 0$$
$$x_1, x_2, x_3 \geq 0$$

The subproblems solved would be

$\max f(x_1, x_2, x_3)$	$\max f(0, x_2, x_3)$
subject to $g(x_1, x_2, x_3) = 0$	subject to $g(0, x_2, x_3) = 0$
$\max f(x_1, 0, x_3)$	$\max f(x_1, x_2, 0)$
subject to $g(x_1, 0, x_3) = 0$	subject to $g(x_1, x_2, 0) = 0$

plus all problems resulting when two variables are set equal to zero, a pair at a time, and, finally, when all variables are set equal to zero, after which the process terminates, if, in fact, the origin satisfies the constraint.

Clearly, this kind of approach is highly undesirable from a computational point of view.

ii. *Necessary Conditions for the General Problem—* *The Kuhn–Tucker Conditions*

We wish to consider the problem without nonnegativity constraints and where constants $b_1, \ldots, b_m$ are left in the constraint expressions

$$\max f(x_1, \ldots, x_n)$$

subject to   (2.40)

$$h_i(x_1, \ldots, x_n) \leq b_i, \quad i = 1, \ldots, m$$

(Clearly, nonnegativity constraints are of this form also, so they are not being prohibited—just not explicitly written.)

We can obtain an equivalent, equality-constrained problem for which we know how to write necessary conditions.

Let $x_1^0, \ldots, x_n^0$ satisfy the constraints of problem (2.40). Then there exist unique *nonnegative* values that we shall write as

$$(z_1^0)^2, \ldots, (z_m^0)^2$$

such that

$$h_i(x_1^0, \ldots, x_n^0) + (z_i^0)^2 = b_i, \quad i = 1, \ldots, m$$

Conversely, if

$$x_1^0, \ldots, x_n^0$$

and nonnegative values

$$(z_1^0)^2, \ldots, (z_m^0)^2$$

satisfy $h_i(x_1^0, \ldots, x_n^0) + (z_i^0)^2 = b_i, \quad i = 1, \ldots, m$

then $h_i(x_1^0, \ldots, x_n^0) \leq b_i, \quad i = 1, \ldots, m$

A simple proof by contradiction will then show that if

$$(x_1^*, \ldots, x_n^*, z_1^*, \ldots, z_m^*)$$

is optimal for

$$\max f(x_1, \ldots, x_n)$$

subject to   (2.41)

$$h_i(x_1, \ldots, x_n) + z_i^2 = b_i, \quad i = 1, \ldots, m$$

then $(x_1^*, \ldots, x_n^*)$ is optimal for problem (2.40). The argument is omitted.

Since problem (2.41) is constrained by equalities, we turn our attention to that problem.

The Lagrangian function for problem (2.41) is

$$F(x_1, \ldots, x_n, z_1, \ldots, z_m, \lambda_1, \ldots, \lambda_m)$$
$$= f(x_1, \ldots, x_n) + \sum_{i=1}^{m} \lambda_i[h_i(x_1, \ldots, x_n) + z_i^2 - b_i]$$

and the necessary conditions are then

$$\frac{\partial F}{\partial x_j} = \frac{\partial f}{\partial x_j} + \sum_{i=1}^{m} \lambda_i \frac{\partial h_i}{\partial x_j} = 0, \quad j = 1, \ldots, n \tag{2.42}$$

$$\frac{\partial F}{\partial z_i} = 2\lambda_i z_i = 0, \quad i = 1, \ldots, m \tag{2.43}$$

$$\frac{\partial F}{\partial \lambda_i} = h_i(x_1, \ldots, x_n) + z_i^2 - b_i = 0, \quad i = 1, \ldots, m \tag{2.44}$$

From equations (2.43) if $\lambda_i \neq 0$, $z_i = 0$, and from equations (2.44) we then obtain $h_i(x_1, \ldots, x_n) = b_i$.

We may combine equations (2.43) and (2.44) and write the equivalent system

$$\lambda_i[h_i(x_1, \ldots, x_n) - b_i] = 0, \quad i = 1, \ldots, m \tag{2.45}$$

We have already seen that if equations (2.43) and (2.44) are satisfied so are equations (2.45). Conversely, it is easily seen that if equations (2.45) are satisfied so are *both* equations (2.43) and (2.44).

Thus an equivalent set of necessary conditions is

$$\frac{\partial f}{\partial x_j} + \sum_{i=1}^{m} \lambda_i \frac{\partial h_i}{\partial x_j} = 0, \quad j = 1, \ldots, n$$
$$\lambda_i[h_i(x_1, \ldots, x_n) - b_i] = 0, \quad i = 1, \ldots, m \tag{2.46}$$
$$h_i(x_1, \ldots, x_n) \leq b_i, \quad i = 1, \ldots, m$$

There is one more important set of relationships that must be added to system (2.46), however. Using the result

$$\frac{\partial z^*}{\partial b_i} = -\lambda_i^*$$

[which, recall, required the strong assumption that $\mathbf{x}^* = (x_1^*, x_2^*, \ldots, x_n^*)$ and $\boldsymbol{\lambda}^* = (\lambda_1^*, \lambda_2^*, \ldots, \lambda_m^*)$ be differentiable functions of $(b_1, \ldots, b_m)$], suppose that $\lambda_k^* > 0$. The implication then is that if $b_1, \ldots, b_{k-1}, b_{k+1}, \ldots, b_m$ are held constant and b_k slightly increased, the optimal objective function value in the new problem created will be smaller than for the given problem. Let us examine this result more closely. Let b_k be increased slightly to b'_k. Certainly, every vector $\mathbf{x}$ satisfying $h_k(\mathbf{x}) \leq b_k$ will also satisfy $h_k(\mathbf{x}) \leq b'_k$, since $b'_k > b_k$. Thus if $\mathbf{x}$ satisfies

$$h_i(\mathbf{x}) \leq b_i, \quad i = 1, \ldots, m$$

$\mathbf{x}$ also satisfies

$$h_i(\mathbf{x}) \leq b_i, \quad i = 1, \ldots, m, i \neq k$$
$$h_k(\mathbf{x}) \leq b'_k$$

Consequently, if x^* is optimal for problem (2.40), $\mathbf{x}^*$ will satisfy the constraints *where b_k is replaced by $b'_k > b_k$*. Thus it is *impossible* that the optimal objective function value in the new problem be *less than* $f(\mathbf{x}^*)$. But this contradicts the result of assuming $\lambda_k > 0$. Therefore, we must have

$$\lambda_i \leq 0, \quad i = 1, \ldots, m$$

Summarizing, if $\mathbf{x}^*$ is an optimal solution to

$$\max f(\mathbf{x})$$

subject to

$$h_i(\mathbf{x}) \leq b_i, \quad i = 1, \ldots, m$$

then there exists $\boldsymbol{\lambda}^*$ such that $\mathbf{x}^*, \boldsymbol{\lambda}^*$ satisfy

$$\frac{\partial f}{\partial x_j} + \sum_{i=1}^{m} \lambda_i \frac{\partial h_i}{\partial x_j} = 0, \quad j = 1, \ldots, n$$
$$\lambda_i[h_i(\mathbf{x}) - b_i] = 0, \quad i = 1, \ldots, m$$
$$h_i(\mathbf{x}) \leq b_i, \quad i = 1, \ldots, m \tag{2.47}$$
$$\lambda_i \leq 0, \quad i = 1, \ldots, m$$

The necessary conditions (2.47) are known as the Kuhn–Tucker conditions after the mathematicians H. Kuhn and A. W. Tucker [5].

a. A MORE GENERAL FORM OF THE KUHN–TUCKER CONDITIONS

Suppose that the rth constraint had actually been

$$h_r(\mathbf{x}) \geq b_r$$

The form (2.40) may be obtained and the problem remains the same if we consider

$$\max f(\mathbf{x})$$

subject to

$$h_i(\mathbf{x}) \le b_i, \quad i = 1, \ldots, m, \quad i \ne r$$
$$-h_r(\mathbf{x}) \le -b_r$$

Following the derivation of the Kuhn–Tucker conditions for problem (2.40), we observe they may be expressed nearly identically for the new problem—the only difference being that $-\lambda_r \le 0$, or $\lambda_r \ge 0$. Also, we know, for equality constraints, that the multipliers are not sign constrained.

Thus we may write necessary conditions for a different form of the general problem

$$\max f(\mathbf{x})$$

subject to

$$
\begin{aligned}
h_i(\mathbf{x}) &\le b_i, \quad i = 1, \ldots, r \\
h_i(\mathbf{x}) &\ge b_i, \quad i = r + 1, \ldots, s \\
h_i(\mathbf{x}) &= b_i, \quad i = s + 1, \ldots, m
\end{aligned}
\tag{2.48}
$$

These conditions are

$$\frac{\partial f(\mathbf{x})}{\partial x_j} + \sum_{i=1}^{m} \lambda_i \frac{\partial h_i(\mathbf{x})}{\partial x_j} = 0, \quad j = 1, \ldots, n$$
$$\lambda_i[h_i(\mathbf{x}) - b_i] = 0, \quad i = 1, \ldots, s$$
$$h_i(\mathbf{x}) \le b_i, \quad i = 1, \ldots, r$$
$$h_i(\mathbf{x}) \ge b_i, \quad i = r + 1, \ldots, s$$
$$h_i(\mathbf{x}) = b_i, \quad i = s + 1, \ldots, m$$
$$\lambda_i \le 0, \quad i = 1, \ldots, r$$
$$\lambda_i \ge 0, \quad i = r + 1, \ldots, s$$

The corresponding conditions for problems in which the minimization of some function is the objective may also be derived (see the problems).

b. COMPUTATION WITH THE KUHN–TUCKER CONDITIONS

For problem-solving purposes, the typical conditions (2.47) are not particularly well suited. The reasons for this are apparent; first, the equations in (2.47) may be nonlinear and, second, one must deal with a set of inequalities as well.

Ordinarily, one first seeks to discover which of the λ_i are negative and which are zero. There are 2^m possible combinations of $\lambda_1, \ldots, \lambda_m$ in terms of the values "negative" and "zero."

Choosing a particular combination, say $\lambda_p, \lambda_q, \ldots, \lambda_z$ negative, all others zero, some of the difficulty is eliminated; for then one can treat

$$\frac{\partial f}{\partial x_j} + \sum_{i=1}^{m} \lambda_i \frac{\partial h_i}{\partial x_j} = 0, \quad j = 1, \ldots, n$$

$$h_i(\mathbf{x}) - b_i = 0, \quad i = p, q, \ldots, z \tag{2.49}$$

which is a system of *equations* in a *smaller* number of variables, and seek to either eliminate possible combinations by contradicting the original constraints or the assumptions that led to the system (2.49), or by finding a solution to system (2.49) that also satisfies conditions (2.47).

Example 2.14

Consider the problem

$$\max \left(\tfrac{1}{3}\right)x + y$$

subject to

$$x - y \leq 0$$
$$-x - y \leq -3$$
$$-x \leq 0 \tag{2.50}$$
$$-y \leq 0$$

for which the Kuhn–Tucker conditions become

$$\tfrac{1}{3} + \lambda_1 - \lambda_2 - \lambda_3 = 0$$
$$1 - \lambda_1 - \lambda_2 - \lambda_4 = 0$$
$$\lambda_1(x - y) = 0$$
$$\lambda_2(-x - y + 3) = 0$$
$$-\lambda_3 x = 0$$
$$-\lambda_4 y = 0$$
$$x - y \leq 0$$
$$-x - y \leq -3 \tag{2.51}$$
$$-x \leq 0$$
$$-y \leq 0$$
$$\lambda_1 \leq 0$$
$$\lambda_2 \leq 0$$
$$\lambda_3 \leq 0$$
$$\lambda_4 \leq 0$$

Suppose first that all multipliers are negative. The system of interest becomes

$$\tfrac{1}{3} + \lambda_1 - \lambda_2 - \lambda_3 = 0$$
$$1 - \lambda_1 - \lambda_2 - \lambda_4 = 0$$
$$x - y = 0$$
$$-x - y + 3 = 0$$
$$x = 0$$
$$y = 0$$
$$x - y \leq 0$$
$$-x - y \leq -3$$
$$-x \leq 0$$
$$-y \leq 0$$

which is clearly inconsistent. Hence there is no solution to the system (2.51) having all multipliers negative.

There are 15 other possibilities for the λ's, but consider

$$\lambda_1 = \lambda_3 = 0 \quad \text{and} \quad \lambda_2, \lambda_4 < 0$$

in which case system (2.51) provides

$$\tfrac{1}{3} - \lambda_2 = 0$$
$$1 - \lambda_2 - \lambda_4 = 0$$
$$-x - y + 3 = 0$$
$$y = 0$$
$$x - y \leq 0$$
$$-x - y \leq -3$$
$$-x \leq 0$$
$$-y \leq 0$$

which is again inconsistent.

In fact, if we pursued the remaining 14 possibilities, we would find a contradiction in every case; the necessary conditions have no solution. The reader should verify geometrically that the objective function can be made infinitely large and the constraints remain satisfied.

Example 2.15

Consider next the problem

$$\max (\tfrac{1}{3})x + y$$

subject to

$$-x + y \leq 0$$
$$x + y \leq 3$$

(2.52)

Now the necessary conditions become

$$\tfrac{1}{3} - \lambda_1 + \lambda_2 = 0$$
$$1 + \lambda_1 + \lambda_2 = 0$$
$$\lambda_1(-x + y) = 0$$
$$\lambda_2(x + y - 3) = 0$$
$$-x + y \leq 0$$
$$x + y \leq 3$$
$$\lambda_1 \leq 0$$
$$\lambda_2 \leq 0$$

(2.53)

(a) $\lambda_1 = 0, \lambda_2 = 0$ provides an inconsistent system of equations.
(b) $\lambda_1 = 0, \lambda_2 < 0$ yields

$$\tfrac{1}{3} + \lambda_2 = 0$$
$$1 + \lambda_2 = 0$$

which is inconsistent.
(c) $\lambda_1 < 0, \lambda_2 = 0$ provides an inconsistent system.
(d) $\lambda_1 < 0, \lambda_2 < 0$ yields the system of equations

$$\tfrac{1}{3} - \lambda_1 + \lambda_2 = 0$$
$$1 + \lambda_1 + \lambda_2 = 0$$
$$-x + y = 0$$
$$x + y - 3 = 0$$

The solution of the first two equation yields

$$\lambda_2 = -\tfrac{2}{3}; \qquad \lambda_1 = -\tfrac{1}{3}$$

The second two equations yield

$$x = y = \tfrac{3}{2}$$

and all relationships of system (2.53) are satisfied.

In this situation, which, owing to the nature of the objective function and the constraints, is a rather special one, the necessary conditions for a relative maximum also suffice for a global maximum; thus $x = y = \tfrac{3}{2}$ is the optimal solution. The reader can verify the solution to the problem geometrically, but we emphasize the general nonsufficiency of the Kuhn–Tucker conditions.

c. ANOTHER COMPUTATIONAL APPROACH

For the problem

$$\max f(\mathbf{x})$$

subject to

$$h_i(\mathbf{x}) \leq b_i, \quad i = 1, \ldots, m \tag{2.54}$$

we can follow a computational procedure that might result in reduction of the total amount of work necessary to locate a global maximum.

As we have seen in examining the Kuhn–Tucker conditions, if $h_i(\mathbf{x}^*) < b_i$, then $\lambda_i^* = 0$, and from observing the necessary conditions, that constraint could have been neglected. This is not to say that the optimal solution to the problem without the constraint is the same as for the problem with the constraint; rather, it says that if the constraint had been neglected x^* would still be found as a point satisfying the necessary conditions.

To illustrate, consider the problem

$$\max f(x)$$

subject to

$$x \leq b$$

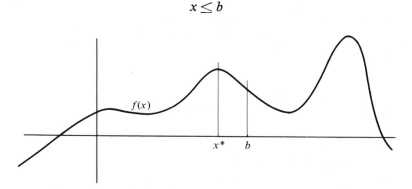

Figure 2.15. A constraint that could be neglected.

where $f(x)$, x^*, and b are as in Figure 2.15.

Obviously, we have $x^* < b$, but if we eliminate the constraint, the point x^* still satisfies the necessary condition.

On the other hand suppose we change b so that our diagram becomes Figure 2.16.

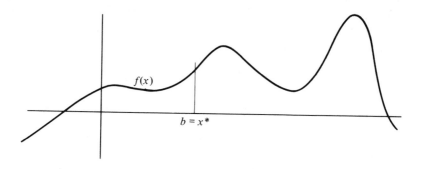

Figure 2.16. A necessary constraint.

Now, the constraint is satisfied as an equality, and if it were neglected, the point x^* would *not* be found. Of course there are special and infrequent instances where x^* satisfies the constraint as an equality, but where neglecting the constraint would still permit x^* to be found (as in Figure 2.17, for example).

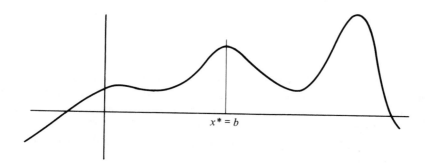

Figure 2.17. Coincidence of constraint constant and relative extremum.

At any rate, of importance here is the fact that satisfaction as a strict inequality implies that the constraint could have been neglected, insofar as finding points which satisfy necessary conditions is concerned. Thus suppose that we approach the given problem (2.54) as follows.

Neglect all the constraints and obtain all points satisfying the resulting necessary conditions. If one or more of these points *satisfy the neglected constraints,* then the one giving the largest value of f is an optimal solution to the given problem. (Why?)

If none of those points satisfies the neglected constraints, it means that if the problem *has* an optimal solution at least one of the constraints must be satisfied as a strict equality (recall that f might become infinite or that the set of solutions satisfying all the constraints might be null).

Thus one is led to solve next each problem from the family

$$\max f(\mathbf{x}) \qquad \max f(\mathbf{x}) \qquad \max f(\mathbf{x})$$
$$, \qquad\qquad , \ldots ,$$
$$\text{subject to } h_1(\mathbf{x}) = b_1 \quad \text{subject to } h_2(\mathbf{x}) = b_2 \qquad \text{subject to } h_m(\mathbf{x}) = b_m$$

Thus one obtains m *sets* of points satisfying the necessary conditions for the respective problems.

If there are some points that satisfy all the neglected constraints as well, the one giving the best objective function value is an optimal solution to the original problem as well (why?), and we stop.

Otherwise, one next treats each distinct member of the family of $\binom{m}{2}$ problems,

$$\max f(\mathbf{x})$$

subject to

$$h_j(\mathbf{x}) = b_j, \quad j = 1, \ldots, m$$
$$h_k(\mathbf{x}) = b_k, \quad k = 1, \ldots, m$$
$$j \neq k$$

because now we know *at least two* of the constraints must be satisfied as strict equalities. If some points are found that satisfy all the constraints, the one giving the best f value is $\mathbf{x}^*$, and we stop; otherwise, we consider all combinations of three constraints as equalities, and so on.

The advantage of the procedure is that we are always able to treat equality-constrained problems; the disadvantage, clearly, is that the potential number of these that may need to be solved can be enormous.

It should also be pointed out that if the given problem has also a set of equality constraints, that is,

$$\max f(\mathbf{x})$$

subject to

$$h_i(\mathbf{x}) \leq b_i, \quad i = 1, \ldots, m$$
$$h_i(\mathbf{x}) = b_i, \quad i = m + 1, \ldots, k$$

then those equality constraints must be included in each of the subproblems solved, the foregoing procedure allowing us some latitude only with inequality constraints.

iii. *Convex and Concave Functions of Several Variables*

The notion of convexity and concavity is quite easily extended to functions of several variables. In fact, extending the definition of convex functions of one variable introduced in Section 2.2-iiia, we say

Definition. A function f is convex over a set X if for every $x_1, x_2 \in X$

$$f(\lambda x_2 + (1 - \lambda)x_1) \leq \lambda f(x_2) + (1 - \lambda)f(x_1)$$

for all $\lambda, 0 \leq \lambda \leq 1$.

Clearly, the only difference between this definition and the corresponding one for functions of one variable is the fact that presently our points lie in a higher-dimensional Euclidean space. Geometrically, our line segments in the plane have become line segments in E^n, and this raises a question about X. In one dimension we considered functions defined on an interval; this guaranteed that

$$\lambda x_2 + (1 - \lambda)x_1$$

was itself part of the region of definition; the definitions in one dimension would be meaningless if the region of definition were other than an interval. Similarly, in n dimensions, X must be such that for any $x_1, x_2 \in X$, $\lambda x_2 + (1 - \lambda)x_1$ lies again in X for $0 \leq \lambda \leq 1$. This property describes a very important class of sets in E^n, *convex sets*. Sets that are convex in the plane are illustrated in Figure 2.18. Examples of nonconvex sets are given in Figure 2.19. A formal definition of convex sets in E^n is the following:

Figure 2.18. Examples of convex sets.

Figure 2.19. Examples of non-convex sets.

Definition. A set $X \subset E^n$ is said to be *convex* if for every $x_1, x_2 \in X$ the line segment joining $\mathbf{x}_1, \mathbf{x}_2$ lies entirely within X.

Thus we speak always of convex, strictly convex, concave, and strictly concave functions *over convex sets*, and we define convex functions as follows.

Definition. A function f is said to be *convex* over the convex set X if for every $\mathbf{x}_1, \mathbf{x}_2 \in X$

$$f(\lambda \mathbf{x}_2 + (1 - \lambda)\mathbf{x}_1) \leq \lambda f(\mathbf{x}_2) + (1 - \lambda)f(\mathbf{x}_1)$$

for all λ, $0 \leq \lambda \leq 1$.

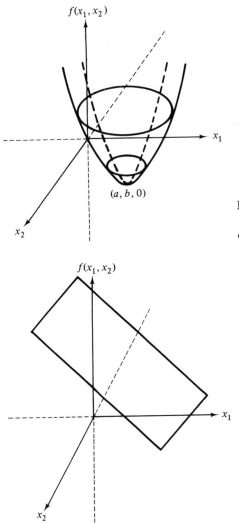

Figure 2.20. $f(x_1, x_2) = (x_1 - a)^2 + (x_2 - b)^2$ is (strictly) convex over any convex subset of E^2.

Figure 2.21. $f(x_1, x_2) = c_1 x_1 + c_2 x_2 + c_3$ is convex and concave over any convex subset of E^2.

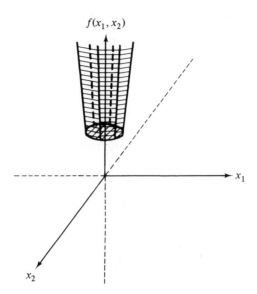

$f(x_1, x_2)$

x_1

x_2

Figure 2.22. $f(x_1, x_2)$ is convex over any convex subset of E^2.

Concave, strictly convex, and strictly concave functions are defined in similar fashion.

The reader will appreciate that, in general, applying these definitions to an arbitrary function to establish, for example, its convexity, is not a simple matter. There are equivalent characterizations of convexity, but these will not be employed here. Examples of convex functions are illustrated in Figures 2.20, 2.21, and 2.22.

As would be expected, Theorem 2.3 carries over directly to n dimensions. The proof of the theorem, in fact, is identical, and we combine that result and the comparable result for concave functions, while proving only the convex portion.

Theorem 2.7

If $f(x_1, x_2, \ldots, x_n)$ is a convex (concave) function on the convex set $X \subset E^n$, and $\mathbf{x}_0$ is a relative minimum (maximum) of f, then $\mathbf{x}_0$ is a global minimum (maximum) of f.

Proof: Let f be convex over the convex set X. Let $\mathbf{x}_0$ be a relative minimum of f over the set X. By definition, then

$$f(\mathbf{x}_0) \leq f(\mathbf{x}), \quad \text{all } \mathbf{x} \in \text{ a neighborhood } N \text{ of } \mathbf{x}_0$$

(If $\mathbf{x}_0$ is a boundary point of X, then the notion of neighborhood is interpreted so as to include only points of X.)

If $\mathbf{x}_0$ is not a global minimum, there is $\mathbf{x}^*$ in X with

$$f(\mathbf{x}^*) < f(\mathbf{x}_0)$$

The line segment joining $\mathbf{x}^*$, $\mathbf{x}_0$ lies wholly within X, by virtue of X's convexity; that is, for $0 \le \lambda \le 1$,

$$\lambda \mathbf{x}^* + (1 - \lambda)\mathbf{x}_0$$

belongs to X, and some points of the segment (infinitely many of them, in fact) lie within the neighborhood N. Thus, for those $x \in N$, we have $x = \lambda'\mathbf{x}^* + (1 - \lambda')\mathbf{x}_0$, for some values of λ'. Since f is convex, then by definition

$$f(\lambda'\mathbf{x}^* + (1 - \lambda')\mathbf{x}_0) \le \lambda'f(\mathbf{x}^*) + (1 - \lambda')f(\mathbf{x}_0)$$

Now, for $\lambda' \ne 0$

$$\lambda'f(\mathbf{x}^*) + (1 - \lambda')f(\mathbf{x}_0) < \lambda'f(\mathbf{x}_0) + (1 - \lambda')f(\mathbf{x}_0) = f(\mathbf{x}_0)$$

so we have

$$f(\lambda'\mathbf{x}^* + (1 - \lambda')\mathbf{x}_0) < f(\mathbf{x}_0)$$
or
$$f(\mathbf{x}) < f(\mathbf{x}_0)$$

which contradicts the fact that $\mathbf{x}_0$ was a relative minimum. Thus no such $\mathbf{x}^*$ can exist, and $\mathbf{x}_0$ is a *global minimum*.

Exercise 2.5

Let f be a concave function over a closed convex set X. Show that the global minimum of f occurs on the boundary of X.

iv. *Sufficiency of the Necessary Conditions—Special Cases*

The reason for the sufficiency of the conditions in Example 2.15 is that we are maximizing a concave function subject to linear constraints. (Minimizing a convex function over linear constraints also provides sufficient conditions, of course.)

Let us formalize and strengthen the previous statement.

Theorem 2.8

If $f(\mathbf{x})$ is concave and $h_i(\mathbf{x})$ is convex for $i = 1, \ldots, m$, then the K–T (Kuhn–Tucker) conditions for the problem,†

---

†Again, any nonnegativity constraints are considered incorporated among the given constraints.

$$\max f(\mathbf{x})$$

subject to

$$h_i(\mathbf{x}) \leq b_i, \quad i = 1, \ldots, m$$

are also sufficient for an optimal solution.

It should be noted that the statement of Theorem 2.8 differs from that of Theorem 2.7. The present statement is that if the K–T conditions are satisfied, a point which does so is immediately globally optimal. Theorem 2.7 makes no mention of the satisfaction of the K–T conditions, but makes a stronger assumption, that a relative extremum is available.

Proof: The theorem is easily established since the saddle point inequalities (2.35) hold globally (which may be shown to be the case under the hypotheses of this theorem) for the function

$$F(x_1, \ldots, x_n, \lambda_1, \ldots, \lambda_m) = f(x_1, \ldots, x_n) + \sum_{i=1}^{m} \lambda_i[h_i(x_1, \ldots, x_n) - b_i]$$

That is, if $\mathbf{x}^*$, $\boldsymbol{\lambda}^*$ satisfy the necessary conditions for this problem, then

$$F(\mathbf{x}, \boldsymbol{\lambda}^*) \leq F(\mathbf{x}^*, \boldsymbol{\lambda}^*) \leq F(\mathbf{x}^*, \boldsymbol{\lambda}) \tag{2.55}$$

for *all* feasible $\mathbf{x}$ and for all $\boldsymbol{\lambda} \leq \mathbf{0}$.

Again, we shall use inequality (2.55) without showing it to hold.

Let $\mathbf{x}$ be any feasible solution to the given problem; then

$$\lambda_i^*(h_i(x_1, \ldots, x_n) - b_i) \geq 0$$

since
$$\lambda_i^* \leq 0$$

Then
$$f(\mathbf{x}) \leq f(\mathbf{x}) + \sum_{i=1}^{m} \lambda_i^*(h_i(\mathbf{x}) - b_i) = F(\mathbf{x}, \boldsymbol{\lambda}^*)$$

Also
$$F(\mathbf{x}^*, \boldsymbol{\lambda}^*) = f(\mathbf{x}^*)$$

so, using the left part of inequality (2.55), we obtain

$$f(\mathbf{x}) \leq f(\mathbf{x}^*)$$

and $\mathbf{x}^*$ is a global maximum.

v. *Generalization of Lagrange Multipliers*

In the present section we discuss some aspects of an optimization technique due to Everett [1].

Consider the problem

$$\max f(x_1, \ldots, x_n)$$

subject to $\qquad h_i(x_1, \ldots, x_n) \le b_i, \quad i = 1, \ldots, m$

Theorem 2.9

Suppose that we *arbitrarily* select m nonpositive values $\lambda_1^0, \ldots, \lambda_m^0$, and construct the function

$$F(\mathbf{x}, \boldsymbol{\lambda}^0) = f(\mathbf{x}) + \sum_{i=1}^{m} \lambda_i^0 h_i(\mathbf{x})$$

Suppose that $\mathbf{x}^*$ is a global maximum for F. Then, under these conditions $\mathbf{x}^*$ is a global maximum for the problem

$$\max f(\mathbf{x})$$

subject to $\qquad h_i(\mathbf{x}) \le h_i(\mathbf{x}^*), \quad i = 1, \ldots, m$

Proof: By assumption,

$$f(\mathbf{x}^*) + \sum_{i=1}^{m} \lambda_i^0 h_i(\mathbf{x}^*) \ge f(\mathbf{x}) + \sum_{i=1}^{m} \lambda_i^0 h_i(\mathbf{x})$$

for all $\mathbf{x}$. That is,

$$f(\mathbf{x}^*) \ge f(\mathbf{x}) + \sum_{i=1}^{m} \lambda_i^0 [h_i(\mathbf{x}) - h_i(\mathbf{x}^*)] \tag{2.56}$$

Since the λ_i^0 are nonpositive, if $h_i(\mathbf{x}) \le h_i(\mathbf{x}^*)$ for all i, then certainly $f(\mathbf{x}^*) \ge f(\mathbf{x})$, since the second term on the right side of (2.56) will then be nonnegative for any $\mathbf{x}$.

Suppose, then, that we select a set of nonpositive λ's, $\lambda_1^0, \lambda_2^0, \ldots, \lambda_m^0$, and solve the *unconstrained* problem

$$\max F(\mathbf{x}, \boldsymbol{\lambda}^0)$$

Should it occur that $h_i(\mathbf{x}^*) = b_i$, $i = 1, \ldots, m$, the *constrained* problem has been solved. Of course, for arbitrary choices of nonpositive $\lambda_1^0, \ldots, \lambda_m^0$, the chances that this will occur are quite remote.

The point of interest, however, and the point which provides a useful optimization technique is that with λ_j fixed, all $j \ne i$, $h_i(\mathbf{x}^*)$ *is a monotonically nondecreasing function of* λ_i.

What this means is that we know in advance in which direction to vary λ_i from its previous value so as to achieve $h_i(\mathbf{x}^*) = b_i$, or at least to attempt to approach that value. For example, suppose that we specify a set of λ's, $\lambda_1^0, \ldots, \lambda_m^0$, obtain $\mathbf{x}^*$ for the corresponding unconstrained problem, and find that

$$h_i(\mathbf{x}^*) < b_i, \qquad [h_i(\mathbf{x}^*) > b_i]$$

Then, holding $\lambda_1, \ldots, \lambda_{i-1}, \lambda_{i+1}, \ldots, \lambda_m$ fixed and increasing [decreasing] λ_i, we obtain a new unconstrained problem whose optimal solution $\mathbf{x}'^*$ causes $h_i(\mathbf{x}'^*)$ to at least not move in the direction *away* from b_i, and, possibly, $h_i(\mathbf{x}'^*)$ will move in the direction toward b_i from $h_i(\mathbf{x}^*)$.

To see what might be possible, suppose that we select a set of λ's, solve the unconstrained problem, and find that $h_k(\mathbf{x}^*) > b_k$. Incidentally, if in cases where $h_i(\mathbf{x}^*) = b_i$, $i \neq k$, we found $h_k(\mathbf{x}^*) > b_k$, then we know the optimal objective function value for the original problem can be no greater than $f(\mathbf{x}^*)$. (Why?)

Then, holding all other λ's fixed, we reduce λ_k and solve the unconstrained problem, obtaining $\mathbf{x}_1^*$. Suppose that we found $h_k(\mathbf{x}^*) > h_k(\mathbf{x}_1^*) > b_k$. We know in advance, by the monotonicity property, that we will *not* have $h_k(\mathbf{x}_1^*) > h_k(\mathbf{x}^*)$.

Seeing that we have moved in the proper direction, assume that we further reduced λ_k, obtained $\mathbf{x}_2^*$ as an optimal solution to the new unconstrained problem, and discovered that

$$h_k(\mathbf{x}^*) > h_k(\mathbf{x}_1^*) > b_k > h_k(\mathbf{x}_2^*)$$

Then we realize we have reduced λ_k too far, and for the next round would *increase* it from the value that gave $\mathbf{x}_2^*$, perhaps interpolating in some fashion between the values that yielded $\mathbf{x}_1^*$ and $\mathbf{x}_2^*$, respectively.

The monotonicity property *always* tells in what direction the λ's should be varied, and it is this property that makes the method work, although, as we shall discuss later, the method is not guaranteed.

Example 2.16

For the moment let us illustrate the procedure. We consider the problem

$$\min x^2 + y^2$$
subject to
$$2x + y \leq 2$$
$$-2x + y \leq 1$$

for which the optimal solution is $(0, 0)$.

Choosing arbitrarily $\lambda_1 = -1$, $\lambda_2 = -2$, we have for the maximization problem

$$F(x, y, \lambda_1, \lambda_2) = -x^2 - y^2 - 2x - y + 4x - 2y$$
$$= -x^2 - y^2 + 2x - 3y$$

and $x^* = 1$, $y^* = -\frac{3}{2}$ is the global maximum of F. Consequently,

$$h_1(x^*, y^*) = 2x^* + y^* = 2 - \frac{3}{2} = \frac{1}{2} < b_1 = 2$$
$$h_2(x^*, y^*) = -2x^* + y^* = -2 - \frac{3}{2} = -\frac{7}{2} < b_2 = 1$$

Let us keep $\lambda_1 = -1$, and increase λ_2 to $\lambda_2 = -1$, obtaining

$$F = -x^2 - y^2 - 2x - y + 2x - y$$
$$= -x^2 - y^2 - 2y$$

whose global maximum is $x_1^* = 0$, $y_1^* = -1$.

Now $h_1(x_1^*, y_1^*) = -1 < b_1 = 2$

and $h_2(x_1^*, y_1^*) = -1 < b_2 = 1$

Let us next increase λ_2 to $\lambda_2 = 0$, maintaining $\lambda_1 = -1$, since we still have $h_2(x_1^*, y_1^*) < b_2$. F becomes

$$F = -x^2 - y^2 - 2x - y$$

and $x_2^* = -1, \qquad y_2^* = -\frac{1}{2}$

$$h_1(x_2^*, y_2^*) = -\frac{3}{2} < b_1$$
$$h_2(x_2^*, y_2^*) = \frac{3}{2} > b_2$$

and we see that we increased λ_2 by too large an amount.

Rather than further adjusting λ_2, let us keep $\lambda_2 = 0$ and increase λ_1 from -1 to $\lambda_1 = 0$, since $h_1(x_2^*, y_2^*) < b_1$.

If we do this, we have

$$F = -x^2 - y^2$$

whose global maximum is $x_3^* = 0$, $y_3^* = 0$. Thus

$$h_1(x_3^*, y_3^*) = 0 < b_1$$
$$h_2(x_3^*, y_3^*) = 0 < b_2$$

Now, we know that $x^* = 0$, $y^* = 0$ is, indeed, the optimal solution to the original problem. Yet we have

$$h_1(0, 0) < b_1$$
$$h_2(0, 0) < b_2 \tag{2.57}$$

On the one hand this should not be disturbing, for Theorem 2.9 was not if and only if; that is, it admits the possibility of optimizing the constrained problem *without* having $h_i(\mathbf{x}^*) = b_i$ for all i. (In fact, we knew in advance that the optimal solution would satisfy the constraints as strict inequalities.) On the other hand, however, one might be disturbed that with $\lambda_1 = \lambda_2 = 0$, which is the *maximum* value they can assume according to Theorem 2.9, we still find, in the inequalities (2.57), that to achieve equality at least one λ needs to be *increased*.

Owing to the simplicity of our problem, further analysis is possible; let us attempt to derive nonpositive λ_1, λ_2 such that

$$x^*(\lambda_1, \lambda_2), \qquad y^*(\lambda_1, \lambda_2)$$

do indeed satisfy the constraints as equalities.

F will have the general form

$$F(x, y, \lambda_1, \lambda_2) = -x^2 - y^2 + x(2\lambda_1 - 2\lambda_2) + y(\lambda_1 + \lambda_2)$$

$$\frac{\partial F}{\partial x} = -2x + 2\lambda_1 - 2\lambda_2$$

$$\frac{\partial F}{\partial y} = -2y + \lambda_1 + \lambda_2$$

Setting these derivatives equal to zero yields the unique solution

$$y = \frac{\lambda_1 + \lambda_2}{2}$$

$$x = \lambda_1 - \lambda_2$$

(Notice that x, y so defined are optimal for the given unconstrained problem regardless of the values of λ_1, λ_2, since the matrix of second partials, the Hessian, is always negative definite.)

To satisfy the given constraints as equalities, we must have

$$2(\lambda_1 - \lambda_2) + \frac{\lambda_1 + \lambda_2}{2} = 2$$

$$-2(\lambda_1 - \lambda_2) + \frac{\lambda_1 + \lambda_2}{2} = 1$$

But this system has also a unique solution, with λ_1 and λ_2 both positive.

Thus in Example 2.16 one observes that there are *no* nonpositive λ_1, λ_2 such that

$$h_1(x^*(\lambda_1, \lambda_2), y^*(\lambda_1, \lambda_2)) = b_1$$

and

$$h_2(x^*(\lambda_1, \lambda_2), y^*(\lambda_1, \lambda_2)) = b_2$$

It is precisely this event that sometimes defeats the method; that is, in some problems there is *no* set of nonpositive λ's that gives $h_i(\mathbf{x}^*) = b_i$, $i = 1, \ldots, m$. And, for the general problem, without equality one cannot conclude that the constrained function has actually been optimized.

a. SUFFICIENCY CONDITION DUE TO EVERETT

Everett was able to obtain a condition that is *sufficient* to guarantee the theoretical convergence of the method: if $f(\mathbf{x}^*)$ is a concave function of b_1, $\ldots$, b_m in a neighborhood of the given constants, then, in theory, the method will work; but we observe that verifying the presence of that property in a given problem is not a simple task.

In the preceding paragraph we spoke of convergence, *in theory*, since a finite number of iterations will still not, in general, provide *exact* equality, even though the $h_i(\mathbf{x}^*)$ may all become quite close to the respective b_i.

Another problem is the following. Suppose that given a set of λ's we cannot solve the unconstrained problem exactly; that is, if $\mathbf{x}^*$ is the true optimal solution, we obtain $\mathbf{x}' \neq \mathbf{x}^*$ as a result, perhaps, of approximation error in the solution procedure. A question, then, is to what extent the corresponding constrained problem is affected by the appoximation in the unconstrained problem.

Toward answering this latter question, Everett proved the following:

Theorem 2.10

If $\mathbf{x}'$ provides an objective function value within ϵ of the true objective function value for the unconstrained problem, then $\mathbf{x}'$ provides a value of f within ϵ of the true maximum value for the problem

$$\max f(\mathbf{x})$$

subject to

$$h_i(\mathbf{x}) \leq h_i(\mathbf{x}'), \quad i = 1, \ldots, m$$

A partial answer to the first question, regarding the fact that one does not achieve exact equality in the constraints, can also be supplied by Theorem 2.10, since we can see that not achieving equality through the choice of the λ's essentially results in approximating a true solution, although the value of ϵ is not obvious in this case.

What must be pointed out as a most attractive property of this method is that absolutely nothing is required of f or the h_i in the way of being well behaved, for example, differentiability or even continuity.

In fact, Everett's original formation of the problem was slightly different. He treated the problem

$$\text{max } f(\mathbf{x})$$

subject to $\qquad\qquad h_i(\mathbf{x}) \leq b_i, \quad i = 1, \ldots, m \qquad\qquad (2.58)$

and $\qquad\qquad\qquad\qquad \mathbf{x} \in S$

where S is an arbitrary subset of E^n.

Now although $\mathbf{x} \in S$ is simply another set of constraints, this particular form was chosen in case there are constraints that do not have an inequality representation. For example, from the nature of a particular problem, it may be that in order for a feasible solution to make sense x_j must be a non-negative integer, $j = 1, \ldots, n$.

Given an arbitrary constrained problem, then, the problem solver has within his control what he wishes to consider as a set S, and this freedom can facilitate problem solution. Integer constraints are very common in practical optimization problems, as we shall see later in the book, and they provide an example where the formulation (2.58) is useful.

2.6. Introduction to the Calculus of Variations

It is the purpose of this section to provide a brief introduction to a class of optimization problems and means for their solution, both different, but in many ways similar, to the preceding portion of the chapter.

Previously, we have been presented a set of points S and a function $f(\mathbf{x})$, $\mathbf{x} \in S$, the problem being to locate $\mathbf{x}^* \in S$ such that $f(\mathbf{x}^*) \leq f(\mathbf{x})$, all $\mathbf{x} \in S$, or $f(\mathbf{x}^*) \geq f(\mathbf{x})$, all $x \in S$. Now we wish to consider S as being a set of *functions*. By a *functional* on S we mean a function $J(\cdot)$ such that to each *function* in S we associate a single real number. Optimization of functionals constitutes the class of problems in the calculus of variations. Some examples of functionals follow.

1. Let S be a set of bounded functions on a real interval $[a, b]$. For $f \in S$, define

$$J(f) = \max_{[a, b]} \{f(x)\}$$

2. Let S be a set of functions that are integrable on $[a, b]$. Define

$$J(f) = \int_a^b f(x) \, dx$$

3. Let S be a set of functions defined on $[a, b]$, each with the property that the arc $y = f(x)$ traced out has a measurable length. Define

$$J(f) = l(f)$$

where $l(f)$ is the length of the arc.

i. *Developing a General Variational Problem*

Given, then, a set of functions S and a *functional* J, a typical problem of the *variational calculus* is

$$\max_{f \in S}\dagger \{J(f)\} \quad \text{or} \quad \min_{f \in S} \{J(f)\}$$

Constrained functional optimization problems are also common. For example, if both J_1 and J_2 are functionals, a problem might be given

$$\max_{f \in S} J_1(f)$$

subject to

$$J_2(f) \leq 0$$

In our brief treatment, however, we shall consider only unconstrained problems.

As in the case of most of this chapter, both the problem class and solution method we shall present are truly classical. Euler, who died in 1783, was the first contributor to this area of study.

Let us examine next what is generally considered the simplest problem of the calculus of variations. There is given

1. An infinite set M, called the admissible set of curves, each of the form $y = f(x)$, $x_0 \leq x \leq x_1$, with x_0, x_1 given. By a curve is meant a continuous function whose first derivative is piecewise continuous.
2. A function F of three variables such that for every $y \in M$ the functional

$$J(y) = \int_{x_0}^{x_1} F(x, y, y') \, dx$$

exists.

Thus, as y ranges over all M, the set of values produced has a greatest lower bound L and a least upper bound U. If there exists $y^* \in M$ such that $J(y^*) = L$ or $J(y^*) = U$, then y^* is said to provide the global minimum or global maximum of J over M, respectively.‡

The problem is to determine all admissible functions $y(x) \in M$ that minimize or maximize $J(y)$.

As in the theory of function optimization, analysis will not provide global extrema, and one is forced to deal with relative extrema.

---

†As in the optimization of functions, it is possible to encounter problems where sup or inf would replace max or min.

‡Achieving the value of the g.l.b. or l.u.b. is clearly equivalent to the existence of a global minimum or maximum, respectively.

It turns out that for purposes of obtaining usable results the statement of the simplest variational problem remains a bit too general, and stronger assumptions on the nature of M and F are required. Hence we pose the more restricted problem. Given

1. a set of admissible curves $y(x)$ each with the following properties:
   (a) $y(x)$ passes through two given points in the plane (x_0, y_0) and (x_1, y_1),
   (b) y may be represented by $y = f(x)$ on $[x_0, x_1]$,
   (c) y is continuous with a continuous first derivative,
   (d) y lies in a fixed region R of the plane
2. a function $F(u_1, u_2, u_3)$ of three variables which is continuous and has continuous partial derivatives of the first *three* orders for all points (u_1, u_2, u_3) such that $(u_1, u_2) \in R$ and such that u_3 is finite,

optimize the functional

$$J(y) = \int_{x_0}^{x_1} F(x, y, y') \, dx$$

ii. *Relative Extrema*

We now define a neighborhood of a curve y and the related concept of a relative extremum in the case of functionals. One defines $\bar{y}$ to be a *relative minimum* if there exists $\epsilon > 0$ such that

$$J(y) \geq J(\bar{y}) \tag{2.59}$$

for every admissible y satisfying

$$|\bar{y} - y| < \epsilon, \quad \text{for } x_0 \leq x \leq x_1 \tag{2.60}$$

A *relative maximum* is defined in obvious fashion.

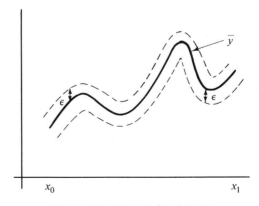

Figure 2.23. An ϵ strip about $\bar{y}$.

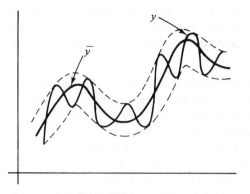

Figure 2.24. An ill-behaved function in the ϵ neighborhood.

Inequality (2.59) defines the concept of relative extremum in expected fashion; it is the nature of the neighborhood as defined by (2.60) that has undergone an interesting change.

A function y that satisfies inequality (2.59) must lie in an ϵ strip about $\bar{y}$, as shown in Figure 2.23. Thus an element y of the neighborhood is "close" to $\bar{y}$ in the sense that $|\bar{y}(x) - y(x)| < \epsilon$ for given x. However, there are functions in the neighborhood of Figure 2.23 that behave quite differently from $\bar{y}$, an example being depicted in Figure 2.24.

a. Necessary Conditions—Derivation

We shall now obtain necessary conditions for a curve $\bar{y}$ to be a relative extremum for the functional J. Since a global extremum must also be a relative extremum, locating all relative extrema will turn up all global extrema.

Suppose, then, that $\bar{y} = f(x)$ is a relative minimum for

$$J(y) = \int_{x_0}^{x_1} F(x, y, y') \, dx$$

Since all admissible y lie in a region R of the plane, it is possible for some $\epsilon > 0$ to construct an ϵ strip about $\bar{y}$ that lies entirely within R. Now let $y_1 \in M$ be an element of the neighborhood constructed. Define a function

$$u = y_1 - \bar{y}$$

which is called the *total variation of $\bar{y}$.*

Since all curves in the neighborhood pass through the points (x_0, y_0) and (x_1, y_1), then

$$u(x_0) = 0 = u(x_1)$$

Also, $|u(x)| < \epsilon$ in $[x_0, x_1]$.

Define, now

$$\Delta J = \int_{x_0}^{x_1} F(x, y_1, y_1') \, dx - \int_{x_0}^{x_1} F(x, \bar{y}, \bar{y}') \, dx$$
$$= \int_{x_0}^{x_1} [F(x, \bar{y} + u, \bar{y}' + u') - F(x, \bar{y}, \bar{y}')] \, dx \qquad (2.61)$$

which is called the *total variation of the integral J.*

Based on our assumption, $\bar{y}'$, y_1', and u' all exist and are continuous. Since $\bar{y}$ is a minimum, we have $\Delta J \geq 0$ if $\epsilon > 0$ is chosen sufficiently small.

The next step is to apply Taylor's formula to the *integrand* of equation (2.61). By virtue of assumption 1(c) and the continuity of $\bar{y}'$, y_1', and u', the foregoing is legitimate. We find

$$F(x, \bar{y} + u, \bar{y}' + u') - F(x, \bar{y}, \bar{y}')$$
$$= F_x(x, \bar{y}, \bar{y}')(x - x) + F_{\bar{y}}(x, \bar{y}, \bar{y}')u + F_{\bar{y}'}(x, \bar{y}, \bar{y}')u' \qquad (2.62)$$
$$+ \text{ higher-order terms}$$

where we have used subscripts on F to indicate partial derivatives.

We next want to consider *special* variations of the form

$$u = \delta g \qquad (2.63)$$

where g is any function of x on $[x_0, x_1]$ that has a continuous first derivative and satisfies the condition $g(x_0) = 0 = g(x_1)$, and δ is a constant sufficiently small to *guarantee*

$$|u(x)| < \epsilon$$

Using equations (2.61) and (2.62), we may write

$$\Delta J = \int_{x_0}^{x_1} [\delta g F_{\bar{y}} + \delta g' F_{\bar{y}'}] \, dx + \text{higher-order terms}$$

Or

$$\Delta J = \delta \left\{ \int_{x_0}^{x_1} [F_{\bar{y}} g + F_{\bar{y}'} g'] \, dx + o(\delta) \right\}^{\dagger} \qquad (2.64)$$

It must be, then, that

$$\int_{x_0}^{x_1} [F_{\bar{y}} g + F_{\bar{y}'} g'] \, dx = 0 \qquad (2.65)$$

for if not, ΔJ could be made negative as well as greater than or equal to zero

---

†The symbolism here may be unfamiliar to the reader. $o(\delta)$ is used to denote a quantity with the property that $\lim_{\delta \to 0} [o(\delta)/\delta] = 0$; that is, $o(\delta)$ goes to zero faster than δ as $\delta \to 0$.

simply by choosing δ *small and* selecting it first to be *positive* then *negative*. But ΔJ must be ≥ 0 by assumption. Integrating the second integrand term in equation (2.65) by parts, we obtain

$$\int_{x_0}^{x_1} F_{y'} g' \, dx = F_{y'} g \Big|_{x_0}^{x_1} - \int_{x_0}^{x_1} g \frac{d}{dx} F_{y'} \, dx \qquad (2.66)$$

Using equation (2.66), equation (2.65) may be written

$$\int_{x_0}^{x_1} g \left(F_y - \frac{d}{dx} F_{y'} \right) dx + F_{y'} g \Big|_{x_0}^{x_1} = 0$$

But since, by assumption, g vanishes at both x_0 and x_1, we deduce that for a global extremum it is necessary that

$$\int_{x_0}^{x_1} g \left(F_y - \frac{d}{dx} F_{y'} \right) dx = 0 \qquad (2.67)$$

It may be shown then that equation (2.67) implies

$$F_y - \frac{d}{dx} F_{y'} = 0 \qquad (2.68)$$

Equation (2.68) is called the Euler equation, and, occasionally, the Euler–Lagrange equation. It is a condition necessary for a relative extremum of J. In the event F is of the form $F(x, y, y', \ldots, y^{(n)})$, equation (2.68) becomes

$$F_y - \frac{d}{dx} F_{y'} + \frac{d^2}{dx^2} F_{y''} - \cdots + (-1)^n \frac{d^n}{dx^n} F_{y^{(n)}} = 0$$

Example 2.17

Consider the variational problem

$$\min \left\{ J(y) = \int_{x_0}^{x_1} F(x, y, y') \, dx = \int_{x_0}^{x_1} [1 + (y')^2]^{1/2} \, dx \right\}$$

Here

$$F_y = 0 \qquad (2.69)$$

and

$$F_{y'} = \frac{y'}{[1 + (y')^2]^{1/2}}$$

Using equation (2.69), equation (2.68) becomes

$$\frac{d}{dx} F_{y'} = 0$$

But performing the differentiation $(d/dx) F_{y'}$ and equating the result to zero, we find

$$y''[1 + (y')^2] = (y')^2 y''$$

or $$y''[1 + (y')^2 - (y')^2] = 0$$

Hence $y'' = 0$. Thus $y' = c$, a constant, and finally $y = cx + b$, b another constant.

Thus a straight line is obtained as the optimal function, and knowledge of the two points through which it must pass uniquely determines c and b.

The reader should note that the objective functional is simply the expression for the arc length of a differentiable function y which passes through (x_0, y_0) and (x_1, y_1). The application of the Euler equation has shown that the least distance between a pair of points in the plane is, indeed, provided by a straight line.

REFERENCES

[1] Everett, H., "Generalized Lagrange Multiplier Method for Solving Problems of Optimum Allocation of Resource," *Operations Research*, Vol. 11, pp. 399–417, 1963.

[2] Gue, R., and M. E. Thomas, *Mathematical Methods of Operations Research*, The Macmillan, Company, New York, 1968.

[3] Hadley, G., *Nonlinear and Dynamic Programming*, Addison-Wesley Publishing Company, Inc., Reading, Mass., 1964.

[4] Hancock, H., *Theory of Maxima and Minima*, Dover Publications, Inc., New York, 1960.

[5] Kuhn, H., and A. Tucker, "Nonlinear Programming," in *Proceedings of the Second Berkeley Symposium on Mathematical Statistics and Probability*, J. Neyman, ed., University of California Press, Berkeley, Calif., 1951.

PROBLEMS

1. For the following functions, locate the critical points and identify each as not a relative extremum, a relative maximum, or a relative minimum.
   (a) $f(x) = x^7 + 5x^6 + 3x^5 - 17x^4 - 16x^3 + 24x^2 + 16x - 16$
   (b) $f(x) = x^{11} - 9x^{10} + 21x^9 + 19x^8 - 93x^7 - 27x^6 + 135x^5 + 81x^4$
   (c) $f(x) = x^9 + 9x^8 + 21x^7 - 31x^6 - 189x^5 - 177x^4 + 247x^3 + 603x^2 + 432x + 108$
   (d) $f(x) = axe^{-(1/x^2)}$, a a constant

2. Suppose that, at x_0, f has a left-hand derivative $f'(x_{0-})$ and a right-hand derivative $f'(x_{0+})$ with

$$f'(x_{0-}) \neq f'(x_{0+})$$

What is a necessary condition that x_0 be a relative maximum? A relative minimum? Combine the two cases into a single statement.

3. Describe the critical points and their types for the functions
   (a) $f(x) = ax \cos x + b \cos x$, for $a, b > 0$
   (b) $f(x) = x \sin \dfrac{1}{x}$, $x \neq 0$
   $f(0) = 0$

4. Find the required global extrema—if they exist—for each of the following problems. If they do not, give suprema or infima if those exist.
   (a) $ax^2 + bx + c$ is to be solved. $b < 0$, $c < 0$. What is the largest value of a such that both roots are positive real numbers? The smallest value?
   (b) (x_1, y_1) is a fixed point in the plane. What is the shortest distance from (x_1, y_1) to the line $ax + by = c$?
   (c) Points a, b are collinear, lying on line l. What is the distance of the minimum-length path from a to b from among those paths from a to b that are perpendicular to l at a.
   (d) Find the minimum and maximum values of the function

   $$f(x) = \frac{1}{x^3 - \frac{5}{4}x^2 + \frac{1}{8}x + \frac{1}{8}}$$

   on the interval $[-1, 1]$.

5. Define f by the following:
   For $x \leq 0$, $f(x) = x$ for x rational
   $= 1$ for x irrational
   For $x > 0$, $f(x) = 1 - x$, x rational
   $= \frac{1}{2}$, x irrational
   What are the relative maxima and minima of f?

6. Theorem 2.2 gives information about the implication of a first nonvanishing derivative of f. What do we conclude if there is *no* such derivative; that is, all vanish?

7. Write the conditions under which the Taylor representation (2.3) is valid.

8. Let f be convex on $[a, b]$. Prove that the global maximum of f occurs at either a or b.

9. The defining inequalities (2.5) and (2.7) are often difficult to apply in order to discover whether a given function f is convex or concave.
   For each of the following functions and property—convexity or concavity —establish whether or not the appropriate inequality holds, and hence whether or not the statement is true.
   (a) e^x is convex on any real interval.
   (b) $\sin x$ is concave on $[0, (\pi/2)]$.
   (c) For $c > 0$ and $0 < p < 1$,

$$f(x) = \frac{c}{1 - p^{2^z}}$$

is convex on $[0, a]$ for any $a > 0$.

(d) x^2 is strictly convex on any interval $[a, b]$.

(e) For $f(x) > 0$ everywhere on $[a, b]$ and $f(x)$ concave on $[a, b]$, $\ln [f(x)]$ is concave on $[a, b]$.

(f) $\ln x$ is concave on $(0, a]$ for any $a > 0$.

10. There are several expressions that are equivalent to inequalities (2.5) or (2.7) for defining convex functions. We ask the reader to establish

(a) if $f''(x) > 0$ on $[a, b]$, then f is strictly convex on $[a, b]$.

(b) if $f''(x) \geq 0$ on $[a, b]$, then f is convex on $[a, b]$.

11. Write the definition of a *differentiable* function of several variables.

12. Apply the determinant form of the sufficient condition for a function $f(x, y)$ and compare the result with that given in an elementary calculus text.

13. For each of the following functions, find *all* the critical points.

(a) $\frac{1}{2}\mathbf{x}^T A \mathbf{x}$, where A is a symmetric real matrix with elements a_{ij}.

(b) $\sum_{j=1}^{n} (x_j - c_j)^2$

$c_1, \ldots, c_n$, real constants.

(c) $\prod_{j=1}^{n} (x_j - c_j)^2$

(d) $f(x, y) = \frac{3}{4}x^2 - 3xy + 3y^2$

(e) $f(x_1, x_2) = \frac{1}{2x_1 x_2} - \frac{2}{x_1^2 x_2} - \frac{4}{x_1 x_2^2}$

(f) $f(x_1, x_2) = x_1 x_2 + \frac{a}{x_1} + \frac{b}{x_2}$, $a, b > 0$; then, $a, b < 0$

(g) $f(x_1, x_2) = \frac{x_1^2}{1 + x_1^2} + x_2$

14. For each function of Problem 13 find a solution to the necessary conditions that provides

(a) the largest function value.

(b) the smallest function value.

15. The matrices whose determinants were important for sufficiency may be considered as being generated from A in the following way, where

$$i_1 = 1, \, i_2 = 2, \ldots, i_j = j, \ldots, i_n = n$$

1. A_1 is the 1×1 array consisting of the element left upon crossing off all rows except i_1 and all columns except i_1.

2. A_2 is the 2×2 array obtained by crossing off all rows except i_1, i_2 *and* all columns except i_1, i_2.

. . .

k. A_k is the $k \times k$ array obtained by crossing off all rows except $i_1, i_2, \ldots, i_k$ and all columns except $i_1, i_2, \ldots, i_k$.

Test the following statement for several real, symmetric matrices A. The signs of the submatrices are the same if $i_1, i_2, \ldots, i_n$ is any ordering of the integers $1, 2, \ldots, n$, that is, if the submatrices are generated with $i_1, \ldots, i_n$ an arbitrary ordering of $1, 2, \ldots, n$.

16. Let $a_1x_1 + a_2x_2 + a_3x_3 = z_0$ define a plane in E^3. Let $\mathbf{p} = (p_1, p_2, p_3)$ be a point in E^3. Find the equation of the line through $\mathbf{p}$ such that the distance from $\mathbf{p}$ to the plane along that line is minimal.

17. Find all critical points of each of the following functions:

    (a) $f(x_1, x_2) = x_1^2 x_2^2 - 2x_1 x_2^2 - 12x_1 x_2 + x_2^2 + 6x_2 x_1^2 + 6x_2 + 9x_1^2 - 18x_1 + 9$

    (b) $f(x_1, x_2, x_3) = (x_1, x_2, x_3) \begin{pmatrix} 1 & -\frac{5}{4} & -1 \\ -\frac{3}{4} & 2 & -1 \\ 2 & 1 & 3 \end{pmatrix} \begin{pmatrix} x_1 \\ x_2 \\ x_3 \end{pmatrix}$

    (c) $f(x_1, x_2, x_3, x_4) = x_1^2 - 2x_1 x_3 + \frac{1}{2}x_1 x_4 + 3x_2 x_4 - 2x_2^2 + x_1 x_4 + x_3^2 - \frac{1}{2}x_4^2 - 2x_4 x_3 - 9x_1 + 8x_2 + 2x_4$

18. For each of the following hypothetical characteristic equations, tell what conclusions are possible regarding the definiteness of the corresponding matrix without approximating any roots.

    (a) $p(x) = x^3 - x^2 - 4x + 4 = 0$
    (b) $p(x) = x^4 - 5x^2 + 4 = 0$
    (c) $p(x) = -x^5 - x^4 + 4x^3 + 4x^2 = 0$
    (d) $p(x) = x^5 - 4x^4 - x^3 + 16x^2 - 12x = 0$
    (e) $p(x) = x^4 - x^3 - x^2 + 1 = 0$

19. In two dimensions, show that line and line segment as *defined* actually are what they should be.

20. For an arbitrary real matrix A of size $n \times n$, with elements a_{ij}, form the matrix B, where

$$b_{ij} = \frac{a_{ij} + a_{ji}}{2}, \quad \text{for } i, j = 1, \ldots, n$$

With respect to quadratic forms, observe and discuss the importance of B relative to A.

21. Draw a simple region of definition S in the plane and illustrate in E^3 that the conclusion of Theorem 2.4 is not valid for extrema on S's boundary.

22. Consider the function

$$f(x, y) = (x - a^2 y^2)(x - b^2 y^2)$$

with a, b given. Has f a minimum at $(0, 0)$? Has f a maximum at the origin among points (x, y) on the parabola

$$x = \frac{a^2 + b^2}{2} \cdot y^2 ?$$

Is the origin a relative extremum among points on a line, $y = cx$, through the origin? Are there any other extrema on $y = cx$?

(The importance of this example of Peano, as cited by Hancock [3] and other authors, is that it provided a counterexample to an intended sufficient condition formulated by Lagrange.)

23.
$$\min \sum_{j=1}^{n} \sqrt{x_j}$$

subject to

$$\sum_{j=1}^{n} x_j = C$$

$$x_j \geq 0$$

(Nonnegativity can be automatically satisfied.)

24. Find the rectangle of largest area that has perimeter $= 24$ units.

25. Find the cylinder of largest volume that has surface area $= 40$ square units.

26. $(1, -2, 3) = \mathbf{y}$ is a point in E^3. Find a point (x_1, x_2, x_3) that will minimize

$$\sum_{i=1}^{3} |y_i - x_i|$$

and that satisfies

$$\sum_{i=1}^{3} x_i = 8$$

27. By looking at past data, a firm's management has decided that if it manufactures x_j units of product $j, j = 1, 2, 3$, during some time period, then a good approximation to its net profit is

$$p(x_1, x_2, x_3) = -\tfrac{1}{2}x_1^2 - 5x_2^2 + 2x_3^2 + 2x_1x_2 + 6x_1 + 5x_2 + 20x_3$$

Each unit of product x_1, x_2, and x_3 requires 2, 1, and 3 units, respectively, of resource 1, and the 1000 units of resource 1 must be utilized. Respective per unit requirements for resource 2 are 1, 3, and 4 units, and all 500 units must be exhausted. Find production quantities so that profit is maximized. Observe whether or not the solution makes sense in the real world. What information is given by the values of the multipliers?

Apply the sufficient condition arising from equation (2.29) to verify that the solution is globally optimal.

28. (a) There are three towns assumed located in a plane, and for an appropriate scaling and some origin their locations are $(-4, 8)$, $(6, 0)$, $(-2, -5)$. A new fire station is to be located to serve the three towns. The towns have respective populations of 4000, 3000, and 6000. It is desired to locate the new station so as to minimize

$$\sum_i \left(\frac{\text{population of town } i}{\text{total population}}\right) \times (\text{distance to station})^2$$

Towns 1 and 2 are ruthlessly competing for a new thumbtack factory,

each wanting the fire station nearer it than its rival; so to compromise, the new station must be precisely equidistant from towns 1 and 2.
Find an optimal location for the fire station.

(b) The thumbtack factory, not wishing to become involved in a feud, decided to locate in town 3. Five years later, the respective populations are 1000, 1000, and 15,000. Is the fire station still optimally located, given that towns 1 and 2 yet demand their distance constraint?
Use the sufficiency test for both subproblems.

29. Comment on the solution of the following problem, where m, n are large. How might it be solved without treating such a large problem?

$$\max f(x_1, \ldots, x_n) = \sum_{k=1}^{r} f_k(\mathbf{x}_k)$$

subject to

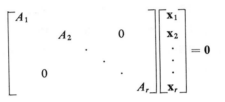

A_k is an $m_k \times n_k$ real matrix, $\mathbf{x}_k$ is an $n_k \times 1$ vector, and

$$\sum_{k=1}^{r} m_k = m \quad \text{and} \quad \sum_{k=1}^{r} n_k = n$$

30. Solve each of the following problems, establish that points found are truly global extrema, state whatever information is given by the multipliers, and verify whatever information is given regarding the constraints.

(a) $$\min x_1^2 + x_2^2 + x_3^2$$

subject to

$$x_1 + \tfrac{3}{4}x_2 - \tfrac{3}{8}x_3 = 4$$
$$\tfrac{8}{3}x_1 + 2x_2 - x_3 = \tfrac{32}{3}$$

(b) $$\min x_1^2 + x_2^2 + 2x_3^2 - 2x_3(x_1 + x_2)$$

subject to

$$x_1 + x_2 \qquad - 2x_4 = 4$$
$$-2x_1 + 3x_2 - x_3 + 4x_4 = 2$$
$$x_2 - \tfrac{1}{3}x_3 \qquad = 2$$

(c) $$\min 4x_1^2 - 4x_1x_2 + 4x_1x_3 + x_2^2 - 2x_2x_3 + x_3^2$$

subject to

$$\tfrac{1}{2}x_1 - 2x_2 \qquad + x_4 = 4$$
$$-\tfrac{5}{3}x_1 + \tfrac{7}{2}x_2 - x_3 - x_4 = -5$$
$$-2x_1 + \tfrac{3}{2}x_2 - x_3 \qquad = -1$$

(d)
$$\min x_1^2 x_2^2 - 2x_1 x_2$$

subject to

$$
\begin{aligned}
2x_1 + \tfrac{1}{2}x_2 - x_3 &= 1 \\
-x_1 \qquad\quad + 2x_3 + x_4 &= 0 \\
2x_2 - x_3 - 3x_4 &= 2
\end{aligned}
$$

(e)
$$\min x_1^2 x_2^2 - 2x_1 x_2$$

subject to

$$
\begin{aligned}
x_1^2 + 2x_2 - x_3 + x_4^2 &= 4 \\
2x_1 x_2 \qquad + x_3 - x_4^2 &= -1 \\
x_1 x_3 - 2x_2^2 \qquad\qquad &= -2
\end{aligned}
$$

31. Derive the LaGrange conditions for one constraint and two variables under the assumption that

$$\frac{\partial g(x_0, y_0)}{\partial x} \neq 0$$

32. The ABC Refuse Disposal Company has a problem. They have been asked by the What Hath God Wrought Iron Company to haul 500,000 cubic feet of waste material across a particular river, since dumping the waste into the river no longer seems the thing to do. To establish a fee, ABC decided to compute the *smallest cost* for which it can do the job, and then request, as a fee, that amount plus some exorbitant fraction thereof.

    ABC plans to construct a rectangular bargelike vessel and to utilize one of its tugboats to haul barge loads of waste across the river. Regardless of the dimensions of the barge, each round trip will cost *$100.00*.

    The bottom of the barge must be constructed of material costing $5 per square foot, while the material for the sides costs $2 per square foot.

    The maximum barge load is equal to the volume so described. The bottom of the barge will have a salvage value equal to one fourth its original cost. Find dimensions of the barge that will minimize the total cost of the operation, subject to the additional constraint that exactly 5000 square feet of material are to be used for construction. Solve the same problem with no additional constraint.

33. Find the dimensions of the rectangular solid of largest volume whose total surface area is 100 square units and whose girth is 20 units. Next, verify whether your solution to the constrained problem gives the proper indication of the effect of decreasing either the surface area or the girth.

34. Construct a max problem of $f(x_1, x_2)$, subject to a single constraint, where the saddle point conditions (2.35) are false for the Lagrangian F.

35. The concepts of convex set and convex function are closely related. Let f be a convex function over all of E^n. Define

$$X = \{\mathbf{x} \mid f(\mathbf{x}) \leq a\}$$

for a given real number a. Show that X is convex. If a is the minimum value of f on X, what is the conclusion?

36. Show that the following characterization of a function f is equivalent to its being convex: The Hessian of f is positive semidefinite.

37. Find all solutions to the necessary conditions for each of the following problems:
    (a) $$\min x_1^2 - 2x_1 + x_2^2 - 2x_2 + 2$$
    subject to
    $$4x_1 - x_2 \geq 4$$
    $$-3x_1 - x_2 \geq 18$$

    (b) $\min x_1^2 + x_2^2 - 2x_1 + 25$ subject to the same constraints as in (a).
    (c) $\max x_1^2 - 2x_1 + x_2^2 - 2x_2 + 2$ subject to the same constraints as in (a).
    (d) $\max x_1^2 + x_2^2 - 4x_1 - 24x_2 + 160$ subject to the same constraints as in (a).
    (e) Obtain solutions to the problems geometrically and compare with the points found. (This may be facilitated by factoring the objective functions.)

38. Putney Products is diversifying and will begin manufacturing artificial flowers in addition to pipe cleaners. It is their guess that during each of the first few months of the expanded product line manufacturing x_1 cartons of pipe cleaners and x_2 cartons of artificial flowers will net them an approximate profit of

$$p(x_1, x_2) = \frac{x_1}{x_2 + 1} + 4x_1 x_2$$

Now, as is well known, the two resources needed for these particular products are wire and fuzz. A carton of pipe cleaners requires 2 pounds of wire and $3\frac{1}{2}$ pounds of fuzz, whereas a carton of artificial flowers requires $\frac{1}{2}$ pound of wire and 10 pounds of fuzz. Storage space is limited at P.P., and 1000 pounds of wire and 1 ton of fuzz may be stored. How many cartons of each product should be made to maximize $p(x_1, x_2)$? (May one *disregard* non-negativity constraints?)

39. Let the probability of a profit on a stock sold at time t_2 after being bought at $t_1 < t_2$ be

$$P = \int_{t_1}^{t_2} \frac{1}{\pi} \frac{\alpha}{\alpha^2 + x^2}\, dx, \quad \alpha > 0$$

The integrand is just the Cauchy density function. (What is the expected value of that distribution?) Assuming that one can see into the future in that fashion, select times t_1, t_2 whose squares sum to no more than 4, and where $t_1 \leq 0$, such that P is maximized.

40. Under the same conditions as Problem 39, maximize

$$E = \int_{t_1}^{t_2} \frac{x}{\pi} \frac{\alpha}{\alpha^2 + x^2}\, dx, \quad \alpha > 0$$

What *is* E?

41. Find the point of the set $S = \{(x, y) | x^2 + y^2 \leq 1\}$ that lies nearest the curve $x^2y - 2x + 3y = 10$ and that lies in the nonnegative quadrant.

42. Solve Problem 41, but neglect nonnegativity.

43. Solve geometrically and by means of the Kuhn–Tucker conditions:

$$\max -(x - 1)^2 - (y - 5)^2$$

subject to

$$-x^2 + y - 4 \leq 0$$
$$-(x - 2)^2 + y - 3 \leq 0$$

44. As in Problem 43, treat the problem

$$\max x - y$$

subject to

$$x + y \leq 1$$
$$x, y \geq 0$$

45. As in Problem 43, treat the problem

$$\max x^2 + y^2$$

subject to

$$x + y \leq 2$$
$$x, y \geq 0$$

46. Use the method of including equality constraints one at a time and solve the following problems:

(a) $$\max u^2 + v^2$$

   subject to

$$u + v \leq 3$$
$$u, v \geq 0$$

(b) $$\max x_1 - 2x_2$$

   subject to

$$-x + y \leq 1$$
$$x + y \leq 5$$
$$\tfrac{2}{3}x + y \leq 4$$
$$x, y \geq 0$$

(c) $$\max x_1^2 + x_2^2 + x_3^2 + 2x_1x_2 + 3x_1x_3$$

   subject to

$$2x_1 + 3x_2 + 6x_3 \leq 24$$
$$0 \leq x_3 \leq 5$$
$$0 \leq x_2 \leq 10$$

47. Solve each of the following problems by the iterative method.

(a)
$$\min xy^2 + \frac{y + 2}{x} + x^2$$

subject to

$$x, y \geq 0$$

(b)
$$\min x^2 + y^2 + 4x + 8y + 20$$

subject to

$$x, y \geq 0$$

(c)
$$\min x_1^2 + x_2^2 + 8x_3^2 + 4x_1 x_2$$

subject to

$$x_1, x_2, x_3 \geq 0$$

(d)
$$\max x_1 - x_2$$

subject to

$$2x_1 + 3x_2 = 6$$

$$x_1, x_2 \geq 0$$

48. Derive the Kuhn–Tucker conditions for the problem

$$\min f(\mathbf{x})$$

subject to

$$h_i(\mathbf{x}) \leq b_i, \quad i = 1, \ldots, m$$

49. Derive the conditions that were presented for problem (2.48) in the text (page 96).

50. Do Problem 49 where the objective is minimization.

51. Consider the problem

$$\max f(x)$$

subject to

$$x \leq b$$

Suppose that f is concave and that $x^* < b$. What then can be said about neglecting the constraint?

52. Let $f(\mathbf{x})$ be defined on a convex set X.
(a) If the set $Y = \{(\mathbf{x}, z) \mid f(\mathbf{x}) \leq z\}$ is convex, is f convex?
(b) Illustrate the set Y in two dimensions for both convex and nonconvex f.

53. A function property of related but lesser importance than convexity is *quasi-convexity*. A function f defined on convex X is said to be quasi-convex if
1. $Y = \{\mathbf{x} \mid f(\mathbf{x}) \leq a\}$ is convex for all real a, or, equivalently,
2. $f(\lambda \mathbf{x}_2 + (1 - \lambda)\mathbf{x}_1) \leq \max [f(\mathbf{x}_2), f(\mathbf{x}_1)]$ for all $0 \leq \lambda \leq 1$ and all $\mathbf{x}_1, \mathbf{x}_2 \in X$.

(a) Show that (1) and (2) are indeed equivalent.

(b) Draw a function that is quasi-convex but not convex.

(c) Give the mathematical expression for a function as in (b). (It may be difficult to find one of these.)

(d) Why might a mathematical programming problem in which all constraints are quasi-convex be of interest?

54. Prove that for text problem (2.40) (page 93) with f concave and the h_i convex, with x_0, λ_0 feasible for the problem, and $F(x, \lambda)$ the Lagrangian

(a) $F(x, \lambda_0)$ is a concave function of feasible x.

(b) $F(x_0, \lambda)$ is a convex function of feasible λ.

55. Using Everett's method, solve the problem

$$\min x_1^2 + x_2^2 - \tfrac{1}{2}x_1 - \tfrac{5}{2}x_2 + \tfrac{13}{8}$$

subject to

$$x_2 - x_1^2 \leq 3$$
$$x_1 + x_2 \leq 5$$

56. Same objective function as Problem 55, but subject to the constraints

$$x_1 \leq 0$$
$$x_1 + x_2 \leq 5$$

57. Same objective function as Problem 55, but with the constraints

$$x_1 + x_2 \geq 5$$
$$-x_1 + x_2 \geq 5$$
$$x_2 \leq 10$$

58. Solve the problem

$$\max -x_1^2 - 4x_2^2 - 4x_1x_2^2 + 32x_1 + 64x_2 - 256$$

subject to

$$-\tfrac{3}{2}x_1^2 + x_2 \geq 2$$
$$-\tfrac{7}{4}x_1 + x_2 \leq 7$$

59. Apply the necessary condition to the following problem. Find the curve $y(x)$ that passes through the points (x_0, y_0), (x_1, y_1) and that minimizes the integral of $y(x)$ over the interval $[x_0, x_1]$. How would you describe the result? Does this problem have a global minimum with the usual assumptions regarding, F, y, and so on?

60. It is desired to find the curve $y(x)$ joining the points $(2, 1)$, $(8, 5)$ that minimizes over $[2, 8]$ the sum of the integrals of $(y'(x))$ and $4y(x)$. Find such a $y(x)$, sketch it, and compute the minimum value of the integral.

61. Write several functionals for which the Euler equation gives no information.

LINEAR PROGRAMMING

3.1. Introduction

In this chapter we shall address ourselves to a particular class of constrained optimization problems. By this time we hope to have convinced our readers that the general constrained optimization problem poses a most difficult task. The general mathematical-programming problem is, in fact, not solvable at the present.

The student has most probably begun to realize, in the course of his studies to date, the importance of the concept of *linearity* in areas of science and engineering. As an elementary example, consider interpolation in a table of logarithms. To find a close approximation to the log of a number not appearing in the table, we *assume* the log function to be linear over a small interval and are thereby able to approximate the desired quantity.

In electrical engineering the study of linear systems is of central importance. In control theory, methods for controlling linear systems are developed in detail. Mathematics itself is composed in no negligible part of studies of linear structures—linear transformations, linear groups, linear differential equations, and systems of linear equations, to name several examples.

This pervasive interest in linearity is not accidental, as we might suspect,

3

and the principal reason for this focus of attention is that what we might term *linear problems* are almost invariably more amenable to solution than are nonlinear problems. In itself, this is not sufficient justification for interest in linear problems, for problems of interest, as we have seen and shall continue to realize, are not usually predisposed to being linear. Thus the development of sophisticated techniques for dealing with linear problems might be challenged on pragmatic grounds. But it is on pragmatic grounds, actually, that we may justify such development; for quite often devices for solving nonlinear problems either do not exist or are so unwieldy as to preclude explicit solution. In such cases it is often expedient, even necessary, to approximate the given problem by one that is solvable or more easily solvable. Thus the substitution of a linear problem for a nonlinear one is a frequent strategy.

That linear problems are not inherently of interest is not an inference that should be drawn from this brief discussion; rather we have attempted to offer some additional motivation for our work in this chapter. A succinct and descriptive assessment is due to Philip Wolfe [7], who said, ". . . linearity represents our compromise between reality and the limitations of our tools for dealing with it."

A principal stimulus for the application of mathematical methods to problems of management, planning, and the like, was World War II. As a result of this emphasis, linear programming formally came into being in 1947, and is, therefore, a quite recent addition to the repertoire of optimization techniques.

G. B. Dantzig is credited with the founding of the theory of and solution methods for linear-programming problems, and in his book [2], Dantzig gives an interesting history of not only the work of his particular group, but also of relevant work prior to 1947. Principal contributions and contributors subsequent to that date are also mentioned, and the reader would find this historical sketch of interest.

The Soviet economist L. V. Kantorovich was apparently the first to formulate, in the late 1930s [4], certain realistic industrial-type problems as linear optimization problems, although he offered little in the way of solution methodology or mathematical analysis of the problems. His work did not receive, even in his own country, the attention it merited.

3.2. The Problem and Its Structure

Toward defining a linear-programming problem, let us first define a mathematical-programming problem, which we might roughly describe as the allocation of some quantity of a resource or resources so as to optimize (maximize or minimize) some function of these allocations, while simultaneously satisfying a set of constraints, themselves functions of the allocations. The constrained optimization problems of Chapter 2 certainly qualify as mathematical-programming problems.

i. *Definitions and Conventions*

As one might suspect, a linear-programming problem is a mathematical-programming problem in which both the function to be optimized, called the *objective function*, and the constraining functions are linear. Therefore, a typical objective function is

$$f(x_1, \ldots, x_n) = c_1 x_1 + c_2 x_2 + \cdots + c_n x_n$$

where c_i, $i = 1, \ldots, n$, are real numbers. A typical constraint takes one of the forms

$$g_i(x_1, \ldots, x_n) = a_{i1} x_1 + \cdots + a_{in} x_n \leq b_i$$
$$g_j(x_1, \ldots, x_n) = a_{j1} x_1 + \cdots + a_{jn} x_n = b_j$$
$$g_k(x_1, \ldots, x_n) = a_{k1} x_1 + \cdots + a_{kn} x_n \geq b_k$$

where $a_{ir}, a_{jr}, a_{kr}, b_i, b_j, b_k, r = 1, \ldots, n$ are real numbers. Although they qualify on the grounds of linearity, constraints of the form

$$a_{i1}x_1 + \cdots + a_{in}x_n < b_i$$
$$a_{j1}x_1 + \cdots + a_{jn}x_n > b_j$$

will not be admissible, for reasons to be described later.

The exclusion of strict inequalities does not constitute any decided restriction on the kinds of real-world problems that we are interested in solving. Optimization problems in business and industry usually involve constraints of the kind "the greatest amount of money that can be spent on this activity is b_i dollars," in which case we expect a $\leq$ constraint, or "the volume of the cargo hold must be at least b_j cubic feet," whereupon a constraint with $\geq$ would likely develop.

Furthermore, since we have decided that our problems will involve tangible resources, we demand that the variables over which the optimization will be performed be nonnegative, although later we shall learn how this restriction may be eliminated.

Thus a typical linear-programming problem becomes

$$\max \text{ or } \min z = c_1 x_1 + c_2 x_2 + \cdots + c_n x_n$$

subject to

$$
\begin{aligned}
a_{i1}x_1 + a_{i2}x_2 + \cdots + a_{in}x_n &\leq b_i, && \text{for } i = 1, \ldots, r_1 \\
a_{i1}x_1 + a_{i2}x_2 + \cdots + a_{in}x_n &\geq b_i, && \text{for } i = r_1 + 1, \ldots, r_2 \\
a_{i1}x_1 + a_{i2}x_2 + \cdots + a_{in}x_n &= b_i, && \text{for } i = r_2 + 1, \ldots, m \\
x_i &\geq 0, && \text{for } i = 1, \ldots, n
\end{aligned}
\tag{3.1}
$$

In matrix-vector notation we write

$$\max \text{ or } \min z = \mathbf{cx} \tag{3.2}$$

subject to

$$A\mathbf{x} (\leq \geq =) \mathbf{b} \tag{3.3}$$

$$\mathbf{x} \geq \mathbf{0} \tag{3.4}$$

where $\mathbf{c} = (c_1, c_2, \ldots, c_n)$.†

---

†Although they may represent costs, profits, etc., the c_j will always be termed costs.

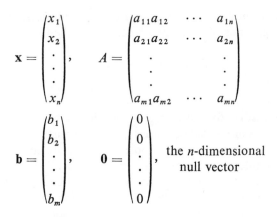

$$\mathbf{x} = \begin{pmatrix} x_1 \\ x_2 \\ \cdot \\ \cdot \\ \cdot \\ x_n \end{pmatrix}, \qquad A = \begin{pmatrix} a_{11}a_{12} & \cdots & a_{1n} \\ a_{21}a_{22} & \cdots & a_{2n} \\ \cdot & & \cdot \\ \cdot & & \cdot \\ \cdot & & \cdot \\ a_{m1}a_{m2} & \cdots & a_{mn} \end{pmatrix}$$

$$\mathbf{b} = \begin{pmatrix} b_1 \\ b_2 \\ \cdot \\ \cdot \\ \cdot \\ b_m \end{pmatrix}, \qquad \mathbf{0} = \begin{pmatrix} 0 \\ 0 \\ \cdot \\ \cdot \\ 0 \end{pmatrix}, \quad \begin{matrix} \text{the } n\text{-dimensional} \\ \text{null vector} \end{matrix}$$

The quantity c_i is known as the *cost* of x_i. The matrix A is known as the *coefficient matrix*, and the vector **b** is known as the *requirements vector*. Columns of A are called *activity vectors*.

By ($\leq \geq =$) we imply that any of and, in fact, all these types of constraints may occur in a given problem.

A vector $\mathbf{x} = (x_1, x_2, \ldots, x_n)$ satisfying expression (3.3) is called a *solution* to the problem. If, in addition, **x** satisfies inequality (3.4), it is called a *feasible solution*.

If $\mathbf{x}^*$ is a feasible solution and $\mathbf{cx}^* \leq (\geq) \mathbf{cx}$ for all feasible solutions **x**, then $\mathbf{x}^*$ is an *optimal* solution to the min (max) problem, if $\mathbf{cx}^*$ is finite.

Thus feasible solutions that produce infinite values of the objective function are not considered optimal. If our linear objective function assumes an infinite value, then, clearly, some of the variables must be infinite. We have, then, the following type of situation—we are attempting, for example, to perform a maximization and discover that we may make the objective function as large as we choose, while at the same time satisfying the constraints, by allowing certain of the variables to become infinite. We have, however, decided that our problems would represent real-world situations, characterized by finite quantities of expendable resources.

If so, the unboundedness mentioned contradicts the underlying nature of the problem, and we must conclude that the existence of an unbounded solution indicates an erroneous or incomplete formulation of the problem.

From our work in classical optimization techniques, the general linear-programming problem appears to be solvable, in principle anyway, by methods previously described. We have observed the shortcomings of these devices and, in the case of a linear-programming problem, a more subtle difficulty, not obvious from the form of the problem, also exists; this will be discussed later in problems involving two variables.

Let us first gain some insight into the geometry of linear-programming problems by considering properties of two- and three-dimensional constraints and objective functions.

ii. *Some Geometry of Linear-Programming Problems*

Consider the straight lines $ax_1 + bx_2 = c_1, ax_1 + bx_2 = c_2, \ldots, ax_1 + bx_2 = c_n$. These are, of course, just a set of parallel lines in E^2. For example, if $c_n > c_{n-1} > \cdots > c_3 > c_2 > c_1$, we might have the family of parallel lines shown in Figure 3.1 or 3.2.

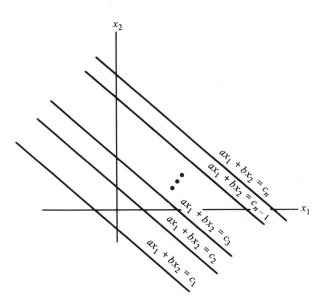

Figure 3.1. A family of lines $ax_1 + bx_2 = c_i$ $(i = 1, \ldots, n)$.

Therefore, in two dimensions, the sets of points satisfying $ax_1 + bx_2 \leq c$ and $ax_1 + bx_2 \geq c$ might be the sets A and B, respectively, both of which include the line $ax_1 + bx_2 = c$ itself† (Figure 3.3).

If in the example in Figure 3.3 we simultaneously demand $x_1 \geq 0$, $x_2 \geq 0$, our sets A, B shrink to those shown in Figure 3.4, that is, to subsets of the nonnegative quadrant of the plane.

Were the constraint $ax_1 + bx_2 = c$, with $x_1, x_2 \geq 0$, then the set of points would be the heavy line segment labeled D in Figure 3.4. Situations may also arise in which there is but a single feasible point, or no feasible point. For

---

†Clearly, the direction of increasing c and that portion of the plane corresponding to $ax_1 + bx_2 \leq c$ or $ax_1 + bx_2 \geq c$ depend upon the actual equation.

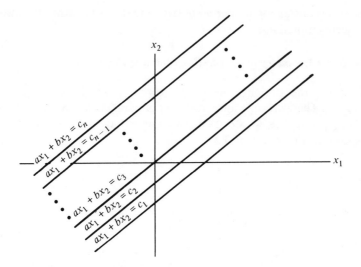

Figure 3.2. A family of lines $ax_1 + bx_2 = c_i$ ($i = 1, \ldots, n$).

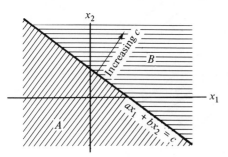

Figure 3.3. Sets of points in the plane satisfying $ax_1 + bx_2 \lesseqqgtr c$.

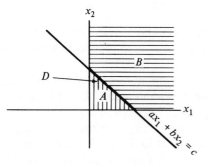

Figure 3.4. Effect of nonnegativity restrictions.

example, in two dimensions the point $(0, 0)$ is the only point satisfying $x_1 + x_2 \leq 0$, $x_1 \geq 0$, $x_2 \geq 0$. There are no points (x_1, x_2) satisfying $x_1 + x_2 \leq -1$, $x_1 \geq 0$, $x_2 \geq 0$.

Since a feasible solution to a linear-programming problem must simultaneously satisfy a number of constraints of the kind mentioned and illustrated, the set of all feasible solutions to a linear-programming problem must therefore consist of the *intersection* of a number of the kinds of sets we have mentioned.

To see this, let us consider another example in two dimensions.

Consider the problem

$$\max z = c_1 x_1 + c_2 x_2$$

subject to

$$a_1 x_1 + a_2 x_2 \leq a_3$$
$$e_1 x_1 + e_2 x_2 \leq e_3$$
$$d_1 x_1 + d_2 x_2 \leq d_3$$
$$x_1, x_2 \geq 0$$

The set of feasible solutions to this problem may be represented by the boundaries and the shaded region in Figure 3.5. The dashed lines represent three of the family of parallel lines representing three distinct values z_0, z_1, z_2 of the objective function, where $z_2 < z_0 < z_1$.

The problem of finding the maximum value of the objective function subject to the constraints may then be considered to be the selection of a real number z_0 and, consequently, a straight line $c_1 x_1 + c_2 x_2 = z_0$ such that $c_1 x_1 + c_2 x_2 = z_0$ intersects the set of feasible solutions, and $c_1 x_1 + c_2 x_2 = z'$ does not intersect that set if $z' > z_0$. All points lying in the intersection, then, are optimal solutions to the problem.

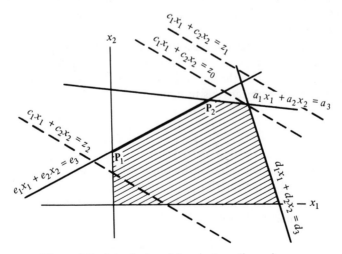

Figure 3.5. A typical problem in two dimensions.

Regarding the example of Figure 3.5, it becomes obvious that z_0 is the maximum value, and the point of intersection of the lines $d_1x_1 + d_2x_2 = d_3$, $a_1x_1 + a_2x_2 = a_3$ is the unique, optimal solution.

Had the family of possible objective functions been parallel to one of the constraining lines, there would have existed an infinite number of optimal solutions, that is, a portion of the boundary of the set of feasible solutions. For example, with respect to Figure 3.5 again, suppose the objective function that we were minimizing had been $e_1x_1 + e_2x_2$, and that e_3 were the minimum value. Then every point on the heavy segment joining P_1 and P_2—including P_1 and P_2—corresponds to an optimal solution.

Figure 3.5 and similar drawings used to illustrate the geometry of linear-programming problems are, of course, nothing more than contour maps such as appeared in Chapter 2. Here, the contours or surfaces of constant function value are simply lines or planes—depending upon the dimension of the illustration.

a. Extrema on Boundaries

It may be observed that if any straight line is chosen as a member of a family of parallel lines corresponding to fixed objective function values and moved parallel to its original position, so as to achieve the optimal value, while maintaining intersection with the feasible solution set, then optimal solutions correspond *only* to boundary points of the feasible set, *never to interior points.*

It may be shown that in a space of arbitrary dimension, optimal solutions to linear-programming problems, if optimal solutions exist, occur only on the boundaries of the sets of feasible solutions. (This is another reason why classical methods of analysis are of little use in solving linear-programming problems. The difficulties accruing to such situations have been discussed in Chapter 2.)

In two or three dimensions this fact is obvious from geometric considerations. In the diagram shown in Figure 3.6, let $\mathbf{x}^* = (x_1^*, x_2^*)$ be assumed optimal, and let the line through it be an optimal line for the problem; that is, suppose that the objective is to maximize

$$c_1x_1 + c_2x_2 = z$$

Then

$$c_1x_1^* + c_2x_2^* = z^*$$

would be the maximum value of the objective function.

Since (x_1^*, x_2^*) is an interior point, there is a neighborhood of it entirely within the feasible set. It is seen that there are feasible points (x_1', x_2') within any such neighborhood with

$$c_1x_1' + c_2x_2' = z' > z^*$$

so $\mathbf{x}^*$ cannot be optimal. It is also clear that the slope of our family of lines

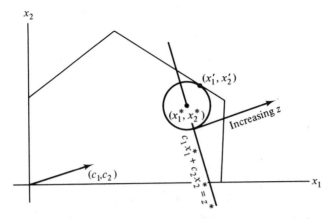

Figure 3.6. Optimal solutions lie only on the boundaries.

(all of which, of course, are perpendicular to the assumed direction of increasing z) and the choice of a maximization as opposed to minimization do not change the result. We ask the reader to provide a proof of the following:

Theorem 3.1

An optimal solution to a linear-programming problem in n variables can only lie on the boundary of the set of feasible solutions. (After proving the result in the fashion indicated above, the reader should observe Problem 8 in Chapter 2.)

Theorem 3.1 illustrates the problem with constraints that are strict inequalities. With all or a portion of the feasible set *boundary* removed, there may be *no* optimal solution.

Beyond three dimensions, however, geometric insight and intuition are of no use in proving statements about linear-programming problems, and the proofs of some of the properties that are so obvious in two or three dimensions are, for n dimensions, somewhat complicated. We wish to emphasize, however, that the insights and understanding gained from observing the geometry of the two- and three-dimensional situations are invaluable in understanding what transpires in n dimensions. (This advice is not restricted to the study of linear programming.)

Suppose that the problem

$$\max z = c_1 x_1 + c_2 x_2$$

subject to

$$a_1 x_1 + a_2 x_2 \leq a_3$$

$$e_1 x_1 + e_2 x_2 \geq e_3$$

$$x_1, x_2 \geq 0$$

has the form shown in Figure 3.7, where the objective function has been plotted for two arbitrary values $z_1, z_2, z_1 < z_2$. In this case it is clear that the objective function may be made arbitrarily large, while the constraints are satisfied. Thus we have an example of a problem with an unbounded solution.

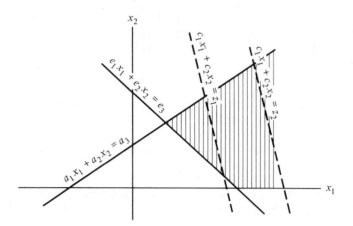

Figure 3.7. A problem with an unbounded solution.

With reference to Figure 3.7, again, if the objective function had been $a_1x_1 + a_2x_2$, the maximum value of the objective function would certainly be finite, a_3 in fact; and in addition to an infinite number of solutions where both x_1, x_2 are finite, there also would exist optimal solutions for which both x_1, x_2 were arbitrarily large.

As another example, consider a set of constraints

$$a_1x_1 + a_2x_2 \leq a_3$$
$$e_1x_1 + e_2x_2 \leq e_3$$
$$x_1, x_2 \geq 0$$

as appearing in Figure 3.8. Here we have a case where there is no feasible solution to the problem.

Our geometric insight may be extended to three dimensions, in which objective functions with fixed values describe planes, sets of feasible solutions are subsets of E^3, and the maximization or minimization process may be viewed as moving a plane through three-dimensional space so as to maintain an intersection of the plane with the three-dimensional feasible solution set.

b. EXTREME POINTS AND CONVEX SETS

We have had an opportunity to observe another theoretically important property of optimal solutions. If there happened to occur a unique optimal

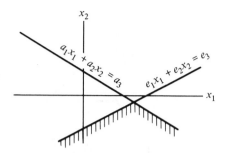

Figure 3.8. A problem with no feasible region.

solution, then a point representing a "corner" of the feasible region provided this solution (see, e.g., Figure 3.5). Furthermore, in case there were an infinite number of optimal solutions, such a "corner" point provided at least one of them (Figure 3.5 again).

This property, too, may be shown to be the rule in n dimensions. What we have imprecisely termed "corners" are called *extreme points* of the set of feasible solutions.

An extreme point may also be described as a point $\mathbf{x}$ of the set such that there do *not* exist distinct points $\mathbf{x}_1, \mathbf{x}_2$, belonging to the set with $\mathbf{x}$ lying on the segment joining $\mathbf{x}_1, \mathbf{x}_2$ and $\mathbf{x} \neq \mathbf{x}_1, \mathbf{x} \neq \mathbf{x}_2$. In other words, $\mathbf{x}$ does not lie *between* two different points of the set.

To express the definition in terms of the parametric form of the line segment joining two points, we are asserting the *nonexistence* of $\mathbf{x}_1 \neq \mathbf{x}_2$ such that

$$\mathbf{x} = \lambda \mathbf{x}_2 + (1 - \lambda)\mathbf{x}_1$$

for some

$$\lambda, 0 < \lambda < 1$$

and the strict inequalities involving λ convey the notion of "between."

In Figure 3.9, for example, $\mathbf{P}_0, \mathbf{P}_1, \mathbf{P}_2, \mathbf{P}_3,$ and $\mathbf{P}_4$ are the only extreme points. $\mathbf{P}_5$ and $\mathbf{P}_6$ are not extreme points.

This definition of extreme point raises a new question, however. Consider the line segment joining any two points of the feasible region in a linear-programming problem. Will the segment lie wholly within the region? It will be noticed that this is the case in every feasible region considered thus far, and it may be proved that the set of feasible solutions to a linear-programming problem is always such that every point on the line segment joining any pair of points in the set lies within the set. Such a set is called *convex*, and convex sets play a most important role in the theory of linear programming. (Convex sets were defined and discussed in Chapter 2.)

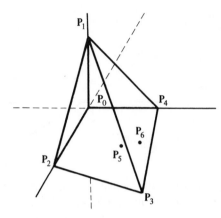

Figure 3.9. An example of extreme points in three dimensions.

c. CONVEXITY OF THE FEASIBLE SET

It is straightforward to establish that the set of feasible solutions to a linear-programming problem is always a convex set. Let $(x_1', x_2', \ldots, x_n') = \mathbf{x}'$ and $(x_1'', x_2'', \ldots, x_n'') = \mathbf{x}''$ be feasible solutions. Consider a constraint

$$a_{i1}x_1 + a_{i2}x_2 + \cdots + a_{in}x_n \leq b_i \tag{3.5}$$

Does

$$\lambda \mathbf{x}'' + (1 - \lambda)\mathbf{x}' = \mathbf{x}$$

satisfy that constraint for $0 \leq \lambda \leq 1$? We note that

$$a_{i1}(\lambda x_1'' + (1 - \lambda)x_1') + a_{i2}(\lambda x_2'' + (1 - \lambda)x_2')$$
$$+ \cdots + a_{in}(\lambda x_n'' + (1 - \lambda)x_n')$$
$$= \lambda(a_{i1}x_1'' + \cdots + a_{in}x_n'') + (1 - \lambda)(a_{i1}x_1' + \cdots + a_{in}x_n')$$
$$= \lambda\xi_1 + (1 - \lambda)\xi_2$$

where

$$\xi_1 = a_{i1}x_1'' + \cdots + a_{in}x_n'' \leq b_i$$
$$\xi_2 = a_{i1}x_1' + \cdots + a_{in}x_n' \leq b_i$$

Consequently,

$$\lambda\xi_1 + (1 - \lambda)\xi_2 \leq b_i$$

and $\mathbf{x}$ satisfies expression (3.5).

In the same way one can show $\mathbf{x}$ to be feasible for $\geq$ type constraints. $\mathbf{x}$ is clearly feasible for equality constraints, and, finally, $\mathbf{x}$ has nonnegative components, of course.

d. Optimal Solutions and Extreme Points

It follows that the set of feasible solutions to a linear-programming problem is a convex set. The importance of convex sets to the theory of linear programming can be seen as follows.

Recalling that our planes are called *hyperplanes* in n dimensions, we have that an optimal hyperplane (if there is one) will have the property of intersecting the feasible set at *boundary points* only; intersection with interior points of the feasible set is forbidden by Theorem 3.1.

Such a hyperplane is called a *supporting* hyperplane or a *support* of X, the feasible set. Consider the example of Figure 3.10 in two dimensions. The hyperplanes $h_1, \ldots, h_5$ are supporting hyperplanes; h_6 is not.

The following theorem is stated without proof.

Theorem 3.2

If X is the set of feasible solutions to a linear-programming problem and h is a supporting hyperplane of X, then h contains an extreme point of X.

The truth of the theorem is well illustrated by Figure 3.10, and it is to be noted that mere convexity of a set does not make the theorem applicable. (The problems at the end of the chapter touch upon that point.) But convexity is certainly necessary for the application of the theorem.

Now, whereas Theorem 3.1 asserts that one may consider only the *boundary* of X when searching for an optimal solution, Theorem 3.2 states that the search may be confined to the *extreme points of X alone*.

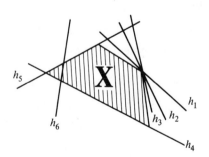

Figure 3.10. Supporting hyperplanes in two dimensions.

e. Introduction to the Computational Procedure

The implication here then is that if we had available a technique for constructing all the extreme points of a feasible set, an optimal solution, if one existed, could be obtained by merely evaluating the objective function at each of the extreme points. Obviously, were the number of extreme points large, such a technique would become impractical; but it will be the case that the computational methods to be described in this chapter are essentially just refinements of this fundamental idea.

The computational procedure to be presented here is known as the *simplex method* and was named by its originators. Simplices are special kinds of convex sets, and in some of the early linear-programming problems treated, the feasible solution sets were simplices—hence, the name. Simplices are sufficiently important constructs—not only in linear programming—that simplectic geometry is itself an area of study in mathematics.

3.3. Solution of Linear-Programming Problems

i. *Slack and Surplus Variables*

Consider once again the linear-programming problem

$$\text{max or min } z = c_1 x_1 + c_2 x_2 + \cdots + c_n x_n$$

subject to the constraints (3.1).

Obtaining even a feasible solution to this problem requires the solution of a system of linear equalities and inequalities. We know that the solution of a system of linear equations is straightforward, at least in principle, whereas, solutions of systems of inequalities are not so. Thus a first approach to obtaining a feasible solution to the original might be the transformation of that system into an equivalent, in some sense, system of equations.

To do so turns out to be a simple task. Let us define a set of *nonnegative* variables

$$x_{n+1}, x_{n+2}, \ldots, x_{n+r_2}$$

and write the system (3.1) as

$$
\begin{aligned}
a_{i1}x_1 + a_{i2}x_2 + \cdots + a_{in}x_n + x_{n+i} &= b_i, &&\text{for } i = 1, \ldots, r_1 \\
a_{i1}x_1 + a_{i2}x_2 + \cdots + a_{in}x_n - x_{n+i} &= b_i, &&\text{for } i = r_1 + 1, \ldots, r_2 \\
a_{i1}x_1 + a_{i2}x_2 + \cdots + a_{in}x_n \phantom{{}- x_{n+i}} &= b_i, &&\text{for } i = r_2 + 1, \ldots, m \\
x_i &\geq 0, &&\text{for } i = 1, 2, \ldots, n + r_2
\end{aligned}
\tag{3.6}
$$

In words, we have *added* a different nonnegative quantity to the left side of each of the $\leq$ constraints to force equality, and equalities have been obtained in place of the $\geq$ constraints by *subtracting* a different nonnegative quantity from those left-hand sides. The original equalities, being already in the desired form, are not modified.

The variables x_{n+i}, $i = 1, \ldots, r_1$, are termed *slack variables*, since they "take up the slack" allowed by the $\leq$. The variables x_{n+j}, $j = r_1 + 1, \ldots, r_2$, are termed *surplus variables*, since they "use up the surplus" allowed by the $\geq$.

Example 3.1

Consider the system

$$
\begin{aligned}
2x_1 - \tfrac{1}{2}x_2 + x_3 + x_4 &\leq 2 \\
x_1 + 2x_2 + 2x_3 - 3x_4 + x_5 &\geq 3 \\
x_1 \quad\quad - x_3 + x_4 - x_5 &\geq \tfrac{2}{3} \\
3x_1 - x_2 \quad\quad + 2x_4 - \tfrac{3}{2}x_5 &= 1 \\
x_i \geq 0, \; i = 1, \ldots, 5
\end{aligned}
$$

Let x_6 be a slack variable, and x_7 and x_8 surplus variables; then the system becomes

$$
\begin{aligned}
2x_1 - \tfrac{1}{2}x_2 + x_3 + x_4 \quad\quad + x_6 \quad\quad &= 2 \\
x_1 + 2x_2 + 2x_3 - 3x_4 + x_5 \quad\quad - x_7 \quad &= 3 \\
x_1 \quad\quad - x_3 + x_4 - x_5 \quad\quad\quad - x_8 &= \tfrac{2}{3} \\
3x_1 - x_2 \quad\quad + 2x_4 - \tfrac{3}{2}x_5 \quad\quad\quad &= 1 \\
x_i \geq 0, \; i = 1, \ldots, 8
\end{aligned}
$$

a. EQUIVALENCE OF THE EQUALITY-CONSTRAINED PROBLEM TO THE GIVEN PROBLEM

In what sense are systems (3.1) and (3.6) equivalent? First, suppose that $x'_1, x'_2, \ldots, x'_n$ satisfy (3.1). Then, clearly, there exists a unique set of non-negative real numbers $x'_{n+1}, x'_{n+2}, \ldots, x'_{n+r_2}$, such that

$$
x'_1, \ldots, x'_n, x'_{n+1}, \ldots, x'_{n+r_2}
$$

satisfy system (3.6). The important thing for us here is that the first n components of a feasible solution to system (3.6) comprise a feasible solution to system (3.1).

Next consider the system (3.6) as constraints on the *original objective function*. That is, given the problem

$$
\text{max or min } z = \mathbf{c}\mathbf{x}
$$

subject to

$$
\begin{aligned}
A\mathbf{x} \, (\leq \geq =) \, \mathbf{b} \\
\mathbf{x} \geq 0
\end{aligned}
$$

(3.7)

we can write a problem

$$\text{max or min } z = \hat{\mathbf{c}}\hat{\mathbf{x}}$$

subject to

$$Q\hat{\mathbf{x}} = \mathbf{b}$$
$$\hat{\mathbf{x}} \geq \mathbf{0}$$

(3.8)

where $\hat{\mathbf{c}} = (c_1, c_2, \ldots, c_n, 0, 0, \ldots, 0)$ (i.e., the costs of slack and surplus variables are made zero).

$$\hat{\mathbf{x}} = \begin{pmatrix} x_1 \\ x_2 \\ \cdot \\ \cdot \\ \cdot \\ x_n \\ x_{n+1} \\ \cdot \\ \cdot \\ \cdot \\ x_{n+r_2} \end{pmatrix}$$

$$Q = \begin{pmatrix}
a_{11} & a_{12} & \cdots & a_{1n} & 1 & 0 & \cdots & 0 & 0 & 0 \\
a_{21} & a_{22} & \cdots & a_{2n} & 0 & 1 & & 0 & 0 & 0 \\
\cdot & & & \cdot & & & & \cdot & \cdot & \cdot \\
\cdot & & & \cdot & & & & \cdot & \cdot & \cdot \\
\cdot & & & \cdot & & & & \cdot & \cdot & \cdot \\
a_{r_1,1} & a_{r_1,2} & \cdots & a_{r_1,n} & 0 & 0 & \cdots & 1 & 0 & 0 \\
a_{r_1+1,1} & a_{r_1+1,2} & \cdots & a_{r_1+1,n} & 0 & 0 & \cdots & 0 & -1 & 0 \\
\cdot & & & \cdot & & & & & 0 & \cdot & \cdot \\
\cdot & & & \cdot & & & & & \cdot & \cdot & 0 \\
a_{r_2,1} & a_{r_2,2} & \cdots & a_{r_2,n} & 0 & 0 & \cdots & 0 & 0 & -1 \\
a_{r_2+1,1} & a_{r_2+1,2} & \cdots & a_{r_2+1,n} & 0 & 0 & \cdots & \cdot & \cdot & 0 \\
\cdot & & & \cdot & & & & & & \cdot \\
\cdot & & & \cdot & & & & & & \cdot \\
a_{m1} & a_{m2} & \cdots & a_{mn} & 0 & 0 & \cdots & \cdot & \cdot & 0
\end{pmatrix}$$

Suppose now that $\hat{\mathbf{x}}^* = (x_1^*, x_2^*, \ldots, x_{n+r_2}^*)$ is an optimal solution to problem (3.8). Then it must be that $\mathbf{x}^* = (x_1^*, \ldots, x_n^*)$ is an optimal solution to problem (3.7). To see this, we need only assume the contrary; that there exist feasible $x_1', x_2', \ldots, x_n'$ such that $z' = \sum_{i=1}^{n} c_i x_i' > \sum_{i=1}^{n} c_i x_i^* = z^*$. But, then, as we saw above we can find $x_{n+1}', x_{n+2}', \ldots, x_{n+r_2}'$ such that $x_1', \ldots, x_{n+r_2}'$ is a feasible solution to problem (3.8); consequently, we would have

$$\sum_{i=1}^{n} c_i x_i' + \sum_{i=n+1}^{n+r_2} 0x_i' > \sum_{i=1}^{n} c_i x_i^* + \sum_{i=n+1}^{n+r_2} 0x_i^*$$

which implies that $\hat{x}^*$ was not optimal for (3.8). The contradiction is established, and we have that the first n components of an optimal solution to (3.8) constitute an optimal solution to (3.7). (A maximization problem has been assumed; the proof for the minimization problem should be obvious.) Now, at least, we may turn our attention to the solution of a system of equations, rather than a system of inequalities and equations. That is, we may assume our linear-programming problems have only equality constraints, excluding those for nonnegativity.

Solutions to systems of equations are discussed in Chapter 1.

b. PREPARATORY REMARKS

Our strategy will be one of obtaining a feasible solution, and then modifying it iteratively until either an optimal solution is found or it is discovered that none exists. It may be the case, as we have seen, that a given system of constraints has no feasible solution; that is, the feasible region is null. The system

$$x_1 + 2x_2 + x_3 \leq 4$$
$$x_1 - x_2 - 2x_3 \leq 2$$
$$2x_1 + x_2 - x_3 \geq 8$$

for example, is such a system, since if the first two relations hold, the third cannot. In large problems, of course, such inconsistencies may not be at all obvious, but it is a convenient property of the computational devices to be developed here that no preliminary investigations along these lines are necessary. The *simplex method* may be applied to such a system, and the lack of feasibility is signaled by the occurrence of certain events in the simplex computation itself.

ii. *Basic Feasible Solutions*

Let us consider a particular class of linear-programming problems and temporarily defer discussion of the general problem.

Consider

$$\text{max or min } z = \mathbf{c}\mathbf{x}$$

subject to

$$A\mathbf{x} \leq \mathbf{b}, \quad \mathbf{b} \geq 0$$
$$\mathbf{x} \geq 0$$

(3.9)

A system of constraining equations may be obtained in this case by the addition of slack variables alone, as we have seen, and we may consider the problem

$$\text{max or min } z = c_1 x_1 + c_2 x_2 + \cdots + c_n x_n$$

subject to

$$
\begin{aligned}
a_{11}x_1 + a_{12}x_2 + \cdots + a_{1n}x_n + x_{n+1} &= b_1 \\
a_{21}x_1 + a_{22}x_2 + \cdots + a_{2n}x_n \quad\quad + x_{n+2} &= b_2 \\
&\vdots \\
a_{m1}x_1 + a_{m2}x_2 + \cdots + a_{mn}x_n \quad\quad\quad + x_{n+m} &= b_m \\
x_1, x_2, \ldots, x_{n+m} &\geq 0
\end{aligned}
\tag{3.10}
$$

Obtaining a feasible solution to this problem is a triviality. The fact that we have exactly one slack variable in each equation and that each slack appears in exactly one equation allows solution by inspection. Certainly, $x_{n+1} = b_1, x_{n+2} = b_2, \ldots, x_{n+m} = b_m, x_1 = x_2 = \cdots = x_n = 0$ is effective. This solution, of course, gives an objective function value of zero and may, indeed, be far from optimal.

In the system (3.10) let P_j be a vector denoting the jth column of the coefficient matrix. Then (3.10) may be written

$$x_1 P_1 + x_2 P_2 + \cdots + x_{n+m} P_{n+m} = b$$

or

$$\sum_{j=1}^{n+m} x_j P_j = b \tag{3.11}$$

For the particular solution found, we have

$$\sum_{j=n+1}^{n+m} x_j P_j = b$$

Since column P_j $(n + 1 \leq j \leq n + m)$ is the $(j - n)$th unit vector in E^m, the set $\{P_{n+1}, P_{n+2}, \ldots, P_{n+m}\}$ forms a *basis* for E^m; consequently, the solution $x_i = 0$, $x_j = b_j$ $(i = 1, \ldots, n$ and $j = n + 1, \ldots, n + m)$ is called a *basic* solution. The vectors P_j (forming a basis) are called the *basis vectors*. The x_j corresponding to the basis vectors are known as *basic variables*. By virtue of the feasibility of the solution, we have found what is called a *basic feasible solution*. [Also, we have that the rank of our coefficient matrix in system (3.10) is m.]

In similar fashion, by selecting *any* m linearly independent columns, setting the variables *not* associated with these columns to zero, and solving the resulting system, we obtain a basic solution, but not necessarily a feasible solution. Further mention of basic solutions is to be found in Chapter 1.

The simplex method deals in basic feasible solutions, and it generates them by changing bases—one vector at a time—having begun with an initial basis.

a. BASIC FEASIBLE SOLUTIONS AND EXTREME POINTS

Before proceeding with the computational aspects of our problem, let us digress briefly upon the importance of basic feasible solutions to linear-programming problems.

We have observed that if an optimal solution exists for a linear-programming problem, then at least one extreme point of the set of feasible solutions will be an optimal solution. It is also possible to establish a correspondence between extreme points and basic feasible solutions; in fact, it may be proved that every extreme point is a basic feasible solution, and that every basic feasible solution corresponds to an extreme point. (The correspondence is not precisely one to one in all cases, but this detail need not be pursued here.) For us, the important consequence is the following. Since an optimal solution, if one exists, must lie among the set of extreme points of the set of feasible solutions, and since every such extreme point constitutes a basic feasible solution, examination of all possible basic feasible solutions must reveal an optimal solution.

For our system of m equations in $n + m$ variables, where the rank of the coefficient matrix has been shown to be m, there are at most $\binom{n+m}{m}$ basic solutions, since there are precisely that many different sets of m columns from the coefficient matrix, or that many sets of n columns, whose associated variables in equation (3.11) will be set to zero. Of course, there will be this number of basic solutions only if every set of m columns is linearly independent; therefore, we say *at most* $\binom{n+m}{m}$ basic solutions, also recalling that not all these need be feasible. It should also be understood that not all the m variables we are *permitting* to assume nonzero values need be nonzero. A basic feasible solution with less than m nonzero variables is termed a *degenerate basic feasible solution*.

The relationship between extreme points and basic feasible solutions may be easily illustrated in two dimensions.

Example 3.2

Consider the system

$$x_1 + x_2 \leq 4$$
$$2x_1 + x_2 \leq 6 \qquad (3.12)$$
$$x_1, x_2 \geq 0$$

Points satisfying these inequalities are represented by the shaded area of Figure 3.11.

The extreme points of the set of feasible solutions are $(0, 0)$, $(0, 4)$, $(2, 2)$, and $(3, 0)$.

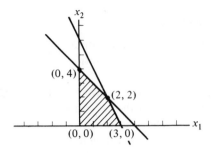

Figure 3.11. Feasible region and extreme points for Example 3.2.

Next consider the system of inequalities obtained by the introduction of slack variables into system (3.12):

$$
\begin{aligned}
x_1 + x_2 + x_3 \quad\quad &= 4 \\
2x_1 + x_2 \quad\quad + x_4 &= 6
\end{aligned}
\tag{3.13}
$$

By setting all possible pairs of variables to zero in system (3.13) and solving the resulting systems, we obtain as basic solutions

$$
\begin{array}{llll}
x_1 = 0, & x_2 = 0, & x_3 = 4, & x_4 = 6 \\
x_1 = 0, & x_2 = 4, & x_3 = 0, & x_4 = 2 \\
x_1 = 0, & x_2 = 6, & x_3 = -2, & x_4 = 0 \\
x_1 = 4, & x_2 = 0, & x_3 = 0, & x_4 = -2 \\
x_1 = 2, & x_2 = 2, & x_3 = 0, & x_4 = 0 \\
x_1 = 3, & x_2 = 0, & x_3 = 1, & x_4 = 0
\end{array}
$$

The extreme points $(0, 0)$, $(0, 4)$, $(2, 2)$, and $(3, 0)$ correspond, respectively, to the basic feasible solutions $(0, 0, 4, 6)$, $(0, 4, 0, 2)$, $(2, 2, 0, 0)$, and $(3, 0, 1, 0)$. Every extreme point is obtained from a basic *feasible* solution to the associated system of equations.

iii. *Computational Techniques*

Let us now choose an example problem of the type (3.9) and begin studying the computational technique in terms of a specific problem. Consider

$$
\max z = 4x_1 + 2x_2 + x_3 + 3x_4
$$

subject to

$$2x_1 + x_2 \qquad\qquad + x_4 \le 4$$
$$2x_1 - x_2 + 2x_3 \qquad\qquad \le \tfrac{3}{2}$$
$$3x_1 + 2x_2 + x_3 + 2x_4 \le 8$$
$$x_1, x_2, x_3, x_4 \ge 0$$

a. Obtaining the Equality-Constrained Problem

We first obtain the new problem

$$\max z = 4x_1 + 2x_2 + x_3 + 3x_4$$

subject to

$$2x_1 + x_2 \qquad\quad + x_4 + x_5 \qquad\qquad\qquad = 4$$
$$2x_1 - x_2 + 2x_3 \qquad\qquad + x_6 \qquad = \tfrac{3}{2} \qquad (3.14)$$
$$3x_1 + 2x_2 + x_3 + 2x_4 \qquad\qquad + x_7 = 8$$
$$x_1, \ldots, x_7 \ge 0$$

An initial basic feasible solution is, then, $x_5 = 4$, $x_6 = \tfrac{3}{2}$, $x_7 = 8$, $x_1 = x_2 = x_3 = x_4 = 0$.

b. Representing the Information

A conventional and useful representation for a linear-programming problem solution is the *tableau*. Let P_j denote the jth column of the coefficient matrix in system (3.14), and let b_i denote the ith component of the requirements vector $\mathbf{b}$. The basis is clearly $\{P_5, P_6, P_7\}$. We may represent the information we have available in tableau form as follows:

Cost Coefficients of the Basic Variables	Basis	P_1	P_2	P_3	P_4	P_5	P_6	P_7	b
0	P_5	2	1	0	1	1	0	0	4
0	P_6	2	−1	2	0	0	1	0	$\tfrac{3}{2}$
0	P_7	3	2	1	2	0	0	1	8

(3.15)

Adjacent to each basis vector we write the cost coefficient of the associated basic variable. For any solution $x_1, x_2, \ldots, x_7$ to the system (3.14), we have $x_1 P_1 + x_2 P_2 + \cdots + x_7 P_7 = \mathbf{b}$, and for the given basic feasible solution, we have

$$x_5 P_5 + x_6 P_6 + x_7 P_7 = \mathbf{b} \qquad (3.16)$$

Note that the basis $\{\mathbf{P}_5, \mathbf{P}_6, \mathbf{P}_7\}$ is the 3×3 identity matrix. Denoting a basis by B, expression (3.16) says

$$B \cdot \begin{pmatrix} x_5 \\ x_6 \\ x_7 \end{pmatrix} = \begin{pmatrix} 4 \\ \frac{3}{2} \\ 8 \end{pmatrix} \tag{3.17}$$

or

$$\begin{pmatrix} x_5 \\ x_6 \\ x_7 \end{pmatrix} = B^{-1} \cdot \begin{pmatrix} 4 \\ \frac{3}{2} \\ 8 \end{pmatrix}$$

But $B^{-1} = I$ for the particular basis here; so we have

$$\begin{pmatrix} x_5 \\ x_6 \\ x_7 \end{pmatrix} = \begin{pmatrix} 4 \\ \frac{3}{2} \\ 8 \end{pmatrix}$$

which is the solution given above.

c. Obtaining a New Basic Feasible Solution

As indicated earlier, our goal is the generation of basic feasible solutions until one is identified as being optimal, or until we discover that no optimal solution exists. Let us defer the question of optimality temporarily and consider the finding of other basic feasible solutions. One strategy then would be to choose a different *basis* and find the appropriate x's to satisfy an equation similar to expression (3.16). We shall seek to change the basis *one vector at a time*, that is, to select a nonbasis vector and replace a basis vector with it.

It may be shown that if $\{\boldsymbol{\beta}_1, \boldsymbol{\beta}_2, \ldots, \boldsymbol{\beta}_k\}$ is a basis and there exists a vector $\boldsymbol{\alpha}$ such that $\boldsymbol{\alpha} = \gamma_1 \boldsymbol{\beta}_1 + \gamma_2 \boldsymbol{\beta}_2 + \cdots + \gamma_k \boldsymbol{\beta}_k$ and $\gamma_r \neq 0$, then $\{\boldsymbol{\beta}_1, \boldsymbol{\beta}_2, \ldots, \boldsymbol{\beta}_{r-1}, \boldsymbol{\alpha}, \boldsymbol{\beta}_{r+1}, \ldots, \boldsymbol{\beta}_k\}$ is also a basis. For example, in terms of tableau (3.15), since $\mathbf{P}_1 = 2\mathbf{P}_5 + 2\mathbf{P}_6 + 3\mathbf{P}_7$, $\{\mathbf{P}_1, \mathbf{P}_6, \mathbf{P}_7\}$ is a basis, as are $\{\mathbf{P}_5, \mathbf{P}_1, \mathbf{P}_7\}$ and $\{\mathbf{P}_5, \mathbf{P}_6, \mathbf{P}_1\}$. Similarly, $\{\mathbf{P}_2, \mathbf{P}_6, \mathbf{P}_7\}$, $\{\mathbf{P}_5, \mathbf{P}_2, \mathbf{P}_7\}$, and $\{\mathbf{P}_5, \mathbf{P}_6, \mathbf{P}_2\}$ are bases. Clearly, there are many such alternatives, but how should we proceed?

The concern is essentially twofold; not all bases provide feasible solutions, and some, even though providing a feasible solution, may provide an objective function value that is worse than one at hand, which seems undesirable for an iterative optimization technique.

Let y_{ij} denote the element in the tableau corresponding to the jth column (beneath $\mathbf{P}_j$) and the ith row (identifying the ith basis vector). For example in (3.15), $y_{11} = 2$, $y_{31} = 3$, $y_{22} = -1$, and $y_{37} = 1$. Suppose that we select some

$\mathbf{P}_j$ not in the basis to enter into the basis. It may be shown that if the rth† basis vector is the one removed, where r is determined by

$$\frac{b_r}{y_{rj}} = \min_{1 \le i \le m} \left\{ \frac{b_i}{y_{ij}} \middle| y_{ij} > 0 \right\}^{\ddagger} = Q(j) \qquad (3.18)$$

then the new basic solution obtained will also be feasible. [The new basic variable will assume the value $Q(j)$, where $Q(j) \ge 0$, since $b_i \ge 0$ for all i.] Furthermore, if this rule is violated, the new basic solution is *guaranteed* to be infeasible.

In the particular example under consideration, i ranges from 1 to 3 and the $\mathbf{P}_j$ eligible for entry at the first iteration are $\mathbf{P}_1$, $\mathbf{P}_2$, $\mathbf{P}_3$, and $\mathbf{P}_4$.

It may also be shown (Problem 3.13) that if the current value of the objective function is z then the new value of the objective function will be, after $\mathbf{P}_j$ replaces the rth basis variable,

$$z' = z + Q(j)(c_j - z_j) \qquad (3.19)$$

where

$$z_j = \sum_i (\text{cost coefficient of } i\text{th basic variable}) \cdot y_{ij} \qquad (3.20)$$

We know that $Q(j)$ is always nonnegative, by its very definition. Therefore, if we are interested in maximizing, we should select a vector $\mathbf{P}_j$ for which $c_j - z_j$ is positive, for then $z' \ge z$. Similarly, if minimization is the goal, a vector $\mathbf{P}_j$ for which $c_j - z_j$ is negative should be entered, since in that case $z' \le z$.

Clearly, then, if we wish to achieve the largest increase or decrease in the objective function value for one basis change, j should be chosen so that

$$|Q(j)(c_j - z_j)| = \max |Q(k)(c_k - z_k)|$$

for all k such that $\mathbf{P}_k$ is not in the basis, and $Q(j)(c_j - z_j) > 0$ for the maximization problem, and $Q(j)(c_j - z_j) < 0$ for the minimization problem. Thus for each possible k such that $\mathbf{P}_k$ is not in the basis, we would compute $Q(k)$—simultaneously determining the vector to leave the basis in case $\mathbf{P}_k$ enters—and select the *pair* of entering and departing vectors so as to achieve maximum improvement in the objective function value.

---

†Order within the basis only is referenced here; that is, the rth basis vector may be $\mathbf{P}_1, \mathbf{P}_5, \mathbf{P}_3$, etc.

‡The notation implies that we examine all quotients b_i/y_{ij} where $y_{ij} > 0$. j is fixed once the vector to enter the basis is chosen. The origins of criterion (3.18) will be investigated later.

In practice, however, this is *not* done; the entering vector P_j is chosen according to

$$|c_j - z_j| = \max|c_k - z_k| \qquad (3.21)$$

for all k such that P_k is not in the basis and $c_k - z_k > 0$ (max problem) or $c_k - z_k < 0$ (min problem). Thus we apparently make a sacrifice in the rapidity of convergence to optimality, since we may not obtain maximum improvement at each step. The sacrifice is actually slight, as the reader may notice if he compares the two methods in the solution to a problem. Furthermore, the obvious saving in computation makes the second alternative more desirable—a single computation for $Q(j)$, say Q, is sufficient.

Therefore, j is first selected to satisfy equation (3.21) and r is next determined to satisfy equation (3.18). The nonbasis vector P_j replaces the rth vector in the present basis, and since

$$Q = \frac{b_r}{y_{rj}} \qquad (3.22)$$

the resulting new objective function value is

$$z' = z + Q(c_j - z_j) \qquad (3.23)$$

Returning to our problem, let us reproduce tableau (3.15), adding the $c_j - z_j$:

c_j	*Basis*	P_1	P_2	P_3	P_4	P_5	P_6	P_7	b	
0	P_5	2	1	0	1	1	0	0	4	
0	P_6	2	-1	2	0	0	1	0	$\frac{3}{2}$	(3.24)
0	P_7	3	2	1	2	0	0	1	8	
	$c_j - z_j$	4	2	1	3	0	0	0		

Since the cost of each basic variable is zero, equation (3.20) shows that $z_j = 0$ for all j, and, in this case, then $c_j - z_j = c_j$. According to criterion (3.21), P_1 should *enter* the basis ($j = 1$); that is, x_1 should become a basic variable. To specify the vector to *leave* the basis (equivalently, the current basic variable that we shall drive to zero), we merely employ expression (3.18), and consider

$$\min\left\{\frac{b_1}{y_{11}}, \frac{b_2}{y_{21}}, \frac{b_3}{y_{31}}\right\} = \min\left\{\frac{4}{2}, \frac{3/2}{2}, \frac{8}{3}\right\}$$

The minimum occurred for $i = 2$, so consequently the second basic vector P_6, which corresponds to $Q = (3/2)/2 = \frac{3}{4}$, is to be removed.

d. TRANSFORMING THE TABLEAU

How is the change accomplished? To answer this question, we shall have to be a little more dogmatic than previously—at least for the present. In tableau (3.24) we find the unit vector $(0, 1, 0)$ under $\mathbf{P}_6$, the vector to be removed. What we wish to do is to construct the same unit vector under $\mathbf{P}_1$, the vector to enter the basis, by performing elementary row operations on the entire array of numbers in the tableau.

Our first operation here is to divide the second row of the array by 2, in order to obtain a 1 in the proper position. Thus our second row becomes

$$1 \quad -\tfrac{1}{2} \quad 1 \quad 0 \quad 0 \quad \tfrac{1}{2} \quad 0 \quad \tfrac{3}{4}$$

We then multiply the new second row by -2 and add the result to the first row, yielding a new first row. In a similar fashion, the new second row is multiplied by -3 and added, term by term, to the third row, resulting in a new third row. $\mathbf{P}_1$ has now replaced $\mathbf{P}_6$ in the basis, and we have

c_j	Basis	$\mathbf{P}_1$	$\mathbf{P}_2$	$\mathbf{P}_3$	$\mathbf{P}_4$	$\mathbf{P}_5$	$\mathbf{P}_6$	$\mathbf{P}_7$	b
0	$\mathbf{P}_5$	0	2	-2	1	1	-1	0	$\tfrac{5}{2}$
4	$\mathbf{P}_1$	1	$-\tfrac{1}{2}$	1	0	0	$\tfrac{1}{2}$	0	$\tfrac{3}{4}$
0	$\mathbf{P}_7$	0	$\tfrac{7}{2}$	-2	2	0	$-\tfrac{3}{2}$	1	$\tfrac{23}{4}$

(3.25)

The new basic feasible solution becomes

$$x_1 = \tfrac{3}{4}, \qquad x_5 = \tfrac{5}{2}, \qquad x_7 = \tfrac{23}{4}$$
$$x_2 = x_3 = x_4 = x_6 = 0$$

with the new basic variable x_1 assuming the value Q.

The reader should note that although we defined the $\mathbf{P}_j$ as columns in (3.15), we have continued to label the columns in that fashion. The numbers y_{ij} beneath a $\mathbf{P}_j$ heading may readily be identified, at any stage, as the real numbers that (uniquely) express the original vector $\mathbf{P}_j$ in terms of the current basis. For example, in tableau (3.25) consider the column labeled $\mathbf{P}_2$. It is

$$\begin{pmatrix} 2 \\ -\tfrac{1}{2} \\ \tfrac{7}{2} \end{pmatrix}$$

$$2\mathbf{P}_5 - \tfrac{1}{2}\mathbf{P}_1 + \tfrac{7}{2}\mathbf{P}_7 = 2\begin{pmatrix} 1 \\ 0 \\ 0 \end{pmatrix} - \tfrac{1}{2}\begin{pmatrix} 2 \\ 2 \\ 3 \end{pmatrix} + \tfrac{7}{2}\begin{pmatrix} 0 \\ 0 \\ 1 \end{pmatrix} = \begin{pmatrix} 1 \\ -1 \\ 2 \end{pmatrix} = \mathbf{P}_2$$

Also, the last column in the tableau identifying the basic variable values ($\frac{5}{2}$, $\frac{3}{4}$, $\frac{23}{4}$) has still been denoted as **b**, although it is different from the original requirements vector **b**.

e. Understanding the Simplex Iteration

The knowledge of the identity of the numbers y_{ij} affords an understanding of the development of equation (3.23). Suppose that at some point we have a basis $B = \{P_1, \ldots, P_m\}$, where the subscripts have been so chosen for notational convenience. If P_j is a vector not in the basis B, then

$$P_j = y_{1j}P_1 + y_{2j}P_2 + \cdots + y_{mj}P_m \qquad (3.26)$$

Furthermore, if our basic feasible solution is $(x_1^0, \ldots, x_m^0)$, we have

$$x_1^0 P_1 + x_2^0 P_2 + \cdots + x_m^0 P_m = b$$

Suppose that P_j is the vector entering the basis, and let x_j be the corresponding new basic variable. Multiplying both sides of expression (3.26) by x_j and adding their difference to this last expression, we obtain

$$x_1^0 P_1 + \cdots + x_m^0 P_m + x_j P_j - x_j(y_{1j}P_1 + \cdots + y_{mj}P_m) = b$$

or

$$(x_1^0 - x_j y_{1j})P_1 + \cdots + (x_m^0 - x_j y_{mj})P_m + x_j P_j = b \qquad (3.27)$$

Clearly, $(x_1^0 - x_j y_{1j}, \ldots, x_m^0 - x_j y_{mj}, x_j)$ is a new solution with a corresponding objective function value.

$$c_1(x_1^0 - x_j y_{1j}) + \cdots + c_m(x_m^0 - x_j y_{mj}) + c_j x_j$$
$$= c_1 x_1^0 + \cdots + c_m x_m^0 - x_j(c_1 y_{1j} + \cdots + c_m y_{mj}) + c_j x_j$$
$$= \sum_{i=1}^{m} c_i x_i^0 + x_j \left(c_j - \sum_{i=1}^{m} c_i y_{ij} \right)$$
$$= z + x_j(c_j - z_j)$$

where

$$z = \sum_{i=1}^{m} c_i x_i^0 \quad \text{and} \quad z_j = \sum_{i=1}^{m} c_i y_{ij}$$

Here we have ignored the fact that one of $\{x_1^0, \ldots, x_m^0\}$ is to become zero, and that x_j must assume a particular value (i.e., Q) in order for a feasible solution to occur when it becomes nonzero. Also, of course, not every $P_j \notin B$ may be brought in; at any rate, the origins of equation (3.23) and consequently the criterion (3.21) should be clearer at this point.

Finally, a feeling for the choice of a departing vector, as summarized in equation (3.22), is available from equation (3.27). The choice of a departing vector must be such that the new variable values must satisfy the nonnegativity conditions, and hence equation (3.22) will accomplish precisely this, as may be observed. Furthermore, if $\mathbf{P}_k$ is the basis vector to be removed, then necessarily $x_k^0 - x_j y_{kj} = 0$. Problems 3.12 and 3.13 also deal with these concepts.

f. INTERPRETING THE $c_j - z_j$

Expression (3.23) assists in conveying a good intuitive interpretation of the numbers $c_j - z_j$; it asserts that for every unit of value x_j is allowed to assume the objective function value will increase by an amount $c_j - z_j$. If we imagine x_j as being zero and begin to increase it continuously, then, of course, if a feasible solution is to be maintained, the other variable values will, in general, vary continuously at the same time. Thus it must be realized that (3.23) reflects these changes in the other variables.

At any rate, $c_j - z_j$ is a very explicit measure of the desirability of allowing x_j to become positive. For this reason, the values $c_j - z_j$ have been called *marginal costs* (profits) or *shadow prices*. (We shall later observe additional uses for these numbers.) Finally, we see that the method of choosing a variable to become basic consists in choosing a variable which has the maximum available *rate* of improving the objective function value.

g. A PROPERTY OF THE TABLEAU

Another interesting property of any simplex tableau is the following: at every stage the columns corresponding to the identity matrix in the initial tableau (our beginning bases will always be identity matrices) will contain B^{-1}, where B is the basis at that stage. In tableau (3.25), for example, the fifth, sixth, and seventh columns correspond to the identity matrix columns in tableau (3.24) and define the matrix

$$A = \begin{pmatrix} 1 & -1 & 0 \\ 0 & \frac{1}{2} & 0 \\ 0 & -\frac{3}{2} & 1 \end{pmatrix}$$

Corresponding to tableau (3.25), the basis is

$$B = (\mathbf{P}_5, \mathbf{P}_1, \mathbf{P}_7) = \begin{pmatrix} 1 & 2 & 0 \\ 0 & 2 & 0 \\ 0 & 3 & 1 \end{pmatrix}$$

Clearly, $AB = BA = I$, which implies $A = B^{-1}$.

The reader is asked to prove the existence of this property in general in the problems.

h. CONTINUATION OF EXAMPLE

The new objective function value in tableau (3.25) is $z = 3$, which could have been calculated, using equation (3.23), in advance. Also, the newly entered variable assumes the value Q. The reader should attempt to remove a vector other than $\mathbf{P}_6$ from the basis in tableau (3.24), given that $\mathbf{P}_1$ will enter, to observe the loss of feasibility that results if a vector other than that determined by condition (3.18) is removed from the basis. In light of condition (3.18), the reader should have asked himself how a departing vector is determined if $y_{ij} \leq 0$ for all i, once the entering vector $\mathbf{P}_j$ has been specified. This question is answered by the following broader result.

It may be shown that if for some $\mathbf{P}_j$ not in the basis $c_j - z_j > 0$ (if maximizing) or $c_j - z_j < 0$ (if minimizing), and $y_{ij} \leq 0$ for all i, then an unbounded solution exists. $\mathbf{P}_j$ need *not* be that vector whose $c_j - z_j$ indicates immediate entry into the basis. It is clear that $c_j - z_j = 0$ for all vectors $\mathbf{P}_j$ in the basis.

To continue, having performed one complete simplex iteration, we would calculate the $c_j - z_j$ in tableau (3.25), determine entering and departing vectors, obtain a new solution, and so forth. But how does the process terminate, if, in fact, it does?

i. TERMINATING THE ITERATIONS AND RECOGNIZING OPTIMALITY

First, from equation (3.23) it is clear that as long as the basic feasible solutions are nondegenerate we obtain an improvement in objective function value, since Q *can* be zero only if some $b_i = 0$. (It should be noted that *degeneracy does not imply that Q will be zero.* Why?)

Let us assume that as we iterate toward an optimal solution we never obtain a degenerate basic feasible solution, and simultaneously we do not encounter the unbounded solution indication. As long as there exists $c_j - z_j > 0$,† we obtain a strict increase in the value of the objective function, which, in turn, implies that a basis (equivalently, a basic feasible solution) is never repeated. We know, however, that there are a finite number of bases (and basic feasible solutions); so strict increases can occur but a finite number

---

†We may, if we wish, confine our attention to either the maximization problem or the minimization problem rather than discussing both. By virtue of the fact that an $\mathbf{x}$ which minimizes $\mathbf{cx}$ will maximize $-\mathbf{cx}$, we may solve any linear-programming problem as a maximization problem or as a minimization problem. When such a choice is made, we shall treat the maximization problem, as we are in this discussion.

of times. Therefore, we must eventually realize $c_j - z_j \leq 0$ for every P_j not in the basis. Thus relation (3.23) would indicate no further feasible improvement. In fact, it may show that $c_j - z_j \leq 0 \,(\geq 0)$ for every P_j not in the basis implies an optimal solution to the maximization (minimization) problem.

Let us return to the current status, as exhibited by tableau (3.25) of our example problem (3.14). The $c_j - z_j$ are found to be 0, 4, −3, 3, 0, −2, 0 for $j = 1, 2, \ldots, 7$. Hence P_2 will enter the basis, and the computation of Q implies that the first basis vector, P_5, must be removed. The next tableau, then, is

c_j	Basis	P_1	P_2	P_3	P_4	P_5	P_6	P_7	b
2	P_2	0	1	−1	$\frac{1}{2}$	$\frac{1}{2}$	$-\frac{1}{2}$	0	$\frac{5}{4}$
4	P_1	1	0	$\frac{1}{2}$	$\frac{1}{4}$	$\frac{1}{4}$	$\frac{1}{4}$	0	$\frac{11}{8}$
0	P_7	0	0	$\frac{3}{2}$	$\frac{1}{4}$	$-\frac{7}{4}$	$\frac{1}{4}$	1	$\frac{11}{8}$
$c_j - z_j$		0	0	1	1	−2	0	0	

Let P_4 enter the basis (P_3 is equally eligible). P_2 must be removed and the tableau becomes

c_j	Basis	P_1	P_2	P_3	P_4	P_5	P_6	P_7	b
3	P_4	0	2	−2	1	1	−1	0	$\frac{5}{2}$
4	P_1	1	$-\frac{1}{2}$	1	0	0	$\frac{1}{2}$	0	$\frac{3}{4}$
0	P_7	0	$-\frac{1}{2}$	2	0	−2	$\frac{1}{2}$	1	$\frac{3}{4}$
$c_j - z_j$		0	−2	3	0	−3	1	0	

An optimal solution is still unavailable; P_3 enters, and P_7 must be removed, whereupon we obtain

c_j	Basis	P_1	P_2	P_3	P_4	P_5	P_6	P_7	b
3	P_4	0	$\frac{3}{2}$	0	1	−1	$-\frac{1}{2}$	1	$\frac{13}{4}$
4	P_1	1	$-\frac{1}{4}$	0	0	1	$\frac{1}{4}$	$-\frac{1}{2}$	$\frac{3}{8}$
1	P_3	0	$-\frac{1}{4}$	1	0	−1	$\frac{1}{4}$	$\frac{1}{2}$	$\frac{3}{8}$
$c_j - z_j$		0	$-\frac{5}{4}$	0	0	0	$\frac{1}{4}$	$-\frac{3}{2}$	

P_6 enters the basis, and we select P_3 for removal (P_1 would do as well), obtaining

c_j	Basis	P_1	P_2	P_3	P_4	P_5	P_6	P_7	b
3	P_4	0	1	2	1	-3	0	2	4
4	P_1	1	0	-1	0	2	0	-1	0
0	P_6	0	-1	4	0	-4	1	2	$\frac{3}{2}$
$c_j - z_j$		0	-1	-1	0	1	0	-2	

P_5 enters the basis, and P_1 must be removed. An optimal tableau is

c_j	Basis	P_1	P_2	P_3	P_4	P_5	P_6	P_7	b
3	P_4	$\frac{3}{2}$	1	$\frac{1}{2}$	1	0	0	$\frac{1}{2}$	4
0	P_5	$\frac{1}{2}$	0	$-\frac{1}{2}$	0	1	0	$-\frac{1}{2}$	0
0	P_6	2	-1	2	0	0	1	0	$\frac{3}{2}$
$c_j - z_j$		$-\frac{1}{2}$	-1	$-\frac{1}{2}$	0	0	0	$-\frac{3}{2}$	

An optimal solution is then $x_4^* = 4$, $x_6^* = \frac{3}{2}$, $x_1^* = x_2^* = x_3^* = x_5^* = x_7^* = 0$, and the maximum objective function value is $z^* = 12$.

j. Degeneracy and Cycling

The importance of the assumption regarding degeneracy should be emphasized. If we admit the possibility of degenerate solutions, then we may not conclude that strict improvement in the objective function value always occurs; consequently, *we admit the possibility of repeating bases and basic feasible solutions.*

To illustrate the potential difficulty, let x_1 be a degenerate basic feasible solution. Suppose we next obtain a basic feasible solution x_2, also degenerate (A degenerate x_1 does *not* imply that the next basic feasible solution obtained will be degenerate. Why?), with $cx_1 = cx_2$, and then obtain $x_3, \ldots, x_k$, all degenerate basic feasible solutions such that $cx_1 = cx_3 = \cdots = cx_k$. Then, suppose that we repeat a basis; that is, the next basic feasible solution obtained is x_1. Will we not iterate through this cycle, again and again, never reaching an optimal solution? Therefore, the question raised by degeneracy is serious, but, happily, not an impediment. First, the only known linear-programming problems that have cycled were a very few problems constructed in attempts to *find* problems of this nature—real-world problems do not seem to cycle. Second, methods have been derived for use in the simplex technique which assure that cycling will not occur, although these will not be discussed here.

Why should a basic feasible solution become degenerate? If we begin with a component of the requirements vector, $b_i = 0$, degeneracy is with us at the outset. But, otherwise, suppose that we have a nondegenerate basic feasible solution and determine that $\mathbf{P}_j$ should next enter the basis. Suppose that when Q is computed we find

$$Q = \min_i \left\{ \frac{b_i}{y_{ij}} \middle| y_{ij} > 0 \right\} = \frac{b_r}{y_{rj}} = \frac{b_s}{y_{sj}} = \cdots = \frac{b_v}{y_{vj}} \qquad (3.28)$$

That is, the minimum value is assumed for more than a single value of i.

The reader should be able to verify that the new b_i obtained by a simplex iteration are

$$b_i' = b_i - Q y_{ik}, \quad i = 1, 2, \ldots, r-1, r+1, \ldots, m$$
$$b_r' = Q$$

when $\mathbf{P}_k$ enters and the rth basis vector is removed, and therefore be able to demonstrate that condition (3.28) implies degeneracy in the next basic feasible solution. [Methods that prevent cycling in any event center about the choice of the vector to leave the basis when (3.28) occurs.]

k. ALTERNATIVE NOTATION

In our tableaux we have labeled basic solutions $\mathbf{b}$, and components thereof $b_1, b_2, \ldots, b_m$, although these symbols were associated with the requirements vector. To denote *basic* feasible solutions and their dependence upon the particular basis B, we shall use $\mathbf{x}(B)$; the ith component of $\mathbf{x}(B)$ will be written $x_i(B)$. It will also be convenient to reference the corresponding cost vector by $\mathbf{c}(B)$, with components $c_i(B)$. In section 3.8 we shall use $\mathbf{x}_B$, $\mathbf{c}_B$ and the like, owing to the greater complexity of the expressions.

l. SIMPLEX MANIPULATIONS IN THE SOLUTION OF SYSTEMS OF EQUATIONS

The manipulations constituting a simplex iteration are by no means motivated by subtleties. All they accomplish, in themselves, is the generation of solutions to systems of equations; they are, after all, just row operations. Consider, for example, the system

$$\begin{aligned} x_1 + 2x_2 + x_3 - x_4 &= 3 \\ x_2 + 2x_3 + x_4 &= 4 \\ x_1 \quad\quad - x_3 + 2x_4 &= 4 \end{aligned} \qquad (3.29)$$

The mechanics of transforming simplex tableaux may be used to find a solution to this system. We first eliminate x_1 from the third equation by adding the negative of row 1 to row 3, getting

$$x_1 + 2x_2 + x_3 - x_4 = 3$$
$$x_2 + 2x_3 + x_4 = 4$$
$$-2x_2 - 2x_3 + 3x_4 = 1$$

Next eliminate x_2 from both the first and third equations to obtain

$$x_1 - 3x_3 - 3x_4 = -5$$
$$x_2 + 2x_3 + x_4 = 4$$
$$2x_3 + 5x_4 = 9$$

We can use the third equation to obtain x_3 or x_4, but suppose that we choose to obtain x_4. We divide that equation by 5, and eliminate x_4 from the first and second equations, row operations again, getting

$$x_1 - \tfrac{9}{5}x_3 = \tfrac{2}{5}$$
$$x_2 + \tfrac{8}{5}x_3 = \tfrac{11}{5} \qquad (3.30)$$
$$\tfrac{2}{5}x_3 + x_4 = \tfrac{9}{5}$$

A solution is then seen to be $x_1 = \tfrac{2}{5}$, $x_2 = \tfrac{11}{5}$, $x_3 = 0$, $x_4 = \tfrac{9}{5}$. The property utilized here is that of the set of solutions to a linear system of equations remains invariant under elementary row operations—a solution to the system (3.30) must be a solution to the system (3.29), and vice versa.

Suppose that we wished a solution with x_4 zero. In the system (3.30), then, we may simply eliminate x_3 from the first and second equations and obtain x_3 from the third equation by letting x_4 be zero. We may write

$$x_1 + \tfrac{9}{2}x_4 = \tfrac{17}{2}$$
$$x_2 - 4x_4 = -5$$
$$x_3 + \tfrac{5}{2}x_4 = \tfrac{9}{2}$$

giving $x_1 = \tfrac{17}{2}$, $x_2 = -5$, $x_3 = \tfrac{9}{2}$, $x_4 = 0$. This solution satisfies system (3.29) as well as system (3.30).

Thus a simplex iteration does nothing more than generate solutions to a system of equations from other solutions. As we have witnessed, however, other considerations are necessary for the retention of feasibility and improvement of objective function value.

3.4. More General Linear-Programming Problems

So far we have discussed the simplex method in some detail, using example (3.14) for illustrative purposes; however, obtaining an initial basic feasible solution requires further discussion. We shall assume that we always transform given problems if necessary to obtain a requirements vector that is nonnegative. Consider now the following minimization problem:

$$\min z = 2x_1 + x_2 - x_3$$

subject to

$$
\begin{aligned}
x_1 + x_2 + x_3 &\leq 3 \\
x_2 + x_3 &\geq 2 \\
x_1 \quad\quad + x_3 &= 1 \\
x_1, x_2, x_3 &\geq 0
\end{aligned}
\tag{3.31}
$$

Converting to the equivalent equality-constrained problem, we have

$$\min z = 2x_1 + x_2 - x_3$$

subject to

$$
\begin{aligned}
x_1 + x_2 + x_3 + x_4 \quad\quad\quad &= 3 \\
x_2 + x_3 \quad\quad - x_5 &= 2 \\
x_1 \quad\quad + x_3 \quad\quad\quad &= 1 \\
x_1, \ldots, x_5 &\geq 0
\end{aligned}
\tag{3.32}
$$

i. *Obtaining an Initial Basic Feasible Solution*

An initial basic feasible solution is no longer obvious. We first note that we do not have exactly one unique variable appearing in each equation. Were the third constraint not present, we would have had such an arrangement. Even then, the initial basic solution would not have been feasible, for we would have had $-x_5 = 2$, or $x_5 = -2$. Clearly, it is the $\geq$ and $=$ constraints that introduce the difficulties. To overcome this hurdle, we *add* a different nonnegative variable, say x_6 and x_7, not appearing elsewhere, in each of the last two equations respectively of problem (3.32). Thus we would obtain

$$x_1 + x_2 + x_3 + x_4 \qquad\qquad\qquad = 3$$
$$x_2 + x_3 \quad - x_5 + x_6 \qquad = 2 \qquad\qquad (3.33)$$
$$x_1 \quad\;\; + x_3 \qquad\qquad\quad + x_7 = 1$$
$$x_1, \ldots, x_7 \geq 0$$

Now, an initial basic solution is easily obtained: $x_1 = x_2 = x_3 = x_5 = 0$, $x_4 = 3$, $x_6 = 2$, $x_7 = 1$. The difficulty, however, is that this solution is not feasible, for it implies

$$x_1 + x_2 + x_3 = 0$$
$$x_2 + x_3 = 0$$
$$x_1 \quad\;\; + x_3 = 0$$

and, therefore, the constraints given in (3.31) are not satisfied, and neither are those of (3.32).

a. A TACTIC—THE M METHOD

What we propose to do, however, is to employ the simplex technique, somehow, and obtain an initial basic feasible solution from our initial basic infeasible solution. What we have done is to *force* an obvious initial basic solution by this artifice, and the variables whose introduction has allowed this are consequently called *artificial variables,* and the columns so obtained in the coefficient matrix are called *artificial vectors.*

From the example, it should be evident that as long as there is a nonzero artificial variable in the solution, the solution is *not* feasible; our objective, then, is to employ the simplex technique and accomplish the reduction of all artificial variables to a zero level (which we know occurs automatically when a corresponding vector leaves the basis). The fact that this end may be achieved by means of the simplex technique itself is a computational convenience, especially in the case of machine computation, since a single routine can treat a problem both before and after feasibility is attained.

To achieve removal of artificials from the basic solution, *one associates with each such variable a cost so undesirable that our technique will drive these variables to zero,* since it is an optimization technique. Thus in a max (min) problem we assign a large negative (positive) cost to each artificial variable. We agree that these costs will be sufficiently large to dominate any expression in which they appear in the computation. We need not assign numbers as these costs, as long as we honor this agreement. Therefore, it is conventional to employ a symbol as this cost, and M or $-M$, is frequently used for the min (max) problem, where M is taken to be positive.

b. APPLYING THE M METHOD

Thus our problem (3.31) becomes

$$\min z = 2x_1 + x_2 - x_3 + Mx_6 + Mx_7$$

subject to

$$
\begin{aligned}
x_1 + x_2 + x_3 + x_4 &= 3 \\
x_2 + x_3 \quad - x_5 + x_6 &= 2 \\
x_1 \quad + x_3 \quad\quad + x_7 &= 1 \\
x_1, \ldots, x_7 &\geq 0
\end{aligned}
$$

Our initial tableau becomes

c_j	Basis	P_1	P_2	P_3	P_4	P_5	P_6	P_7	b	
0	P_4	1	1	1	1	0	0	0	3	
M	P_6	0	1	1	0	-1	1	0	2	(3.34)
M	P_7	1	0	1	0	0	0	1	1	
$c_j - z_j$		$2-M$	$1-M$	$-1-2M$	0	M	0	0		

P_3 should come into the basis, since $-1 - 2M$ is the most negative $c_j - z_j$ available, by virtue of M's dominance. P_7, then, will leave the basis, yielding the tableau

c_j	Basis	P_1	P_2	P_3	P_4	P_5	P_6	P_7	b
0	P_4	0	1	0	1	0	0	-1	2
M	P_6	-1	1	0	0	-1	1	-1	1
-1	P_3	1	0	1	0	0	0	1	1
$c_j - z_j$		$3+M$	$1-M$	0	0	M	0	$1+2M$	

Here we find that P_2 should enter and that P_6 should be removed, yielding the tableau

c_j	Basis	P_1	P_2	P_3	P_4	P_5	P_6	P_7	b	
0	P_4	1	0	0	1	1	-1	0	1	
1	P_2	-1	1	0	0	-1	1	-1	1	(3.35)
-1	P_3	1	0	1	0	0	0	1	1	

In tableau (3.35) both artificial variables are zero, and we find $x_4 = 1$, $x_2 = 1$, $x_3 = 1$, $x_1 = x_5 = 0$, which is a feasible solution for (3.32); this then is selected as our initial basic feasible solution. The P_6 and P_7 columns may be omitted from future tableaux, since at this point a feasible solution for problem (3.32) has been obtained, and the entry of P_6 and P_7 would never improve the value of the objective function. The corresponding columns may, in fact, be removed as soon as the artificial vectors leave the basis. If one were interested in maintaining B^{-1} in the tableau, no columns would ever be dropped, of course.

c. Interpreting Conditions Related to Artificial Variables

It is by no means the case that all the artificial vectors will be removed before a nonartificial vector comes out. In tableau (3.34), for example, if b_1 were any real number less than 1, P_4 would be the first vector removed. A valid question, then, is whether or not all the artificial variables can be reduced to zero. It is sometimes the case that they cannot.

Suppose that we proceed in a problem containing artificial variables and eventually satisfy the optimality conditions, with at least one nonzero artificial variable in the basic solution. Let us observe that there is really nothing artificial about the problem that contains artificial variables—it is a quite legitimate linear-programming problem in its own right. It is "artificial" only with respect to the problem of interest. Thus, under the conditions cited, we have obtained an optimal solution to the problem containing artificial variables. If there existed a feasible solution in which all artificial variables were zero for this artificial problem (equivalent to the existence of a *feasible* solution for the problem given), then the objective function value would certainly be improved, owing to the nature of the costs of the artificial variables. But, by assumption, the optimality criteria are satisfied, so no such solution can exist for the artificial problem; consequently, *no feasible solution can exist in the original problem*.

The nonfeasibility that occurs when an artificial variable is nonzero may be depicted geometrically in a simple instance. We imagine that the problem is subject to the single constraint $x_1 + x_2 = c$ to which we add an artificial variable x_3. It is clear from Figure 3.12 that a point (x_1, x_2, x_3) is a feasible solution to the original constraint if and only if $x_3 = 0$. (Nothing was said about the lack of a feasible solution to a problem constrained by $Ax \leq b$, $x \geq 0$ when $b \geq 0$. Why?)

As long as an artificial variable is positive, the condition that normally signals an unbounded solution can no longer be interpreted as indicating an unbounded solution in the *given problem*, since a *feasible* solution has not, at that point, been obtained for the given problem. Thus the solution process would not be terminated in such a situation, but would be continued.

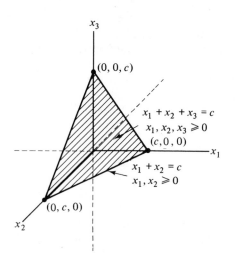

Figure 3.12. Geometric interpretation of adding artificial variable x_3.

Example 3.3

$$\min z = 2x_1 + x_2 + 3x_3 - x_4$$

subject to

$$
\begin{aligned}
2x_1 \quad\quad - x_2 \quad\quad + x_4 &\leq 2 \\
-x_1 \quad + 2x_2 - x_3 \quad\quad\quad &= 4 \\
3x_1 \quad\quad\quad - x_3 + 2x_4 &\geq 8 \\
x_1, x_2, x_3, x_4 &\geq 0
\end{aligned}
$$

has no feasible solution. It should be solved to observe the condition mentioned above.

Solutions satisfying the optimality criteria may occur with artificials in the basis at a zero level. Such a solution is, of course, a bona fide optimal solution, but further analysis of this condition, which we shall not undertake, can determine whether or not any of the original constraints were redundant and, in fact, which ones had this property.

ii. *Multiple Optimal Solutions*

When an optimal solution to a linear-programming problem is available, there may be vectors $\mathbf{P}_j$ not in the basis for which $c_j - z_j = 0$. We know that if $\mathbf{P}_j$ may be brought into the basis ($y_{ij} > 0$ for at least one i) we obtain a different, but optimal, basic feasible solution. In such cases, alternative optimal solutions may be generated.

Let us assume we have found k distinct optimal basic feasible solutions to a problem in which the objective function is $c_1 x_1 + \cdots + c_n x_n$ and the equality constraints obtained were $A\mathbf{x} = \mathbf{b}$. Let our k solutions be $\{\mathbf{x}^{(1)}, \mathbf{x}^{(2)}, \ldots, \mathbf{x}^{(k)}\}$

Let

$$\mathbf{x} = d_1 \mathbf{x}^{(1)} + d_2 \mathbf{x}^{(2)} + \cdots + d_k \mathbf{x}^{(k)}$$

where

$$\sum_{i=1}^{k} d_i = 1, \quad d_i \geq 0, i = 1, 2, \ldots, k$$

The reader should show that $\mathbf{x}$ is both feasible and optimal, and thereby conclude that the existence of two or more optimal basic feasible solutions implies the existence of an infinite number of optimal solutions. (In fact, two or more distinct optimal solutions—basic or not—imply this infinity of optimal solutions, as the proof will show).

In Figure 3.13 we have attempted to depict a situation in which an infinite number of optimal solutions exist. An entire edge of the set of feasible solutions corresponds to optimal solutions. It is also easy to see that entire faces of the feasible set could be optimal solutions. $\mathbf{x}'$ and $\mathbf{x}''$ are two distinct optimal basic solutions (also extreme points of the feasible set), while the line segment joining them presents the infinity of optimal solutions.

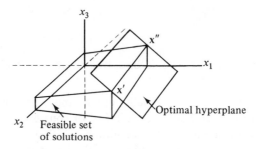

Figure 3.13. An optimal hyperplane illustrating multiple optimal solutions.

If, with the optimality criteria satisfied, one finds for $\mathbf{P}_j$ nonbasic, $c_j - z_j = 0$ and $y_{ij} \leq 0$ for all i, we realize another situation mentioned earlier in the chapter; there are optimal solutions (finite objective function value) *in which some of the variables are infinitely large*. Geometrically, the feasible set is not bounded from above, and continues to intersect the optimal hyperplane infinitely far out.

3.5. Further Geometric Aspects

i. *Simplex Method in the Space of Feasible Solutions*

It is instructive to view a simple example of the simplex technique and the geometry of the solution process simultaneously. Consider the maximization problem

$$\max z = 3x_1 + 2x_2 + x_3$$

subject to

$$2x_1 + x_2 + 2x_3 \leq 4 \qquad\qquad (3.36)$$
$$x_1 + x_2 \qquad\quad \leq 1$$
$$x_1, x_2, x_3 \geq 0$$

By setting all possible pairs of variables to zero in $2x_1 + x_2 + 2x_3 = 4$, we determine three points lying on the constraining plane, and therefore determine it uniquely. These points and the nonnegativity restrictions give the three-dimensional shaded figure in Figure 3.14.

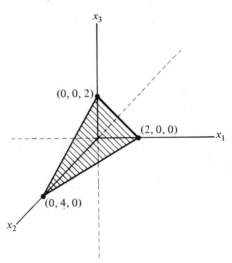

Figure 3.14. Solution space to problem (3.36) with only the constraint $2x_1 + x_2 + 2x_3 \leq 4$.

When we consider the constraint $x_1 + x_2 \leq 1$, then the feasible region is reduced to the three-dimensional shaded figure in Figure 3.15, whose six extreme points have been labeled.

To solve the problem, we first obtain the equivalent problem

$$\max z = 3x_1 + 2x_2 + x_3$$

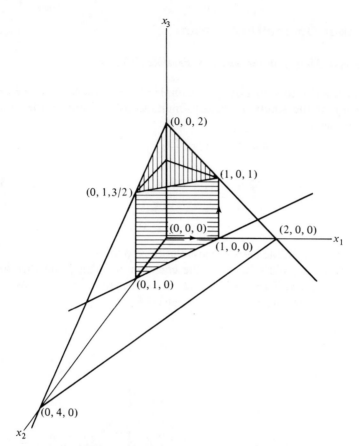

Figure 3.15. Solution space to problem (3.36).

subject to

$$2x_1 + x_2 + 2x_3 + x_4 \qquad\qquad = 4 \qquad\qquad (3.37)$$
$$x_1 + x_2 \qquad\qquad\qquad + x_5 = 1$$
$$x_1, \ldots, x_5 \geq 0$$

The initial tableau is

c_j	Basis	P_1	P_2	P_3	P_4	P_5	b
0	P_4	2	1	2	1	0	4
0	P_5	1	1	0	0	1	1
	$c_j - z_j$	3	2	1	0	0	

Our initial basic feasible solution corresponds to the extreme point $(0, 0, 0)$. $\mathbf{P}_1$ is to come in, and $\mathbf{P}_5$ leaves the basis, giving

c_j	Basis	$\mathbf{P}_1$	$\mathbf{P}_2$	$\mathbf{P}_3$	$\mathbf{P}_4$	$\mathbf{P}_5$	b
0	$\mathbf{P}_4$	0	-1	2	1	-2	2
3	$\mathbf{P}_1$	1	1	0	0	1	1
	$c_j - z_j$	0	-1	1	0	-3	

The solution here, $x_1 = 1$, $x_4 = 2$, $x_2 = x_3 = x_5 = 0$, corresponds to the extreme point $(1, 0, 0)$. $\mathbf{P}_3$ is next to enter and $\mathbf{P}_4$ is removed, giving

c_j	Basis	$\mathbf{P}_1$	$\mathbf{P}_2$	$\mathbf{P}_3$	$\mathbf{P}_4$	$\mathbf{P}_5$	b
1	$\mathbf{P}_3$	0	$-\frac{1}{2}$	1	$\frac{1}{2}$	-1	1
3	$\mathbf{P}_1$	1	1	0	0	1	1
	$c_j - z_j$	0	$-\frac{1}{2}$	0	$-\frac{1}{2}$	-2	

So $x_1 = 1$, $x_3 = 1$, $x_2 = 0$ is an optimal solution to the original problem, and the extreme point is $(1, 0, 1)$.

Figure 3.15 illustrates the manner in which we have moved about the boundary of the feasible region, from extreme point to extreme point as indicated by the arrows. Notice that the simplex technique has always proceeded along what we might call "edges" of the set, never across one of its faces. This is the rule with the simplex technique, and while we could exhibit a precise definition of an "edge," intuition will suffice for our purpose here. Edges of the figure we have been examining are the line segments connecting the following pairs of extreme points; $(0, 0, 2)$ and $(0, 1, \frac{3}{2})$; $(0, 0, 2)$ and $(1, 0, 1)$; $(0, 1, \frac{3}{2})$ and $(0, 1, 0)$; $(1, 0, 1)$ and $(1, 0, 0)$; $(0, 1, 0)$ and $(1, 0, 0)$; $(0, 0, 0)$ and $(0, 1, 0)$; $(0, 0, 0)$ and $(1, 0, 0)$; $(0, 0, 2)$ and $(0, 0, 0)$; $(0, 1, \frac{3}{2})$ and $(1, 0, 1)$. A line segment connecting $(0, 1, 0)$ and $(1, 0, 1)$ is not an edge, and this, too, satisfies our intuitive notions.

Two extreme points that are connected by an edge are termed *adjacent*. Since we have decided that the simplex technique always acts to produce movement from extreme point to extreme point along edges, we may say that it always proceeds to an adjacent extreme point in its search for an optimal extreme point. (Except that in the presence of degeneracy, a simplex iteration may keep it at the same extreme point.)

ii. *Simplex Method in the Space of Requirements and Activity Vectors*

It is instructive to view the simplex procedure from a different point of view. Consider the system

$$2x_1 + x_2 = a_1$$

$$x_1 - x_2 = a_2$$

or $$x_1 \begin{pmatrix} 2 \\ 1 \end{pmatrix} + x_2 \begin{pmatrix} 1 \\ -1 \end{pmatrix} = \begin{pmatrix} a_1 \\ a_2 \end{pmatrix}, \quad a_1, a_2 \geq 0$$

or $$x_1 \mathbf{P}_1 + x_2 \mathbf{P}_2 = \begin{pmatrix} a_1 \\ a_2 \end{pmatrix}$$

Figure 3.16 shows a shaded region that has been *generated* by the vectors $\mathbf{P}_1$ and $\mathbf{P}_2$. This process of generation could be described mathematically and extended to higher dimensions, but two dimensions will suffice for our purposes, and the determination of the cone-shaped region should be clear for arbitrary $\mathbf{P}_1, \mathbf{P}_2$.

A few examples should convince the reader of the fact that the existence of a feasible solution to

$$2x_1 + x_2 = a_1$$

$$x_1 - x_2 = a_2$$

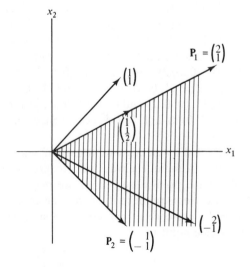

Figure 3.16. The cone generated by a particular basis.

is equivalent to the vector $\begin{pmatrix} a_1 \\ a_2 \end{pmatrix}$ lying *in the cone-shaped region*. For example,

$$2x_1 + x_2 = 2$$
$$x_1 - x_2 = -1$$

has a unique solution $x_1 = \frac{1}{3}$, $x_2 = \frac{4}{3}$, which is certainly feasible. The system

$$2x_1 + x_2 = 1$$
$$x_1 - x_2 = 1$$

has a unique solution, $x_1 = \frac{2}{3}$, $x_2 = -\frac{1}{3}$, which is not feasible. Finally, the system

$$2x_1 + x_2 = 1$$
$$x_1 - x_2 = \frac{1}{2}$$

has a unique solution $x_1 = \frac{1}{2}$, $x_2 = 0$

The various requirements vectors used here are also shown in Figure 3.16, and their relationship to our cone-shaped region is obvious. Our last solution $(\frac{1}{2}, 0)$ is degenerate; this might lead us to suspect a relationship between the requirements vector's lying on the *boundary* or *face* of the figure and degeneracy. (The relationship is obvious in two dimensions, anyway.) That such a relationship exists in n dimensions may be proved algebraically, but will not be discussed here.

An example of a cone in three dimensions *generated* by three vectors $\mathbf{P}_i, \mathbf{P}_j, \mathbf{P}_k$, constituting a basis in E^3, is illustrated in Figure 3.17.

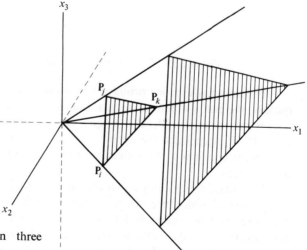

Figure 3.17. A cone in three dimensions.

To return to our original problem (3.37), the system of equations is

$$2x_1 + x_2 + 2x_3 + x_4 \qquad = 4$$
$$x_1 + x_2 \qquad\qquad + x_5 = 1$$

A feasible solution to the system is a vector $(x_1, x_2, \ldots, x_5)$, $x_j \geq 0$, $j = 1, \ldots, 5$, such that

$$x_1\binom{2}{1} + x_2\binom{1}{1} + x_3\binom{2}{0} + x_4\binom{1}{0} + x_5\binom{0}{1} = \binom{4}{1}$$

Figure 3.18 illustrates geometrically each of the column vectors $\mathbf{P}_1 = \binom{2}{1}$, $\mathbf{P}_2 = \binom{1}{1}$, $\mathbf{P}_3 = \binom{2}{0}$, $\mathbf{P}_4 = \binom{1}{0}$, $\mathbf{P}_5 = \binom{0}{1}$, and the requirements vector $\mathbf{b} = \binom{4}{1}$.

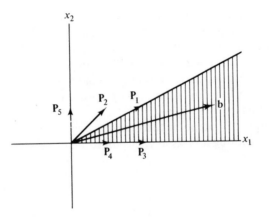

Figure 3.18. Cone of feasible solutions for problem (3.37).

The first basis in our problem was an identity matrix, as usual; consequently, the entire nonnegative quadrant was our first cone and $\mathbf{b}$ lay within it. Our next basis was $\{\mathbf{P}_4, \mathbf{P}_1\}$, whose cone is shaded in Figure 3.18. Our final basis $\{\mathbf{P}_3, \mathbf{P}_1\}$ generated the same cone as did $\{\mathbf{P}_4, \mathbf{P}_1\}$; consequently, the requirements vector was maintained within the cone generated by each pair of basis vectors.

Of course, maintaining a feasible solution is effected algebraically in our computational procedure, when we determine the vector to leave the basis, but we see now precisely which bases provide feasible solutions and which do not. $\{\mathbf{P}_5, \mathbf{P}_2\}$, $\{\mathbf{P}_5, \mathbf{P}_1\}$, and $\{\mathbf{P}_2, \mathbf{P}_1\}$ are all the bases that do not provide a basic feasible solution. $\{\mathbf{P}_4, \mathbf{P}_3\}$, of course, is not a basis.

3.6. Dismissing Nonnegativity Requirements

i. *Nonnegativity Requirements—Discussion*

It is not difficult to conceive of linear-programming problems in which the variables are not all restricted to nonnegative values. We would like to be able to solve such problems. Clearly, however, our techniques were constructed to deal with nonnegative solutions alone.

The reader might ask himself what is the real significance of demanding nonnegative variables. True, we have always developed our procedures around the requirement of nonnegativity, but where did our solution procedure really require it, if it did?

The fact that an optimal solution will only lie on the boundary of the feasible set is certainly independent of the signs of the variables. The reader can remove nonnegativity restrictions and convince himself of this fact for sample problems in two or three variables.

What motivated our computational approach was the fact that if an optimal solution exists then at least one extreme point would generate it, or, equivalently, every optimal hyperplane contains an extreme point of the feasible set. By removing nonnegativity, the reader will find that the previous statement is *not* always true; there are in fact sets that have no extreme points. It is not so much nonnegative variables that are missed, but rather a condition implied by nonnegativity: the feasible set is bounded on one side; that is, it cannot extend infinitely far in directions with negative components.

By the preceding discussion we mean only to point out that nonnegative restrictions are of greater importance in linear programming than merely to reflect realistic problems.

ii. *Solving Problems with Unrestricted Variables*

To solve problems with variables not sign restricted, one has two basic choices—he may either modify the theory to obtain the desired procedures, or he may modify his problem so as to make existing procedures applicable and, of course, obtain an *equivalent* problem.

The dilemma is easily resolved, and the simplest strategy is to transform the problem into an equivalent problem in nonnegative variables, rather than modify the computational methods. A simple substitution serves to effect the transformation.

Suppose that x_j, $j \in J$, is a variable unrestricted in sign, where J is a (possibly empty) subset of $1, \ldots, n$, n being the number of decision variables; that is, these x's will be allowed to assume any value, positive or negative, so as to maximize or minimize the objective function involved.

Let us write $x_j = x_{j1} - x_{j2}, j \in J$, where we require x_{j1}, x_{j2} to be non-negative, and substitute accordingly in the given problem, thus obtaining a new problem. For example, given

$$\max z = x_1 + 2x_2 - x_3$$

subject to

$$2x_1 + x_2 + x_3 \leq 3 \tag{3.38}$$
$$x_1 \qquad - x_3 = 2$$
$$x_1, x_2 \geq 0, \qquad x_3 \text{ unrestricted}$$

we write $x_3 = x_{31} - x_{32}$, and obtain the new problem

$$\max z = x_1 + 2x_2 - x_{31} + x_{32}$$

subject to

$$2x_1 + x_2 + x_{31} - x_{32} \leq 3 \tag{3.39}$$
$$x_1 \qquad - x_{31} + x_{32} = 2$$
$$x_1, x_2, x_{31}, x_{32} \geq 0$$

It is quite simple to show that an optimal solution to problem (3.39) yields an optimal solution to problem (3.38) upon specifying x_3 in terms of x_{31}, x_{32}, as defined. This proof is left as a problem. Furthermore, depending upon the relative magnitudes of x_{31} and x_{32}, an optimal x_3 may indeed be positive, negative, or zero, since it is the difference of two nonnegative values.

The possible outcomes, however, may be further narrowed. In problem (3.39), for instance, we obtain a column $\mathbf{P}_{31}$ and a column $\mathbf{P}_{32}$ from the substitution, and each is but the negative of the other. Consequently, any set containing $\mathbf{P}_{31}$ and $\mathbf{P}_{32}$ cannot be linearly independent, so x_{31} and x_{32} cannot be simultaneously basic variables. Thus, at any stage in the solution of problem (3.39), optimal solutions included, one (or both) of x_{31}, x_{32} is zero, and as a result, we have either

$$x_3 = x_{31} > 0$$
$$x_3 = 0$$
$$x_3 = -x_{32} < 0$$

The ability to solve the more general problem has here been obtained at the expense of having to treat a larger problem. The reader should also observe that simple substitutions allow problems in which $x_j \leq 0$ constraints exist to be amenable to the simplex technique.

3.7. Another Procedure for Artificial Variables

i. *Two-Phase Method*

Other methods are available for handling artificial variables. The *two-phase method* is commonly used for mechanized solution, and we describe it below.

In the first phase, surprisingly known as phase 1, all costs associated with nonartificial variables are set equal to zero, and every artificial variable is assigned a cost of -1. The simplex method is then used to *maximize* the contrived objective function. The goal is a feasible solution for the original problem, so phase 1 may halt as soon as an objective function value of zero is achieved for the revised problem, whether the optimality criteria for the original problem are satisfied or not. In case the maximum objective function value for the revised problem is *negative*, there *is no feasible solution to the initial problem.*

Let us suppose that the phase 1 objective function has a maximum value of zero, in which case there is a second phase, or phase 2. Now, in the optimal phase 1 tableau, *restore the cost coefficients of the nonartificial variables to their original values, cross off any nonbasic artificial columns, and assign a cost of zero to any basic artificial variables. Then recompute the $c_j - z_j$ and apply the simplex method to solve this transformed problem as usual.*

Phase 1 provides a feasible solution; phase 2 proceeds to the solution of the original problem. Notice that phase 2 offers *no* cost incentive for artificial variables to remain zero. Any that were basic at the end of phase 1 remain basic at the outset of phase 2, and begin with value zero. But, with cost zero, an artificial variable might well become positive along with some other variable. Obviously, whenever that happens, the solution becomes infeasible, so one must guard against it. As always, if an artificial vector becomes nonbasic it may be scratched permanently.

Let us illustrate the method on a sample problem, observe an occurrence of an artificial variable surreptitiously becoming positive, and then formulate a strategy for preventing that.

ii. *Overcoming the Problem of a Basic Artificial Variable Being Positive in Phase 2*

To see how the problem may arise, we note that if $\mathbf{P}_k$ enters the basis and the rth basic vector is removed, the new value of the ith basic variable will be

$$x_i(B') = x_i(B) - \frac{x_r(B)}{y_{rk}} \cdot y_{ik}$$

Consider any i for which $x_i(B)$ is artificial. It is zero, but if $x_r(B)/y_{rk} > 0$ and if $y_{ik} < 0$, $x_i(B')$ will be positive; that is, an artificial variable will have become positive, and thus the solution will have become infeasible.

The foregoing also indicates a procedure for avoiding the occurrence of positive artificials. If for any i corresponding to an artificial basis vector, we have $y_{ik} > 0$, there can be no problem (Why?). But if for all artificial i, $y_{ik} \leq 0$, and one, say y_{qk} is negative, then $x_q(B')$ could become positive, so one may enter $\mathbf{P}_k$ and remove the qth basis vector. For that choice $Q = 0$, so the variable values are unchanged. Although there is no improvement in objective function value, wasted effort in terms of optimizing, nevertheless, the method is quite effective in removing a troublesome vector from the basis and preventing an infeasible solution. [If $y_{ik} = 0$, $x_i(B')$ is zero.] An example contrived for our purposes will illustrate the procedure.

Example 3.4

$$\min x_1 + 2x_2 - x_3 - x_4 + 2x_5$$

subject to

$$
\begin{aligned}
x_1 + x_2 - 2x_3 + x_4 + 2x_5 &\geq 4 \\
x_2 + x_3 \quad\quad + 2x_5 &\geq 4 \\
2x_1 \quad\quad - x_3 - x_4 + x_5 &\geq 2 \\
-x_1 + 2x_2 \quad\quad + 2x_4 + x_5 &\geq 2 \\
x_1, \ldots, x_5 &\geq 0
\end{aligned}
$$

Adding surplus variables $x_6, \ldots, x_9$ and artificial variables $x_{10}, \ldots, x_{13}$, the phase 1 problem is

$$\max z = -x_{10} - x_{11} - x_{12} - x_{13}$$

subject to

$$
\begin{aligned}
x_1 + x_2 - 2x_3 + x_4 + 2x_5 - x_6 \quad\quad\quad\quad + x_{10} \quad\quad\quad\quad &= 4 \\
x_2 + x_3 \quad + 2x_5 \quad - x_7 \quad\quad\quad\quad + x_{11} \quad\quad\quad &= 4 \\
2x_1 \quad - x_3 - x_4 + x_5 \quad\quad - x_8 \quad\quad\quad\quad + x_{12} \quad &= 2 \\
-x_1 + 2x_2 \quad\quad + 2x_4 + x_5 \quad\quad\quad - x_9 \quad\quad\quad\quad + x_{13} &= 2 \\
x_1, \ldots, x_{13} &\geq 0
\end{aligned}
$$

The initial tableau appears as

c_j	Basis	1	2	3	4	5	6	7	8	9	10	11	12	13	b
								$\mathbf{P}_j$							
-1	$\mathbf{P}_{10}$	1	1	-2	1	2	-1	0	0	0	1	0	0	0	4
-1	$\mathbf{P}_{11}$	0	1	1	0	2	0	-1	0	0	0	1	0	0	4
-1	$\mathbf{P}_{12}$	2	0	-1	-1	1	0	0	-1	0	0	0	1	0	2
-1	$\mathbf{P}_{13}$	-1	2	0	2	1	0	0	0	-1	0	0	0	1	2
$c_j - z_j$		2	4	-2	2	6	-1	-1	-1	-1	0	0	0	0	

$\mathbf{P}_5$ becomes basic, and with a four-way tie for departing vector, let the fourth basis vector, $\mathbf{P}_{13}$, be removed. The new tableau is

c_j	Basis	1	2	3	4	5	6	7	8	9	10	11	12	b
								$\mathbf{P}_j$						
-1	$\mathbf{P}_{10}$	3	-3	-2	-3	0	-1	0	0	2	1	0	0	0
-1	$\mathbf{P}_{11}$	2	-3	1	-4	0	0	-1	0	2	0	1	0	0
-1	$\mathbf{P}_{12}$	3	-2	-1	-3	0	0	0	-1	1	0	0	1	0
0	$\mathbf{P}_5$	-1	2	0	2	1	0	0	0	-1	0	0	0	2

The objective function value is zero, so we have an optimal phase 1 solution, with three zero-level artificial variables basic. The $\mathbf{P}_{13}$ column has been crossed off. Affixing the original costs and setting all artificial costs to zero, the initial phase 2 tableau, with $c_j - z_j$ values recomputed, appears then as

c_j	Basis	1	2	3	4	5	6	7	8	9	10	11	12	b
								$\mathbf{P}_j$						
0	$\mathbf{P}_{10}$	3	-3	-2	-3	0	-1	0	0	2	1	0	0	0
0	$\mathbf{P}_{11}$	2	-3	1	-4	0	0	-1	0	2	0	1	0	0
0	$\mathbf{P}_{12}$	3	-2	-1	-3	0	0	0	-1	1	0	0	1	0
2	$\mathbf{P}_5$	-1	2	0	2	1	0	0	0	-1	0	0	0	2
$c_j - z_j$		3	-2	-1	-5	0	0	0	0	2	0	0	0	

$\mathbf{P}_4$ should be the entering vector, and according to the usual rule, $\mathbf{P}_5$, the fourth basis column, should be removed—in fact $\mathbf{P}_5$ *must* be removed according to that rule. But if this is done, it is clear that all three of x_{10}, x_{11},

x_{12} will become positive. Thus it is essential that one of P_{10}, P_{11}, P_{12} be removed this time. Suppose c_3 had been -10, so that P_3 should enter. There are negative y_{i3} for $i = 1$ and $i = 3$, but Q is zero, so no values change and feasibility is not lost.

The reader should continue and complete the solution of the present example.

3.8. Duality

We have acknowledged the importance of linearity in numerous disciplines. Another concept of frequent interest is that of duality. In the study of circuits, the idea of a circuit that is the "dual" of a given circuit is common; in graph theory, one often studies the dual of a given graph; in linear algebra, we have the idea of dual space. In all these cases the dual entity and the given entity share a strong relationship of one kind or other. Linear programming also admits the concept of the dual, and it is our intention to introduce the reader to several of the more basic concepts of duality in linear programming.

i. *Dual of a Linear-Programming Problem*

Let us consider the problem

$$\max c_1 x_1 + \cdots + c_n x_n$$

subject to

$$a_{11}x_1 + a_{12}x_2 + \cdots + a_{1n}x_n \leq b_1$$
$$a_{21}x_1 + a_{22}x_2 + \cdots + a_{2n}x_n \leq b_2 \qquad (3.40)$$
$$\vdots \qquad\qquad \vdots \quad \vdots$$
$$a_{m1}x_1 + a_{m2}x_2 + \cdots + a_{mn}x_n \leq b_m$$
$$x_1, \ldots, x_n \geq 0$$

The dual linear-programming problem is *defined* to be

$$\min b_1 y_1 + b_2 y_2 + \cdots + b_m y_m$$

subject to

$$a_{11}y_1 + a_{21}y_2 + \cdots + a_{m1}y_m \geq c_1$$
$$a_{12}y_2 + a_{22}y_2 + \cdots + a_{m2}y_m \geq c_2 \qquad (3.41)$$

$$\cdot \qquad \qquad \cdot \quad \cdot$$
$$\cdot \qquad \qquad \cdot \quad \cdot$$
$$\cdot \qquad \qquad \cdot \quad \cdot$$

$$a_{1n}y_1 + a_{2n}y_2 + \cdots + a_{mn}y_m \geq c_n$$
$$y_1, \ldots, y_m \geq 0$$

The original maximization problem, (3.40), will be called the *primal* problem.

It is actually but a matter of choice which of problems (3.40) or (3.41) is called the primal or the dual, for problem (3.41) may be written

$$\max -b_1 y_1 - b_2 y_2 - \cdots - b_m y_m$$

subject to

$$-a_{11}y_1 - a_{21}y_2 - \cdots - a_{m1}y_m \leq -c_1$$
$$\cdot \qquad \qquad \cdot \quad \cdot$$
$$\cdot \qquad \qquad \cdot \quad \cdot$$
$$\cdot \qquad \qquad \cdot \quad \cdot$$
$$-a_{1n}y_1 - a_{2n}y_2 - \cdots - a_{mn}y_m \leq -c_n$$
$$y_1, \ldots, y_m \geq 0$$

The dual of this problem, as we have seen, is defined to be

$$\min -c_1 x_1 - c_2 x_2 - \cdots - c_n x_n$$

subject to

$$-a_{11}x_1 - \cdots - a_{1n}x_n \geq -b_1$$
$$-a_{21}x_1 - \cdots - a_{2n}x_n \geq -b_2$$
$$\cdot \qquad \qquad \cdot \quad \cdot$$
$$\cdot \qquad \qquad \cdot \quad \cdot$$
$$\cdot \qquad \qquad \cdot \quad \cdot$$
$$-a_{m1}x_1 - \cdots - a_{mn}x_n \geq -b_m$$
$$x_1, \ldots, x_n \geq 0$$

But this last problem is exactly problem (3.40), so we have that the dual of the dual is the primal.

Consider the relationships now between the two problems. The cost vector for the primal is the requirements vector for the dual; the requirements vector for the primal is the cost vector for the dual; the constraint columns of the primal are the constraint rows for the dual, and vice versa. In matrix-vector notation, we may write the primal as

$$\max \mathbf{cx}$$

subject to

$$A\mathbf{x} \leq \mathbf{b}$$
$$\mathbf{x} \geq \mathbf{0}$$

and the dual as

$$\min \mathbf{b}^T\mathbf{y}$$

subject to

$$A^T\mathbf{y} \geq \mathbf{c}^T$$
$$\mathbf{y} \geq \mathbf{0}$$

where $\mathbf{y} = \begin{pmatrix} y_1 \\ \cdot \\ \cdot \\ \cdot \\ y_m \end{pmatrix}$ and $\mathbf{c}$, $\mathbf{x}$, A, and $\mathbf{b}$ are as previously defined in Section 3.2-i.

The superscript T indicates the transpose.

ii. *Observing Relationships Between Primal and Dual*

Consider problem (3.36), which is

$$\max z = 3x_1 + 2x_2 + x_3$$

subject to

$$
\begin{array}{ll}
2x_1 + x_2 + 2x_3 \leq 4 & \\
x_1 + x_2 \qquad\quad \leq 1 & \qquad (3.42) \\
x_1, x_2, x_3 \geq 0 &
\end{array}
$$

The final tableau giving an optimal solution was found to be

c_j	Basis	P_1	P_2	P_3	P_4	P_5	b	
1	P_3	0	$-\frac{1}{2}$	1	$\frac{1}{2}$	-1	1	
3	P_1	1	1	0	0	1	1	(3.43)
	$c_j - z_j$	0	$-\frac{1}{2}$	0	$-\frac{1}{2}$	-2		

Writing the dual of (3.42), we get

$$\min 4y_1 + y_2$$

subject to

$$2y_1 + y_2 \geq 3$$
$$y_1 + y_2 \geq 2$$
$$2y_1 \qquad \geq 1$$
$$y_1, y_2 \geq 0$$

(3.44)

The addition of surplus variables then gives

$$\min 4y_1 + y_2$$

subject to

$$2y_1 + y_2 - y_3 \qquad\qquad = 3$$
$$y_1 + y_2 \qquad - y_4 \qquad = 2$$
$$2y_1 \qquad\qquad - y_5 = 1$$
$$y_1, \ldots, y_5 \geq 0$$

(3.45)

a. SIMPLEX ITERATIONS WITH INFEASIBLE SOLUTIONS

Let us solve problem (3.45) and also use the opportunity to show that we may use the simplex technique with one or more components of the requirements vector negative. Multiplying each of the constraints of (3.45) by -1 gives us the identity matrix as a basis, which we required, but simultaneously produces the negative requirements vector. We have, then,

$$\min 4y_1 + y_2$$

subject to

$$-2y_1 - y_2 + y_3 \qquad\qquad = -3$$
$$-y_1 - y_2 \qquad + y_4 \qquad = -2$$
$$-2y_1 \qquad\qquad + y_5 = -1$$
$$y_1, \ldots, y_5 \geq 0$$

The initial tableau for this problem is

c_j	Basis	P_1	P_2	P_3	P_4	P_5	b
0	P_3	-2	-1	1	0	0	-3
0	P_4	-1	-1	0	1	0	-2
0	P_5	-2	0	0	0	1	-1

Our first goal is obtaining a feasible solution; however, examining the $c_j - z_j$ will be of no use. Therefore, we ask if there are vectors we could insert into the basis and ones which we could remove in order that a feasible solution be obtained. (Notice that the usual criterion for removal is also, in general, not useful in such an instance.)

In the particularly fortuitous instance at hand, one iteration suffices to provide a basic feasible solution; but, in general, this is not the case. Here, inserting P_1 and removing P_4 is effective, and the basic feasible solution $y_3 = 1, y_1 = 2, y_5 = 3, y_2 = y_4 = 0$ is obtained, as shown next.

c_j	Basis	P_1	P_2	P_3	P_4	P_5	b
0	P_3	0	1	1	-2	0	1
4	P_1	1	1	0	-1	0	2
0	P_5	0	2	0	-2	1	3
	$c_j - z_j$	0	-3	0	4	0	

To proceed to an optimal solution, we must insert P_2, which causes P_3 to be removed, resulting in the following tableau:

c_j	Basis	P_1	P_2	P_3	P_4	P_5	b
1	P_2	0	1	1	-2	0	1
4	P_1	1	0	-1	1	0	1
0	P_5	0	0	-2	2	1	1
	$c_j - z_j$	0	0	3	-2	0	

Finally, we insert P_4, remove P_5, and obtain

c_j	Basis	P_1	P_2	P_3	P_4	P_5	b	
1	P_2	0	1	-1	0	1	2	
4	P_1	1	0	0	0	$-\frac{1}{2}$	$\frac{1}{2}$	(3.46)
0	P_4	0	0	-1	1	$\frac{1}{2}$	$\frac{1}{2}$	
	$c_j - z_j$	0	0	1	0	1		

iii. *Results*

Tableaux (3.43) and (3.46) are then optimal tableaux for the primal and dual problems, respectively, and a number of interesting relationships between the two are visible. First, the optimal objective function value of the primal is 4, as is that of the dual. Second, the $c_j - z_j$ for the slack columns of the primal are just the negatives of *a set of optimal dual variables*. The first primal slack column corresponds to the first dual variable; the second primal slack column to the second dual variable. Finally, the $c_j - z_j$ for the surplus columns of the dual are *precisely* the values of the optimal primal variables. Here the signs have *not* been inverted.

iv. *Alternative Solution*

Let us keep these relationships in mind and observe that problem (3.44) may also be solved using artificial variables. Thus we may write

$$\min 4y_1 + y_2 + My_6 + My_7 + My_8$$

subject to

$$
\begin{aligned}
2y_1 + y_2 - y_3 \qquad\qquad + y_6 \qquad\qquad &= 3 \\
y_1 + y_2 \qquad - y_4 \qquad\qquad + y_7 \qquad &= 2 \\
2y_1 \qquad\qquad - y_5 \qquad\qquad + y_8 &= 1 \\
y_1, \ldots, y_8 &\geq 0
\end{aligned}
$$

The successive tableaux iterating toward an optimal solution are

c_j	Basis	P_1	P_2	P_3	P_4	P_5	P_6	P_7	P_8	b
M	P_6	2	1	-1	0	0	1	0	0	3
M	P_7	1	1	0	-1	0	0	1	0	2
M	P_8	2	0	0	0	-1	0	0	1	1

c_j	Basis	P_1	P_2	P_3	P_4	P_5	P_6	P_7	P_8	b
M	P_6	0	1	-1	0	1	1	0	-1	2
M	P_7	0	1	0	-1	$\frac{1}{2}$	0	1	$-\frac{1}{2}$	$\frac{3}{2}$
4	P_1	1	0	0	0	$-\frac{1}{2}$	0	0	$\frac{1}{2}$	$\frac{1}{2}$

c_j	Basis	P_1	P_2	P_3	P_4	P_5	P_6	P_7	P_8	b
M	P_6	0	0	-1	1	$\frac{1}{2}$	1	-1	$-\frac{1}{2}$	$\frac{1}{2}$
1	P_2	0	1	0	-1	$\frac{1}{2}$	0	1	$-\frac{1}{2}$	$\frac{3}{2}$
4	P_1	1	0	0	0	$-\frac{1}{2}$	0	0	$\frac{1}{2}$	$\frac{1}{2}$

c_j	Basis	P_1	P_2	P_3	P_4	P_5	P_6	P_7	P_8	b
0	P_4	0	0	-1	1	$\frac{1}{2}$	1	-1	$-\frac{1}{2}$	$\frac{1}{2}$
1	P_2	0	1	-1	0	1	1	0	-1	2
4	P_1	1	0	0	0	$-\frac{1}{2}$	0	0	$\frac{1}{2}$	$\frac{1}{2}$
$c_j - z_j$		0	0	1	0	1	$M-1$	M	$M-1$	

No longer is an optimal primal solution to be found as the negatives of the $c_j - z_j$ associated with the slack or surplus columns. But let us recall that when $z_j - c_j$ did provide an optimal primal solution it was actually the case that the z_j for the columns corresponding to the slack variables comprised the solution, since $c_j = 0$ for slack variables. Furthermore, as could be shown, an optimal primal solution found is not peculiar to slack or surplus vectors, but to columns corresponding to the *initial identity matrix*, which may, of course, also contain artificial columns.

In the preceding example, the z_j for the columns corresponding to the initial identity matrix, P_6, P_7, P_8, are 1, 0, 1, respectively, which is an optimal primal solution.

v. *Actual Relationships Between Primal and Dual*

The facts that underlie these results are the following: First, if the primal (dual) problem has an optimal solution, then so does the dual (primal); and, in fact, their optimal objective functions have the same value. Next, if one of the problems is solved by the simplex method, then the optimal solution to the other problem appears as the z_j values for the columns corresponding to the *initial identity matrix* in the optimal tableau. But, if the kth constraint in the problem solved is multiplied by -1, then the *negative* of an optimal value for the dual variable is obtained. (This explains the sign anomaly experienced in Section 3.8-iii.)

Certainly, a given primal or dual problem need not have an optimal solution, and *all solution possibilities* in a (primal, dual) pair are as follows:

Table 3.1

Primal	Dual
Optimal solution	Optimal solution
Unbounded solution	No feasible solution
No feasible solution	Unbounded solution
No feasible solution	No feasible solution

vi. *Writing Duals for More General Problems*

The reader will note that our selection of a primal problem was restricted to the class

$$\max \mathbf{cx}$$

subject to

$$A\mathbf{x} \leq \mathbf{b} \tag{3.47}$$

$$\mathbf{x} \geq \mathbf{0}$$

Duality theory is not at all so limited, and the dual is easy to derive for a given problem having any general set of constraints, as well as unrestricted variables. We should notice that any set of constraints may be cast into the constraint form in problem (3.47). For example, the constraint

$$a_{i1} + \cdots + a_{in}x_n = b_i$$

is completely equivalent to the pair

$$a_{i1}x_1 + \cdots + a_{in}x_n \geq b_i$$
$$a_{i1}x_1 + \cdots + a_{in}x_n \leq b_i$$

which is, in turn, equivalent to

$$-a_{i1}x_1 - \cdots - a_{in}x_n \leq -b_i$$
$$a_{i1}x_1 + \cdots + a_{in}x_n \leq b_i$$

Example 3.5

The set of constraints

$$2x_1 + x_2 - 2x_3 + x_4 \leq 3$$
$$x_1 - x_2 + x_3 + 2x_4 \geq 4$$
$$x_1 + 3x_2 - x_3 = 2$$

becomes

$$2x_1 + x_2 - 2x_3 + x_4 \leq 3$$
$$-x_1 + x_2 - x_3 - 2x_4 \leq -4$$
$$x_1 + 3x_2 - x_3 \leq 2$$
$$-x_1 - 3x_2 + x_3 \leq -2$$

We have also seen that a simple transformation allows a problem having unrestricted or negative variables to be expressed as a problem having nonnegative variables. Thus it is possible to transform any given problem into the form (3.47) and write the dual of the problem obtained. Then the dual may be reduced to a more compact form.

Example 3.6

The dual of

$$\max 2x_1 + x_2 + x_3 - x_4$$

subject to

$$x_1 - x_2 + 2x_3 + 2x_4 \leq 3$$
$$2x_1 + 2x_2 - x_3 = 4$$
$$x_1 - 2x_2 + 3x_3 + 4x_4 \geq 5$$
$$x_1, x_2, x_3 \geq 0, \quad x_4 \text{ unrestricted}$$

is, omitting the intermediate steps,

$$\min 3y_1 + 4y_2 + 5y_3$$

subject to

$$y_1 + 2y_2 + y_3 \geq 2$$
$$-y_1 + 2y_2 - 2y_3 \geq 1$$
$$2y_1 - y_2 + 3y_3 \geq 1$$
$$2y_1 + 4y_3 = -1$$
$$y_1 \geq 0$$
$$y_2 \text{ unrestricted}$$
$$y_3 \leq 0$$

The student will be asked to derive some further primal–dual relationships of this nature in the problems at the end of the chapter.

vii. *An Application*

For our purposes here, the importance of duality lies in the ability to solve both the primal and the dual problems by solving just one of the two. Consider then a given problem in which we have more constraints than variables. Assuming the system has a feasible solution, we shall have redundant constraints, but identification of the redundant constraints is, in general, a nontrivial task. It should be pointed out that when one attempts to describe a system or process mathematically, and begins observing relationships between the variables of the system, it is not at all unusual that redundant relationships are formulated. Suppose then that we treat a system $A\mathbf{x} \leq \mathbf{b}$ of m inequalities in n variables, with $m > n$. The addition of slack variables gives a system of m equations in $m + n$ variables. The addition of surplus variables in the dual gives a system of n equations in $m + n$ variables. Thus, owing to the relationship $m > n$, there is more computational effort involved in transforming a tableau in the primal than in the dual. Since we solve both problems by solving one, we minimize the computational effort involved in obtaining an optimal solution by attacking the dual in this situation. We illustrate this discussion by an example.

Example 3.7

By solving the dual, solve the problem

$$\max -x_1 + 2x_2 + 2x_3$$

subject to

$$x_1 + x_2 + x_3 \geq 4$$
$$2x_1 + x_2 \qquad \leq 4$$
$$x_2 + x_3 \leq 3$$
$$x_1 \qquad\qquad \geq 1$$
$$x_1, x_2, x_3 \geq 0$$

We consider the primal problem to be

$$\max -x_1 + 2x_2 + 2x_3$$

subject to

$$-x_1 - x_2 - x_3 \leq -4$$
$$2x_1 + x_2 \qquad \leq 4$$
$$x_2 + x_3 \leq 3$$
$$-x_1 \qquad\qquad \leq -1$$
$$x_1, x_2, x_3 \geq 0$$

The corresponding dual problem is

$$\min -4y_1 + 4y_2 + 3y_3 - y_4$$

subject to

$$-y_1 + 2y_2 \qquad - y_4 \geq -1$$
$$-y_1 + y_2 + y_3 \qquad \geq 2$$
$$-y_1 \qquad + y_3 \qquad \geq 2$$
$$y_1, y_2, y_3, y_4 \geq 0$$

Let us multiply the first constraint by -1, and treat the problem

$$\min\ -4y_1 + 4y_2 + 3y_3 - y_4 + My_7 + My_9$$

subject to

$$
\begin{aligned}
y_1 - 2y_2 \quad\;\; + y_4 + y_5 \qquad\qquad\qquad &= 1 \\
-y_1 + y_2 + y_3 \qquad\quad\; - y_6 + y_7 \qquad\quad &= 2 \\
-y_1 \qquad\;\; + y_3 \qquad\qquad\qquad - y_8 + y_9 &= 2 \\
y_1, \ldots, y_9 &\geq 0
\end{aligned}
$$

The first tableau is

c_j	Basis	1	2	3	4	P_j 5	6	7	8	9	b
0	P_5	1	-2	0	1	1	0	0	0	0	1
M	P_7	-1	1	1	0	0	-1	1	0	0	2
M	P_9	-1	0	1	0	0	0	0	-1	1	2
$c_j - z_j$		$2M-4$	$4-M$	$3-2M$	-1	0	M	0	M	0	

The reader may verify that the final tableau is

c_j	Basis	1	2	3	4	P_j 5	6	7	8	9	b
-4	P_1	1	-2	0	1	1	0	0	0	0	1
3	P_3	0	-2	1	1	1	0	0	-1	1	3
0	P_6	0	-1	0	0	0	1	-1	-1	1	0
$c_j - z_j$		0	2	0	0	1	0	M	3	$M-3$	

The z_j corresponding to the columns containing the initial identity matrix, the fifth, seventh, and ninth columns, are -1, 0, and 3. Recalling that the first dual constraint was multiplied by -1, the negative of an optimal first primal variable has been obtained. An optimal primal solution is then $x_1 = 1$, $x_2 = 0$, $x_3 = 3$.

Example 3.8

Let us consider a primal problem that has an unbounded solution.

$$\max 3x_1 + 2x_2$$

subject to

$$2x_1 - x_2 \leq 4$$
$$x_1 + x_2 \geq 2$$
$$x_1, x_2 \geq 0$$

whose dual is

$$\min 4y_1 - 2y_2$$

subject to

$$2y_1 - y_2 \geq 3$$
$$-y_1 - y_2 \geq 2$$
$$y_1, y_2 \geq 0$$

The feasible region in the dual problem is therefore null. The set of *solutions* to the system is represented by the shaded area in Figure 3.19, and no element of this region is feasible. This is the result we expect, from Table 3.1.

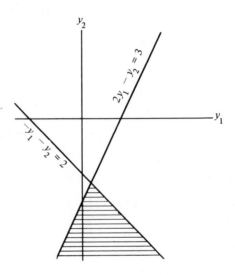

Figure 3.19. Set of feasible solutions to the dual problem of Example 3.8.

viii. *Some Dual Theorems*

The foregoing remarks regarding possible relationships between primal–dual pairs are all rather easily established.

Let us consider the pair of problems

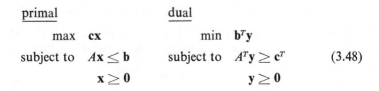

primal	dual	
max cx	min $\mathbf{b}^T\mathbf{y}$	
subject to $A\mathbf{x} \leq \mathbf{b}$	subject to $A^T\mathbf{y} \geq \mathbf{c}^T$	(3.48)
$\mathbf{x} \geq 0$	$\mathbf{y} \geq 0$	

We prove the following:

Lemma 3.1

If $\mathbf{x}$, $\mathbf{y}$ are feasible solutions to their respective problems, then

$$\mathbf{cx} \leq \mathbf{b}^T \mathbf{y}$$

That is, for any pair $\mathbf{x}$, $\mathbf{y}$ of feasible solutions, the primal objective function value is no greater than that of the dual.

Proof: By assumption, $\mathbf{x}$ is feasible, so

$$A\mathbf{x} \leq \mathbf{b}$$

By virtue of $\mathbf{y}$'s nonnegativity, it must be that

$$\mathbf{y}^T(A\mathbf{x}) \leq \mathbf{y}^T \mathbf{b} \tag{3.49}$$

Furthermore, we have by assumption that

$$A^T \mathbf{y} \geq \mathbf{c}^T$$

and since $\mathbf{x}$ is nonnegative

$$(\mathbf{x}^T)A^T \mathbf{y} \geq \mathbf{x}^T \mathbf{c}^T \tag{3.50}$$

But
$$(\mathbf{x}^T)A^T \mathbf{y} = (A\mathbf{x})^T \mathbf{y} = \mathbf{y}^T A\mathbf{x}$$
$$\mathbf{y}^T \mathbf{b} = \mathbf{b}^T \mathbf{y}$$

and
$$\mathbf{x}^T \mathbf{c}^T = \mathbf{cx}$$

So expressions (3.49) and (3.50) give

$$\mathbf{b}^T \mathbf{y} \geq \mathbf{cx}$$

which proves the lemma.

Theorem 3.3

For a feasible primal–dual pair $\mathbf{x}^*$, $\mathbf{y}^*$, we have

$$\mathbf{cx}^* = \mathbf{b}^T \mathbf{y}^* \tag{3.51}$$

if and only if $\mathbf{x}^*$, $\mathbf{y}^*$ are *optimal* solutions for their respective problems.

Proof: First, suppose that we have a feasible pair $\mathbf{x}^*$, $\mathbf{y}^*$ satisfying equation (3.51). For any feasible $\mathbf{x}$ we have, by Lemma 3.1, that

$$\mathbf{b}^T \mathbf{y}^* \geq \mathbf{cx}$$

But
$$\mathbf{c}\mathbf{x}^* = \mathbf{b}^T\mathbf{y}^*$$
so
$$\mathbf{c}\mathbf{x}^* \geq \mathbf{c}\mathbf{x}$$

which implies that $\mathbf{x}^*$ is primal optimal.

Using Lemma 3.1 in the same fashion, for any dual-feasible $\mathbf{y}$, we have

$$\mathbf{b}^T\mathbf{y} \geq \mathbf{c}\mathbf{x}^* = \mathbf{b}^T\mathbf{y}^*$$

which establishes $\mathbf{y}^*$'s optimality.

To complete the proof, observe that if either of $\mathbf{x}^*$, $\mathbf{y}^*$ is assumed non-optimal Lemma 3.1 is contradicted. Hence the theorem is proved.

The question of existence, of course, has not yet been resolved; that is, despite Theorem 3.3 we still have *not* shown the following theorem.

Theorem 3.4

If one of the primal–dual pair (3.48) has an optimal solution, then so does the other.

Proof: Consider the dual problem

$$\min \mathbf{b}^T\mathbf{y}$$

subject to

$$A^T\mathbf{y} \geq \mathbf{c}^T$$
$$\mathbf{y} \geq \mathbf{0}$$

Suppose that we add a surplus variable to each constraint, and multiply all dual constraints by -1. We obtain the problem

$$\min \mathbf{b}^T\mathbf{y}$$

subject to

$$-A^T\mathbf{y} + I_n\mathbf{s} = -\mathbf{c}^T$$
$$\mathbf{y} \geq \mathbf{0}, \qquad \mathbf{s} \geq \mathbf{0}$$

where $\mathbf{s}$ is an n vector of surplus variables and I_n is the $n \times n$ identity matrix. The identity matrix that we desire as an initial basis is present, but to get it we may have created negative elements in the dual requirements vector. Suppose that we are able to overcome that difficulty (we have previously illustrated that it is possible to utilize simplex manipulations and begin problem solving under such circumstances), and that we eventually obtain an optimal basis B, an optimal basic feasible solution $\mathbf{y}_B$, and a cost vector $\mathbf{b}_B$ corresponding to $\mathbf{y}_B$.

Since the optimality criteria are satisfied, we have our (equivalent of $c_j - z_j$) $b_j - z_j \geq 0$ for $j = 1, \ldots, m, \ldots, m + n$. The z_j in this case are the numbers

$$\mathbf{b}_B^T B^{-1}(-\mathbf{a}_j^T), \quad j = 1, \ldots, m$$

where $\mathbf{a}_j^T$ is the jth column of A^T, and the numbers

$$\mathbf{b}_B^T B^{-1}\mathbf{e}_i, \quad i = 1, \ldots, n$$

where $\mathbf{e}_i$ is the ith unit vector. The costs for variables associated with the identity matrix columns are zero, so if the corresponding $b_j - z_j \geq 0$, it is just that

$$-\mathbf{b}_B^T B^{-1}\mathbf{e}_i \geq 0, \quad i = 1, \ldots, n$$

or

$$-\mathbf{b}_B^T B^{-1}I_n \geq 0$$

or

$$-\mathbf{b}_B^T B^{-1} \geq 0 \qquad (3.52)$$

We also have

$$b_j - \mathbf{b}_B^T B^{-1}(-\mathbf{a}_j^T) \geq 0, \quad j = 1, \ldots, m$$

or

$$-\mathbf{b}_B^T B^{-1}\mathbf{a}_j^T \leq b_j, \quad j = 1, \ldots, m$$

or

$$(-\mathbf{b}_B^T B^{-1})A^T \leq \mathbf{b}^T$$

or

$$A(-\mathbf{b}_B^T B^{-1})^T \leq \mathbf{b} \qquad (3.53)$$

Thus by expressions (3.52) and (3.53)

$$(-\mathbf{b}_B^T B^{-1})^T$$

is a *feasible* solution for the primal problem and we have

$$\mathbf{y}_B = B^{-1}(-\mathbf{c}^T)$$

Hence the optimal dual objective function value is

$$\mathbf{b}_B^T\mathbf{y}_B = \mathbf{b}_B^T(B^{-1}(-\mathbf{c}^T)) = \mathbf{c}(-\mathbf{b}_B^T B^{-1})^T$$

which is equal to the primal objective function value for the primal-feasible solution. Consequently, by Theorem 3.3, $-\mathbf{b}_B^T B^{-1}$ is an optimal solution to the primal.

We have thus shown that if the dual has an optimal solution, so also does the primal. What we have shown suffices to prove the more general theorem, since the dual of the dual is the primal. However, one may employ the same constructive-type proof, assuming an optimal primal solution, and show $\mathbf{c}_B B^{-1}$ to be dual optimal, where B is an optimal primal basis.

In our proof the effect of multiplying the dual constraints by -1 is clear. Whereas a set of optimal primal variables would have been $\mathbf{b}_B^T B^{-1}$, they are, after the multiplication, $-\mathbf{b}_B^T B^{-1}$.

To establish the relationship between an unbounded solution in one problem and the absence of a feasible solution in the other, we first observe that if a linear-programming problem has a feasible solution then it has a feasible solution in which every variable is finite. This is a direct consequence of the fact that all constants in the constraints are finite. Suppose then that the primal problem has an unbounded solution. By Lemma 3.1, every dual-feasible solution must produce an infinite value for the dual objective function, which implies that the dual has no feasible solution with every variable finite. However, this last result contradicts our prior observation so that the dual must have *no* feasible solutions.

Again, since the dual of the dual is the primal, if the dual has an unbounded solution, it follows that the primal has no feasible solution. The final possibility, that neither of the pair has a feasible solution, may be illustrated by example, and Problem 3.42 asks for such a pair.

a. COMPLEMENTARY SLACKNESS

There are other interesting relationships between primal–dual pairs. In the present section, we make mention of one of these since the results will be of use in dealing with some special linear-programming problems in later chapters. Consider the pair of problems (3.48), and suppose that the primal has been solved with B being an optimal basis, $\mathbf{x}_B$ an optimal basic solution, and $\mathbf{c}_B$ as before. Theorem 3.5 is proved.

Theorem 3.5

If x_j is basic in $\mathbf{x}_B$, the corresponding optimal dual solution $\mathbf{c}_B B^{-1}$ is such that it satisfies the jth dual constraint as an equality.

Proof: To see this, we let $\mathbf{a}_j$ be the primal activity vector associated with x_j; since x_j is basic, we have

$$c_j - z_j = c_j - \mathbf{c}_B B^{-1} \mathbf{a}_j = 0$$

or
$$\mathbf{c}_B B^{-1} \mathbf{a}_j = c_j \tag{3.54}$$

But equation (3.54) merely states that the jth dual constraint is satisfied as an equation, which is what we wished to establish.

Stronger statements than the preceding are possible. For example, suppose the following primal problem is solved with one slack variable having been added into each constraint:

$$\max \mathbf{cx}$$

subject to

$$Ax + I_m x_s = b \qquad (3.55)$$
$$x, x_s \geq 0$$

where x_s is the vector of slack variables. Suppose that an optimal solution to the primal is (x^*, x_s^*). Let y^* be an optimal solution to the dual of (3.55). Premultiplying each term in the constraint equations of (3.55) by $(y^*)^T$, we have

$$(y^*)^T A x^* + (y^*)^T x_s^* = (y^*)^T b \qquad (3.56)$$

From the proof of Lemma 3.1, inequalities (3.49) and (3.50) in particular

$$(y^*)^T b \geq (y^*)^T A x^* \geq c x^* \qquad (3.57)$$

From Theorem 3.3

$$c x^* = (y^*)^T b \qquad (3.58)$$

Thus we obtain, by comparing expressions (3.57) and (3.58),

$$(y^*)^T A x^* = (y^*)^T b \qquad (3.59)$$

Equations (3.56) and (3.59) then imply

$$(y^*)^T x_s^* = 0$$

But $y^* \geq 0$ and $x_s^* \geq 0$; hence if the kth component of one is positive, then the kth component of the other is zero. That is, if the kth primal slack variable is positive for any *optimal* primal solution, then the kth dual variable is zero for *every* optimal dual solution; and if in *any* optimal dual solution we have the kth variable positive, then the kth primal slack variable will be zero in *every* optimal primal solution. Note that it was *not* necessary to restrict our attention to basic solutions.

Suppose now that we add surplus variables only to the dual and obtain an optimal solution (y^*, y_s^*) for that problem. We have then

$$A^T y^* - I_n y_s^* = c^T \qquad (3.60)$$

and left-multiplying equation (3.60) by $(x^*)^T$, we obtain

$$(x^*)^T A^T y^* - (x^*)^T y_s^* = (x^*)^T c^T$$

As before, we find

$$(x^*)^T y_s^* = 0$$

The nonnegativity of the primal and dual optimal solutions yields the result that if the kth primal variable is positive in *any* optimal solution then the kth dual surplus variable is zero in *every* optimal dual solution, and so on.

ix. *Derivation of a Dual Problem*

Despite the properties exhibited by primal–dual pairs, it may yet seem that the notion of duality has arisen rather unnaturally and arbitrarily. Let us see that dual problems arise quite naturally. Consider the problem

$$\max c_1 x_1 + \cdots + c_n x_n$$

subject to

$$
\begin{aligned}
a_{11}x_1 + a_{12}x_2 + \cdots + a_{1n}x_n &\leq b_1 \\
a_{21}x_1 + a_{22}x_2 + \cdots + a_{2n}x_n &\leq b_2 \\
&\;\;\vdots \\
a_{m1}x_1 + a_{m2}x_2 + \cdots + a_{mn}x_n &\leq b_m \\
x_j \geq 0, \quad j = 1, \ldots, n
\end{aligned}
\tag{3.61}
$$

Let us consider the equivalent equality-constrained problem, and apply the Kuhn–Tucker conditions to it.

$$\max c_1 x_1 + \cdots + c_n x_n$$

subject to

$$
\begin{aligned}
a_{11}x_1 + a_{12}x_2 + \cdots + a_{1n}x_n + u_1^2 &= b_1 \\
a_{21}x_1 + a_{22}x_2 + \cdots + a_{2n}x_n + u_2^2 &= b_2 \\
&\;\;\vdots \\
a_{m1}x_1 + a_{m2}x_2 + \cdots + a_{mn}x_n + u_m^2 &= b_m \\
x_1 - v_1^2 &= 0 \\
x_2 - v_2^2 &= 0 \\
&\;\;\vdots \\
x_n - v_n^2 &= 0
\end{aligned}
\tag{3.62}
$$

The Lagrangian function for problem (3.62) is

$$F(x_1, \ldots, x_n, \lambda_1, \ldots, \lambda_m, u_1, \ldots, u_m, v_1, \ldots, v_n, \gamma_1, \ldots, \gamma_n)$$

$$= \sum_{j=1}^{n} c_j x_j + \lambda_1 (\sum_{j=1}^{n} a_{1j} x_j + u_1^2 - b_1) + \cdots$$

$$+ \lambda_m (\sum_{j=1}^{n} a_{mj} x_j + u_m^2 - b_m) + \gamma_1(x_1 - v_1^2) + \cdots$$

$$+ \gamma_n(x_n - v_n^2) \tag{3.63}$$

$$= - \sum_{i=1}^{m} \lambda_i b_i + (c_1 + \sum_{i=1}^{m} a_{i1}\lambda_i + \gamma_1)x_1 + \cdots$$

$$+ (c_n + \sum_{i=1}^{m} a_{in}\lambda_i + \gamma_n)x_n + \lambda_1 u_1^2 + \cdots + \lambda_m u_m^2$$

$$- \gamma_1 v_1^2 - \cdots - \gamma_n v_n^2$$

We know that for a relative maximum

$$\frac{\partial F}{\partial x_j} = c_j + \sum_{i=1}^{m} a_{ij}\lambda_i + \gamma_j = 0, \quad j = 1, \ldots, n$$

We also require

$$\frac{\partial F}{\partial u_i} = 2\lambda_i u_i = 0, \quad i = 1, \ldots, m$$

Since $\lambda_i u_i = 0$, $\lambda_i u_i^2 = 0$; consequently,

$$\lambda_i(b_i - a_{i1}x_1 - \cdots - a_{in}x_n) = 0, \quad i = 1, \ldots, m$$

Similarly, it must be that

$$\gamma_j v_j^2 = 0, \quad j = 1, \ldots, n$$

or

$$\gamma_j x_j = 0$$

We know that $\lambda_i \leq 0$, $i = 1, \ldots, m$, and $\gamma_j \geq 0$, $j = 1, \ldots, n$, so

$$\gamma_j = -c_j - \sum_{i=1}^{m} a_{ij}\lambda_i \geq 0 \tag{3.64}$$

Suppose that $(\mathbf{x}^*, \boldsymbol{\lambda}^*)$ is optimal for the unconstrained maximization of F. Our foregoing invocation of necessary (and, in this problem, sufficient) conditions shows that on the right side of equation (3.63) all terms vanish save the first. Thus we have that

$$-\sum_{i=1}^{m} \lambda_i^* b_i = \sum_{j=1}^{n} c_j x_j^* \tag{3.65}$$

Now, for any feasible **x** we have

$$\sum_{j=1}^{n} a_{ij}x_j \le b_i, \quad i = 1, \ldots, m$$

and since

$$-\lambda_i \ge 0$$

$$-\lambda_i b_i \ge -\lambda_i \sum_{j=1}^{n} a_{ij}x_j, \quad i = 1, \ldots, m$$

Summing the above inequality over all possible values of i, we have

$$-\sum_{i=1}^{m} \lambda_i b_i \ge \sum_{i=1}^{m} -\lambda_i \sum_{j=1}^{n} a_{ij}x_j = \sum_{j=1}^{n} x_j \sum_{i=1}^{m} -\lambda_i a_{ij} \tag{3.66}$$

But using equation (3.64), we have

$$\sum_{j=1}^{n} x_j \sum_{i=1}^{m} -\lambda_i a_{ij} \ge \sum_{j=1}^{n} c_j x_j \tag{3.67}$$

Inequalities (3.66) and (3.67) then show that for arbitrary feasible **x**, $\boldsymbol{\lambda}$

$$-\sum_{i=1}^{m} \lambda_i b_i \ge \sum_{j=1}^{n} c_j x_j$$

In particular,

$$-\sum_{i=1}^{m} \lambda_i b_i \ge \sum_{j=1}^{n} c_j x_j^* \tag{3.68}$$

for any feasible $\boldsymbol{\lambda}$.

From equation (3.65) $\boldsymbol{\lambda}^*$ provides equality in (3.68); therefore, $\boldsymbol{\lambda}^*$ is an optimal solution to the problem

$$\min \sum_{i=1}^{m} -\lambda_i b_i$$

subject to

$$-\sum_{i=1}^{m} a_{ij}\lambda_i \ge c_j, \quad j = 1, \ldots, n \tag{3.69}$$

$$\lambda_i \le 0, \quad i = 1, \ldots, m$$

With the substitution $y_i = -\lambda_i$, $i = 1, \ldots, m$, problem (3.69) may be written

$$\min \sum_{i=1}^{m} b_i y_i$$

subject to

$$\sum_{i=1}^{m} a_{ij} y_i \geq c_j, \quad j = 1, \ldots, n$$

$$y_i \geq 0, \quad i = 1, \ldots, m$$

which is precisely what was *defined* as the "dual" of problem (3.61).

x. Concluding Remarks on Uses of Duality

Some elementary implications of duality in linear programming have been presented, along with some application. The importance of duality is not at all confined to computational aspects of linear programming; the theory of dual programs contributes heavily to other algorithms for solving problems in many areas of mathematical programming, especially linear programming, and the student pursuing operations research studies will encounter these theories frequently.

3.9. Formulation of Linear-Programming Problems

An important aspect of utilizing linear programming is the ability to formulate given problems as linear-programming problems. Unfortunately, problem formulation cannot be reduced to a set of rules or algorithms, and proficiency in this area must be gained through practice and the application of rational thought, while familiarity with the problem areas is indispensable.

Let us consider a number of elementary and typical example problems and their formulation as linear-programming problems. It should be understood, however, that the problem solver will not, in general, find himself presented with a neat list of all pertinent data; more likely such data will have to be observed and gathered, perhaps from a wide variety of sources. The student should attempt to obtain mathematical formulations of problems that exist about him. Practice of this sort is quite constructive.

Example 3.9

Suppose that we have m locations which serve as points of origin, such as warehouses, for some commodity. Suppose that there are n points of destination, such as retail stores, to which the commodity is to be shipped from the m origins. The cost of shipping one unit of the commodity from origin i to destination j is given as c_{ij}, $i = 1, \ldots, m$, $j = 1, \ldots, n$.

There are a_i units of the commodity available at origin i, $i = 1, \ldots, m$; destination j requires at least b_j units of the commodity, $j = 1, \ldots, n$. In order that satisfaction of the demands be possible, we assume $\sum_{i=1}^{m} a_i \geq \sum_{j=1}^{n} b_j$; in words, the amount available is sufficiently large to satisfy the demands.

The problem then is to determine a shipping strategy, that is, determine how many units should be shipped from origin i to destination j, for all i, j, so as to minimize the total cost, while observing all origin limitations and destination requirements.

Formulation

The number of variables whose optimal values must be determined is mn, since each origin may ship to each destination. A concise and *meaningful* notation then is obtained by letting

$$x_{ij} = \text{number of units to be shipped from origin } i \text{ to destination } j$$

The cost of such a strategy, our present objective function, is

$$c_{11}x_{11} + c_{12}x_{12} + \cdots + c_{1n}x_{1n} + c_{21}x_{21} + \cdots + c_{2n}x_{2n}$$
$$+ \cdots + c_{m1}x_{m1} + \cdots + c_{mn}x_{mn}$$
$$= \sum_{i=1}^{m} \sum_{j=1}^{n} c_{ij}x_{ij}$$

In accordance with the limitation on availability at origin i, we must have

$$x_{i1} + x_{i2} + \cdots + x_{in} \leq a_i, \quad i = 1, \ldots, m$$

which simply says that we cannot ship more than a_i units from origin i. Similarly, satisfaction of requirements at destination j requires

$$x_{1j} + x_{2j} + \cdots + x_{mj} \geq b_j, \quad j = 1, \ldots, n$$

Our problem is mathematically stated, then, as follows:

$$\min \sum_{i=1}^{m} \sum_{j=1}^{n} c_{ij}x_{ij}$$

subject to

$$\sum_{j=1}^{n} x_{ij} \leq a_i, \quad i = 1, \ldots, m$$

$$\sum_{i=1}^{m} x_{ij} \geq b_j, \quad j = 1, \ldots, n$$

$$x_{ij} \geq 0$$

This is an example of a type of problem known as a transportation problem. Transportation problems will be treated in greater detail in another chapter. For these problems, computational simplicities are available, and if the reader will construct an initial tableau for a small problem of this kind, he will observe certain structural properties responsible for these simplicities.

Example 3.10

A small firm manufactures m different products. Each product requires the use of n machines $M_1, \ldots, M_n$. The amount of time, in hours, required of M_j for the production of a single unit of product $i(i = 1, \ldots, m)$ is t_{ij}, $i, \ldots, m, j = 1, \ldots, n$. M_j is available a maximum of T_j hours each week, and 1 hour of time on M_j costs c'_j dollars. If one unit of product i sells for c_i dollars, determine the number of units of each product to manufacture in a given week so that the total profit is maximized.

Formulation

Since we are interested in specifying a number for each product, it seems that our variables should be $x_1, x_2, \ldots, x_m$. Machine j will be used

$$x_1 t_{1j} + x_2 t_{2j} + \cdots + x_m t_{mj} \quad \text{hours}$$

so we are constrained by

$$\sum_{i=1}^{m} x_i t_{ij} \leq T_j, \quad j = 1, \ldots, n$$

Our objective function, the net profit, is

$$z = \sum_{i=1}^{m} c_i x_i - \sum_{j=1}^{n} c'_j \left(\sum_{i=1}^{m} x_i t_{ij} \right)$$

The first term is the amount realized from selling the products (it has been assumed that everything manufactured is sold), and the second, the cost of operating the machines.

This simple problem raises a very pertinent question. The implication is that we shall manufacture an integral number of units of each product, but the simplex technique is not at all aware of this. If, therefore, an optimal solution advises the production of a nonintegral number of one or more products, what are we to do? It is possible, of course, that generating some alternative optimal solutions, if some exist, would produce one with integer variable values, but we cannot rely upon this. Obviously, rounding the numbers involved, or some such tactic, will produce a set of integer values. Of course, we may violate the constraints upon doing so; but even if we do

not, there may be integer solutions that provide *better* objective function
values than does our rounded solution.

Figure 3.20 illustrates such a situation. The point $\mathbf{R}_2 = (2.75, 1.40)$ is the
extreme point which provides the maximum, z^*, for $ax_1 + bx_2$, the objective
function. Rounding this solution to $\mathbf{R}_3 = (3, 1)$ gives an objective function
value of $z_2 < z^*$. But the point $\mathbf{R}_1 = (2, 2)$ provides a value z_1, of the objective function, such that $z_2 < z_1 < z^*$.

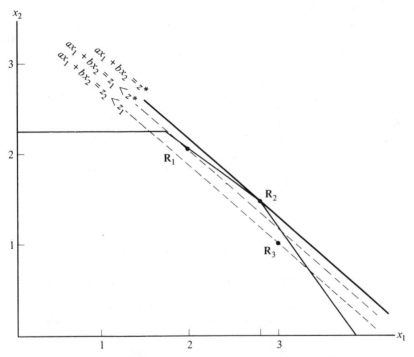

Figure 3.20. Effect of rounding to satisfy integer constraints.

Had the optimal hyperplane contained both $\mathbf{R}_1$ and $\mathbf{R}_2$, then an integer
solution would have given the optimal value z^*, but would not be available
in a basic feasible solution, since $\mathbf{R}_1$ is not an extreme point of the set of
feasible solutions. Here, then, the difference between the objective function
value for the rounded solution and the maximum attainable value of an integer solution is even greater.

It should also be understood that the magnitude of such differences may
be anything but negligible. It is the case that computational devices, which
among others are variations on the simplex technique, for optimizing linear
objective functions subject to linear constraints, and subject to integer con-

straints on some or all variables, exist, but they will not be treated in the present textbook.

Example 3.11

Let us consider an example similar to Example 3.10. The m products require r different resources in their manufacture. The ith product requires s_{ij} units of the jth resource, and there are S_j units of resource j available. Each unit of S_j unused costs c_j'' in storage and handling costs.

Suppose now that our firm manufactures products P_1, P_2, P_3, and P_4. Each product requires the use of one machine of type I, one of type II, and one of type III. There are three machines of type I, A_1, A_2, and A_3; two of type II, B_1 and B_2; and three of type III, C_1, C_2, and C_3.

Times required for the production of one unit of the various products on the various machines are as follows:

		I			*II*			*III*
	A_1	A_2	A_3	B_1	B_2	C_1	C_2	C_3
P_1	$\frac{1}{2}$	1	$\frac{1}{4}$	$\frac{1}{4}$	$\frac{1}{4}$	$\frac{3}{4}$	$\frac{1}{4}$	1
P_2	1	$\frac{2}{3}$	$\frac{1}{2}$	$\frac{3}{4}$	1	$\frac{1}{4}$	$\frac{1}{2}$	$\frac{1}{4}$
P_3	$\frac{1}{4}$	$\frac{1}{4}$	$\frac{1}{4}$	$\frac{1}{2}$	$\frac{1}{4}$	$\frac{1}{3}$	2	$\frac{1}{4}$
P_4	1	$\frac{1}{4}$	$\frac{1}{4}$	$\frac{1}{4}$	$\frac{1}{2}$	$\frac{3}{4}$	$\frac{1}{4}$	1

Machine costs per hour and availabilities in hours per week are given by

	A_1	A_2	A_3	B_1	B_2	C_1	C_2	C_3
Cost	3.5	4	5	4	4.5	4	3	6
Time limits	35	30	45	48	40	35	35	40

Machines A_1, A_2, B_2, C_1, and C_2 must be utilized at least 25, 25, 30, 20, and 20 hours per week, respectively.

The selling prices of the products are 5, 4, 6, and 3 dollars, respectively. Determine an optimal production schedule.

Formulation

Here, it is necessary to determine not only how much of each product to manufacture, but also how much should be produced on the various machines.

The formulation of the objective function and the specification of the constraints are straightforward and are left for the reader.

REFERENCES

[1] CHARNES, A., and W. COOPER, *Management Models and Industrial Applications of Linear Programming*, Vol. I, John Wiley & Sons, Inc., New York, 1961.

[2] DANTZIG, G., *Linear Programming and Extensions*, Princeton University Press, Princeton, N.J., 1963.

[3] HADLEY, G., *Linear Programming*, Addison-Wesley Publishing Company, Inc., Reading, Mass., 1962.

[4] KANTORVICH, L. V., "Mathematical Methods in the Organization and Planning of Production," 1939. A translation appears in *Management Science*, Vol. 6, p. 366, 1960.

[5] SIMMONARD, M., *Linear Programming*, Prentice-Hall, Inc., Englewood Cliffs, N.J., 1966.

[6] SMYTHE, W., and L. JOHNSON, *Introduction to Linear Programming with Applications*, Prentice-Hall, Inc., Englewood Cliffs, N.J., 1966.

[7] WOLFE, P., "Recent Developments in Nonlinear Programming", in *Advances in Computers*, Vol. 3, Academic Press, Inc., New York, 1962.

PROBLEMS

1. For each of the following linear-programming problems, by making sketches of the set of feasible solutions and several members of the family of objective function "lines," decide whether there is a unique optimal solution, an infinite number of optimal solutions, an unbounded solution, an optimal solution with some infinite components, or no feasible solution. When one exists, find an optimal solution and the associated objective function value.

(a)   max $x + 2y$

subject to

$$3x + y \geq 2$$
$$-\tfrac{1}{2}x + y \leq 1$$
$$-\tfrac{1}{4}x + y \geq \tfrac{3}{4}$$
$$x, y \geq 0$$

(b) min $x + 3y$ subject to the same constraints as in (a).

(c) $$\max 6y + 4x$$

subject to

$$y - x \leq 1$$
$$3y + 2x \leq 12$$
$$y + 4x \leq 20$$
$$y, x \geq 0$$

(d) $$\min 2x_1 - x_2$$

subject to

$$\tfrac{1}{3}x_1 + x_2 \leq \tfrac{4}{3}$$
$$-x_1 + x_2 \geq 1$$
$$\tfrac{5}{2}x_1 + x_2 \geq 10$$
$$x_1, x_2 \geq 0$$

(e) $$\max -x_1 + \tfrac{3}{4}x_2$$

subject to

$$x_1 + x_2 \geq 2$$
$$-2x_1 + x_2 \leq -1$$
$$-\tfrac{4}{3}x_1 + x_2 \leq 2$$

2. In two dimensions, prove that an optimal solution *cannot* occur at an interior point of the feasible set. Can you generalize the proof to n dimensions?

3. Let X be the set of feasible solutions to any linear-programming problem in two dimensions, and suppose that X is bounded. What geometric term describes X? (Null sets and sets of a single point will also be considered to satisfy the definition.)

4. Using the approach of Chapter 2 in defining line segments joining given pairs of points, select a particular feasible set $X \subset E^2$ with several extreme points, say $x_1, x_2, \ldots, x_k$, where X is bounded, and show
   (a) That every $x \in X$ is a convex combination of $x_1, x_2, \ldots, x_k$; that is,

$$\mathbf{x} = \sum_{i=1}^{k} \alpha_i \mathbf{x}_i$$

where $\alpha_i \geq 0$, $i = 1, \ldots, k$, and

$$\sum_{i=1}^{k} \alpha_i = 1$$

   (b) That the set of *all* convex combinations of $x_1, \ldots, x_k$ is *exactly* the set X. Can you generalize these results to n dimensions?

5. Sketch several sets that represent feasible solution sets for linear-programming problems in two dimensions. For each set X, construct several lines as the line L, which has the properties
   (a) $L \cap X \neq \emptyset$
   (b) $L \cap X_I = \emptyset$
   where X_I is the set of interior points of X.
        What is the relationship between these lines L and Problem 2?

6. In Problem 5, can any such lines be constructed that do *not* contain an extreme point of X? How does that result relate to Problem 2?

7. Consider the problem
$$\max -2x_1 + 3x_2$$
   subject to
$$-x_1 + \tfrac{3}{2}x_2 \geq -3$$
$$4x_1 - 6x_2 \geq -11$$

   (a) Is Theorem 3.2 true for this problem?
   (b) Add the usual nonnegativity constraints and then answer (a).
   (c) Could you characterize the differences between the respective sets of points in terms other than nonnegativity?
   (d) For the given problem, is Theorem 3.1 satisfied?

8. (a) For each of the following systems of equations, find a basic solution, using elementary row operations (Chapter 1).

   (1) $2x_1 - x_2 + x_3 = 4$
         $x_1 + x_2 + 3x_3 = 6$

   (2) $x_1 + 2x_2 - \tfrac{1}{2}x_3 + 4x_4 = 6$
      $-2x_1 - x_2 + 2x_3 \quad\quad\quad = -2$
           $x_2 - x_3 \quad\quad\quad = 0$

   (3) $4x_1 - x_2 + 2x_3 = 2$
        $x_1 + 2x_2 - x_3 = 0$
        $x_1 \quad\quad + 3x_3 = 4$

   (4) $4x_1 - x_2 + 2x_3 = 2$
        $3x_1 - x_2 - x_3 = -2$
        $x_1 \quad\quad + 3x_3 = 4$

   (5) $4x_1 - x_2 + 2x_3 = 1$
        $3x_1 - x_2 - x_3 = -2$
        $x_1 \quad\quad + 3x_3 = 4$

   (b) Does every consistent system of linear equations have a basic solution? Explain.
   (c) Write a system of equations $A\mathbf{x} = \mathbf{b}$, where
$$r(A, \mathbf{b}) \neq r(A)$$

9. Given the system of inequalities $A\mathbf{x} \gtreqless \mathbf{b}$ with A $m \times n$, add the appropriate slack and surplus variables, obtaining the system

$$A'\mathbf{x}' = \mathbf{b}$$

   (a) What is the rank of A'?
   (b) For the above systems of inequalities, select an A with $r(A) \neq r(A')$. [Can we have $r(A') < r(A)$?] Using row operations, obtain a basic solution to the system $A'\mathbf{x}' = \mathbf{b}$. Having done that, is it possible to deduce that $r(A)$ was not m? How?

10. For each of the systems in Problem 8a, describe the set of all feasible solutions —in terms of some set of variables that may be arbitrarily specified.

11. Suppose that we find *two* bases $B \neq B'$, which give rise to two basic feasible solutions, $\mathbf{x}(B)$ and $\mathbf{x}(B')$ for $A\mathbf{x} = \mathbf{b}$. Strictly speaking, $\mathbf{x}(B)$ is *different* from $\mathbf{x}(B')$ because B is different from B'. However, in some cases it may be that $\mathbf{x}(B)$ and $\mathbf{x}(B')$ correspond to the *same* point in space.
   (a) What property *must* $\mathbf{x}(B)$ and $\mathbf{x}(B')$ have in order that this occur?
   (b) Construct a system $A\mathbf{x} = \mathbf{b}$ and find B, B', $\mathbf{x}(B)$, and $\mathbf{x}(B')$ such that it does occur.

12. Relation (3.18) was given as a rule by which to decide which vector in a basis can be replaced by $\mathbf{P}_j$ not in that basis so that the *new* basic solution remains feasible. The relation is quite simple to derive, but here we ask the reader only to show that (3.18) is valid.
   Let a basis be $B = \{\boldsymbol{\beta}_1, \boldsymbol{\beta}_2, \ldots, \boldsymbol{\beta}_m\}$ with corresponding basic feasible solution $\{b_1, b_2, \ldots, b_m\}$. Then

$$b_1\boldsymbol{\beta}_1 + b_2\boldsymbol{\beta}_2 + \cdots + b_m\boldsymbol{\beta}_m = \mathbf{b} \tag{1}$$

   Suppose that $\mathbf{P}_j \notin B$. $\mathbf{P}_j$ has a unique expression in terms of B:

$$\mathbf{P}_j = y_{1j}\boldsymbol{\beta}_1 + y_{2j}\boldsymbol{\beta}_2 + \cdots + y_{mj}\boldsymbol{\beta}_m \tag{2}$$

   The scalar y_{rj} must be nonzero in order that

$$\{\boldsymbol{\beta}_1, \ldots, \boldsymbol{\beta}_{r-1}, \mathbf{P}_j, \boldsymbol{\beta}_{r+1}, \ldots, \boldsymbol{\beta}_m\} = B'$$

   be a linearly independent set.
   Thus (2) may be solved for $\boldsymbol{\beta}_r$. Substitute that expression into (1), and combine terms, obtaining expressions for the values of the basic variables corresponding to B'. (Expressions for the new variables as functions of Q can also be obtained by expressing the $\mathbf{b}$ column results of the simplex iteration.) Then show that if relation (3.18) is used to choose $\boldsymbol{\beta}_r$, all variable values remain nonnegative.

13. Relation (3.19) or (3.23) is also simple to obtain. Having obtained the new variable values in Problem 12, the objective function value for the new solution is available.

Denoting the costs of the basic variables corresponding to B by

$$c_i(B), \quad i = 1, \ldots, m$$

and basic solution components by

$$x_i(B), \quad i = 1, \ldots, m$$

the new objective function value is

$$z' = \sum_{i=1}^{m} c_i(B')x_i(B') \tag{1}$$

With the $x_i(B')$ known from Problem 12, relation (3.19) follows by rearranging terms from (1).

14. The reasons for modifying the tableau in the way stated are also easy to grasp.

For the basis B, the y_{ij} are uniquely determined, of course. When $\mathbf{P}_j$ replaces $\mathbf{\beta}_r$, these y_{ij} also change, and expressions for the new ones are easily obtained in a manner analogous to obtaining the $x_i(B')$ in Problem 12.

Derive expressions for these new y_{ij}, and verify that the methods for modifying the tableau accomplish *exactly* the obtaining of the new y_{ij} and $x_i(B')$.

15. The terminology "simplex method" is derived from the fact that among the first linear-programming problems to be solved by such methods the feasible sets were often *simplices*. A *simplex* in E^n is defined as the convex polyhedron generated in the sense of Problem 4 by some $n + 1$ points, not all of which lie on a hyperplane in E^n. What does every simplex look like for $n = 2$? For $n = 3$?

16. Using the simplex method, solve each of the following linear-programming problems.

(a)   max $2x_1 + x_2 + x_3 - x_4$

     subject to

$$
\begin{aligned}
x_1 - x_2 - x_3 + x_4 &\leq 1 \\
-x_1 \quad\quad + 2x_3 + 2x_4 &\leq 2 \\
x_1 - x_2 \quad\quad + 3x_4 &\leq 2 \\
x_1, x_2, x_3, x_4 &\geq 0
\end{aligned}
$$

(b)   max $\frac{1}{2}x_{11} + 2x_{12} + x_{13} + 3x_{21} + \frac{1}{4}x_{22} + 2x_{23}$

     subject to

$$
\begin{aligned}
x_{11} + x_{12} + x_{13} &\leq 2 \\
x_{21} + x_{22} + x_{23} &\leq 3 \\
x_{11} + x_{21} &\leq 1 \\
x_{12} + x_{22} &\leq 3 \\
x_{13} + x_{23} &\leq 1 \\
x_{ij} \geq 0, \ i = 1, 2, \ j &= 1, 2, 3
\end{aligned}
$$

(c) min $-x_{11} - 2x_{12} - \frac{1}{2}x_{13} - 2x_{21} - 3x_{22} - \frac{1}{2}x_{23} - x_{31} - x_{32} - 4x_{33}$
subject to

$$x_{11} + x_{12} + x_{13} \le 1$$
$$x_{21} + x_{22} + x_{23} \le 1$$
$$x_{31} + x_{32} + x_{33} \le 1$$
$$x_{11} + x_{21} + x_{31} \le 1$$
$$x_{21} + x_{22} + x_{32} \le 1$$
$$x_{31} + x_{32} + x_{33} \le 1$$
$$x_{ij} \ge 0, \quad \text{all } i, j$$

(For parts (b) and (c), could any statements about *optimal* values of slack variables have been made *before* solving the problem?)

(d) What is the largest value of x_2 for which a point (x_1, x_2, x_3, x_4) satisfies

$$x_1 - x_2 + 2x_3 + x_4 \le 6$$
$$2x_1 + x_2 - x_3 + 3x_4 \le 8$$
$$-x_1 + x_2 - 3x_3 + x_4 \le 5$$
$$x_1, x_2, x_3, x_3, x_4 \ge 0$$

(e) Let X be all points satisfying

$$x_1 + x_2 - 2x_3 - 4x_4 \le 0$$
$$x_1 \quad + 2x_3 + x_4 \le 10$$
$$-x_1 + x_2 \quad + 2x_4 \le 8$$
$$x_1, x_2, x_3, x_4 \ge 0$$

For $\mathbf{x}', \mathbf{x}'' \in X$, define

$$d(\mathbf{x}', \mathbf{x}'') = \sum_{j=1}^{4} (x_i' - x_i'')$$

Find a pair of points in X that maximize d and give the maximum value of d.

(f) min $-2x_1 + x_2 + 2x_3 + x_4 - 3x_5$
subject to

$$2x_1 - x_2 + 3x_3 \quad - x_5 \le 4$$
$$x_2 + 2x_3 + x_4 \quad \le 6$$
$$-x_1 \quad + x_3 - x_4 + x_5 \le 2$$
$$x_1, x_2, x_3, x_4 \ge 0$$

(g) Among those **x** satisfying

$$4x_1 - x_2 + x_3 - x_4 \leq 1$$
$$x_1 + x_2 + x_4 \leq 3$$
$$2x_2 - x_3 + 2x_4 \leq 2$$

is there one for which the function $x_1 - 2x_2 + 2x_3 - 3x_4$ assumes the value $\frac{3}{4}$? The value 3? Explain.

(h) A much-used linear-programming example problem is the diet problem, in which one attempts to select a diet from among a few given foods that satisfies various requirements of food value, calories, vitamin content, and so forth, and that minimizes cost on a daily basis. We ask the reader to solve such a problem using the foods and totally hypothetical information in the following table. The unit for the data is all per ounce.

Available Food	Cost	Calories	Protein Content	Vitamin A	Carbohydrates	Taste, Appearance, Esthetics, etc.
Rum	15	25	8	1	10	9
Bananas	10	20	9	2	9	5
Escargot	20	24	9	4	9	9
Crow	4	4	10	10	3	1
Watermelon	2	2	2	8	1	4

Select the amount of each food to be included in the person's diet so as to minimize the cost while satisfying the following constraints:
No more than 2000 calories must be consumed. Total vitamin A content must be less than total esthetics. Twice the total protein consumed must be less than $\frac{3}{2}$ the total carbohydrates plus 240. The sum of carbohydrates and vitamin A must be less than 400 units.

17. The quantity $c_j - z_j$ has been likened to a partial derivative. Discuss that interpretation, comparing and contrasting the two.

18. True or false? Explain.
   (a) If a basic feasible solution is degenerate, the next one will also be.
   (b) If from one iteration to the next there is no change in variable values, the former solution was degenerate.

19. Re-solve Problem 3.16d by utilizing the maximum-improvement-at-each-stage procedure for selecting entering/departing vector pairs. For that problem, give a quantitative comparison of the computation required by the two methods.

20. A one-to-one correspondence between the extreme points corresponding to a system of linear equations and equalities and those corresponding to the system of equations obtained from the former by the addition of slack and

surplus variables was illustrated (p. 149). Can you *prove* this correspondence in general?

21. (a) Let $\mathbf{P}_{i_1}, \mathbf{P}_{i_2}, \ldots, \mathbf{P}_{i_k}$ be artificial vectors in a basis at a time when the optimality criteria are satisfied—and with the corresponding artificial variables equal to zero. Suppose, for $\mathbf{P}_j$ not in the basis, that we find

$$y_{i_q j} \neq 0, \quad \text{for } q \in \{1, \ldots, k\}$$

Show that $\mathbf{P}_j$ may replace $\mathbf{P}_{i_q}$, and that the new basic feasible solution is also optimal.

   (b) Show that with the same artificial vectors basic, and with zero variable values, if for *all* $\mathbf{P}_j$ not in the basis we have $y_{ij} = 0$ for all $i = i_1, i_2, \ldots, i_k$ then we had k constraints redundant among the original m constraints. (Actually, it is possible to determine which ones *were* redundant. As it happens, under the given conditions, if the artificial vectors—which are unit vectors of course—have the 1's in the $p_1, p_2, \ldots, p_k$ components, respectively, then the $(p_1)^{\text{st}}, (p_2)^{\text{nd}}, \ldots, (p_k)^{\text{th}}$ constraints were redundant. Look back at optimal solutions to some of the problems that required artificial variables and observe the validity of this result.)

22. Suppose one solves a linear-programming problem that had no artificial variables. How might one recognize from the results of the simplex method that there were redundant constraints, or can one?

23. Suppose, with a feasible solution available, that the unbounded solution indication appears. Show, by means of an example, that one cannot simply ignore the indication and have it go away—in other words, that it is persistent. Establish whether it is apparent at *every* iteration, whether it goes, to return at some later time, or what.

24. The simplex method, by design, operates so as to never produce a worse objective function value during an iteration. There are cases in which it is possible to locate an optimal solution in fewer iterations if one took a "loss" during one or more iterations along the way. Produce a sketch in two or three dimensions that illustrates this possibility.

25. Solve the following problem and see if there is any indication of redundant constraints. Can you generalize?

$$\max x_1 - 2x_2 + x_3 - 4x_4 + 3x_5$$

subject to

$$-2x_1 \qquad + x_3 \qquad + x_5 \leq 4$$
$$x_1 + 2x_2 \qquad + 3x_4 + 2x_5 = 6$$
$$3x_1 + 2x_2 - x_3 + 3x_4 + \ x_5 \leq 2$$
$$x_1, x_2, x_3, x_4, x_5 \geq 0$$

26. A company wishes to market four different blends of gasoline. They have available three different types of blending stock, in limited quantities. The

specifications of the blends and the profit per gallon are as follows:

Blend	Stock A(%)	Stock B(%)	Stock C(%)	Profit
1	40	20	40	.10
2	33	33	34	.11
3	50	50	0	.14
4	10	60	30	.25

Blending stock availability is

Stock A	15,000 gallons
Stock B	14,000 gallons
Stock C	20,000 gallons

Formulate this problem as a linear-programming problem and find a solution that maximizes profit.

27. A company distributes two products, and it can either buy them from another company or it can make them in its own plant. The costs of each alternative and the production rates are as follows:

	Product A	Product B
Make	$1.00/unit	$1.70/unit
Buy	$1.20/unit	$1.50/unit
Production rate	3 units/hour	5 units/hour

The company must have at least 100 units of A and 200 units of B each week. There are 40 hours of productive time per week and idle time costs $2.50 per hour. Furthermore, no more than 60 units of A can be made each week, and no more than 120 units of B can be made each week. Also no more than 130 units of B can be bought per week.

State this problem as a linear-programming problem with cost minimization the objective, and then solve the problem.

28. This problem is an application of linear programming to production scheduling. It is desired to develop a production schedule for one product for the next 4 months. The following data are given:

Month	Demand (units)
1	5000
2	6000
3	4000
4	4000

The inventory at the beginning of the period is 600 units, and the inventory at the end of the period must be 500 units. Production during the present month (month 0) was 4000 units.

The unit cost of increasing production from one period to the next is $.020. The unit cost of decreasing production from one period to the next is $.018. The cost of carrying a unit of inventory from one period to the next is $.030.

State this problem as a linear-programming problem, and find an optimal production schedule for the company.

29. State this problem as a linear-programming problem.

At the end of the week the following situation arises in a railroad network:

Station	Number of Cars on Hand
1	47
2	82
3	31
4	29
5	66

Station	Number of Cars Needed
1	28
4	36
6	79
7	68

The mileage matrix for the network is

	1	4	6	7
1	0	176	49	76
2	213	72	149	68
3	39	132	105	163
4	91	0	63	82
5	34	76	92	132

Furthermore, there is no train from station 4 to station 7, and none from station 3 to station 1.

Reallocate the cars so as to minimize total mileage for all cars.

30. Suppose, for some simplex tableau, we find that row i is zero for all nonbasic columns P_j. Then show
(a) Row i is zero *everywhere* except for one column (which column?).
(b) Row i will *never* change.
(c) The ith basis vector will *never* be removed.

31. Utilizing the simplex method, with artificial variables whenever necessary, solve each of the following linear-programming problems (use M method).

(a) $\max x_1 - 2x_2 + 3x_3 - 4x_4$

subject to

$$x_1 + x_2 - \tfrac{1}{2}x_3 + 3x_4 = 5$$
$$x_2 - \tfrac{3}{2}x_3 + x_4 \geq 3$$
$$x_1 + 2x_3 + 2x_4 \geq 4$$
$$x_1, x_2, x_3, x_4 \geq 0$$

(b) $\min 3x_1 - 4x_2 + x_3 + \tfrac{1}{2}x_4$

subject to

$$x_2 + 4x_3 + 2x_4 \leq 10$$
$$2x_1 - \tfrac{3}{2}x_2 - \tfrac{3}{2}x_3 + x_4 \geq -4$$
$$x_1 + 2x_2 - x_3 + 2x_4 \geq 2$$
$$x_1, x_2, x_3, x_4 \geq 0$$

(c) $\max \tfrac{3}{2}x_1 - \tfrac{3}{4}x_2 + 2x_3 + \tfrac{5}{2}x_4$

subject to

$$-x_1 - 2x_2 + x_3 - 2x_4 \leq -2$$
$$2x_1 + x_2 + x_4 \leq 6$$
$$-4x_1 - x_2 - x_3 + 2x_4 \leq -1$$
$$x_1, x_2, x_3, x_4 \geq 0$$

(Also preserve the inverse of an optimal basis.)

(d) Re-solve part (c) *without* using artificial variables.

32. Show an example where the unbounded solution indication appears, but, because the solution is infeasible at that time, is not a true indication.

33. Solve each problem in Problem 31 using the two-phase method.

34. Return to Problem 31(a) and (b), and approach it with the methods of Chapter 2. What, if any, difficulties are encountered, and which procedure seems most effective on the two sample problems?

35. Suppose that $B = (\boldsymbol{\beta}_1, \boldsymbol{\beta}_2, \ldots, \boldsymbol{\beta}_m)$ is a basis and that we have B^{-1} available. $\boldsymbol{\beta}_r$ is about to be replaced by $\mathbf{P}_k$, and we would like to compute $(B')^{-1}$, where $B' = (\boldsymbol{\beta}_1, \ldots, \boldsymbol{\beta}_{r-1}, \mathbf{P}_k, \boldsymbol{\beta}_{r+1}, \ldots, \boldsymbol{\beta}_m)$.

Derive a method for computing $(B')^{-1}$ that, rather than the inversion of a new matrix, expresses $(B')^{-1}$ as a *matrix product* involving B^{-1}.

36. The following problems are to be formulated and solved.

(a) There are five kinds of products that can be transported in a cargo plane whose hold can accommodate 300,000 pounds and 40,000 cubic feet. The weights, volumes, and selling prices of each product are as follows:

Product	Weight	Volume	Selling Price ($)
1	20	10	3
2	30	20	2
3	50	15	5
4	20	50	5
5	35	30	4

In addition, there is a cost of fueling the plane, and this is $1 per pound of cargo. Find a cargo that will maximize profits, fractional numbers for product types being acceptable.

(b) Locate in the plane E^2 the vertices of the rectangle of maximum perimeter that satisfies the following conditions.
(1) Its sides are parallel to the coordinate axes.
(2) The shorter side must have length at least two units.
(3) The area of the rectangle may be at most 23 square units.
(4) The length of the portion of the perimeter that lies in the quadrant of the plane where both coordinates are negative must be at least 3 units and no more than 11 units.

37. You have carefully gathered data and optimized a particular operation, which you had formulated as a linear-programming problem, using that data. At that point it is discovered that the cost of the variable x_j should not have been the c_j you used; rather $c'_j \neq c_j$. What should you do? Make the change and re-solve the problem?

38. For the problem

$$\text{max } \mathbf{cx}$$

subject to

$$A\mathbf{x} \overset{\leq}{\underset{\geq}{=}} \mathbf{b}$$

$$\mathbf{x} \geq \mathbf{0}$$

derive the dual in the fashion of the text's derivation for the problem with constraints $A\mathbf{x} \leq \mathbf{b}$.

39. Write the dual problem for each of the following, and write it in a form as compact as possible.
(a)   max $x_1 - 2x_2 + 2x_3 + x_4 - x_5$
subject to

$$x_1 + 2x_2 + 3x_3 \quad\quad + 2x_5 \leq 6$$
$$x_2 - x_3 + x_4 + x_5 = 7$$
$$-x_1 \quad\quad + 2x_3 - x_4 + x_5 = 5$$
$$x_1, x_3, x_5 \geq 0, \quad x_2, x_4 \text{ unrestricted}$$

(b)
$$\max \sum_{i=1}^{m} \sum_{j=1}^{n} c_{ij} x_{ij}$$

subject to

$$\sum_{j=1}^{n} x_{ij} = a_i, \quad i = 1, \ldots, m$$

$$\sum_{i=1}^{m} x_{ij} = b_j, \quad j = 1, \ldots, n$$

$$x_{ij} \geq 0, \quad \text{all } i, j$$

(c)
$$\min x_1 - 2x_2 + x_3 - x_4 - 3x_5$$

subject to

$$
\begin{aligned}
x_1 - 2x_2 + \tfrac{3}{2}x_3 \qquad\qquad + x_5 &\geq 4 \\
-x_1 + 2x_2 \qquad + 2x_4 + 2x_5 &\geq -3 \\
x_2 - x_3 + 3x_4 + x_5 &= 2 \\
2x_1 \qquad + 2x_3 \qquad - 3x_5 &\leq 4 \\
x_1 \leq 0, x_2, \ldots, x_5 &\geq 0
\end{aligned}
$$

(d)
$$\max \sum_{i=1}^{n} \sum_{j=1}^{n} c_{ij} x_{ij}$$

subject to

$$\sum_{i=1}^{n} x_{ij} = 1, \quad j = 1, \ldots, n$$

$$\sum_{j=1}^{n} x_{ij} = 1, \quad i = 1, \ldots, n$$

$$x_{ij} \geq 0, \quad \text{all } i, j$$

(e)
$$\max 4x_1 + 2x_2 + x_3 - 2x_4 + x_5 - x_6$$

subject to

$$
\begin{aligned}
3x_1 \qquad - x_3 + 2x_4 \qquad + x_6 &\geq 3 \\
2x_2 + x_3 \qquad + x_5 - x_6 &= 6 \\
x_1 - x_2 \qquad + x_4 + 2x_5 + 2x_6 &\leq 4 \\
x_2 - x_3 + 3x_4 \qquad - x_6 &= 3 \\
x_1 \qquad + 2x_3 - x_4 - x_5 + x_6 &\leq 1 \\
x_1, x_3, x_5 \text{ unrestricted}, x_2 &\leq 0 \\
x_4, x_6 &\geq 0
\end{aligned}
$$

40. Consider the development leading to Theorem 3.3. Prove the theorem for the primal problem max $\mathbf{cx}$, subject to $A\mathbf{x} = \mathbf{b}$, $\mathbf{x} \geq \mathbf{0}$.

41. Proceed along the lines of the constructive proof of Theorem 3.4 and show that if max $\mathbf{cx}$, subject to $A\mathbf{x} = \mathbf{b}$, $\mathbf{x} \geq \mathbf{0}$ has an optimal solution, then so does its dual.

42. Construct a primal–dual pair such that neither has a feasible solution.

43. Can you construct a linear-programming problem of the form max $\mathbf{cx}$, subject to $A\mathbf{x} \leq \mathbf{b}$, $\mathbf{x} \geq \mathbf{0}$, such that the dual of the problem is the same as the problem itself? If you can, how would you characterize the matrix A?

44. A most important result in linear inequalities and linear programming is the Farkas lemma. It states that if $\mathbf{b}^T\mathbf{x} \geq 0$ for every $\mathbf{x}$ satisfying

$$A^T\mathbf{x} \geq \mathbf{0}$$

then the system

$$A\mathbf{y} = \mathbf{b}, \quad \mathbf{y} \geq \mathbf{0}$$

has a solution.

The Farkas lemma has applications in nonlinear programming and can be used to derive the duality theory of linear programming in elegant fashion.

What we ask here, however, is that the reader use what he knows about duality theory to prove the Farkas lemma.

45. Find optimal solutions (or determine that none exist) for each of the following problems by solving their respective dual problems.

(a)
$$\max -x_1 + 2x_2 + x_3$$
subject to
$$\begin{aligned}
2x_1 + x_2 + x_3 &\leq 20 \\
-x_1 \quad\quad + x_3 &\leq 5 \\
x_1 - 2x_2 \quad\quad &\leq -1 \\
x_2 + 2x_3 &\leq 6 \\
x_1, x_2, x_3 &\geq 0
\end{aligned}$$

(b)
$$\min 2x_1 + x_2 - 3x_3 + 4x_4$$
subject to
$$\begin{aligned}
x_1 - 2x_2 + x_3 \quad\quad &\leq 3 \\
-2x_1 \quad\quad - x_3 - x_4 &\leq 6 \\
3x_1 - x_2 + 2x_3 + 2x_4 &\geq 2 \\
x_2 \quad\quad + x_4 &= 4
\end{aligned}$$

(c)
$$\max 3x_1 + 2x_2 - \tfrac{5}{2}x_3 - x_4 + 2x_5$$
subject to
$$\begin{aligned}
-2x_1 + x_2 \quad\quad + 2x_4 + x_5 &\geq 10 \\
2x_2 + x_3 - 3x_4 + 2x_5 &\leq 8 \\
x_1 + x_2 \quad\quad + x_4 - 2x_5 &\leq 5 \\
x_1, x_2, x_3, x_4 \geq 0, \; x_5 \text{ unrestricted}
\end{aligned}$$

(d) With c_{ij}, a_i, and b_j given by

$$
C = \begin{pmatrix} 2 & 2 & 1 & 0 \\ 4 & 2 & -1 & 1 \\ 7 & 3 & 4 & 2 \\ 4 & 7 & 2 & 6 \\ 5 & 9 & 4 & 2 \end{pmatrix}, \quad a = \begin{pmatrix} 4 \\ 5 \\ 6 \\ 2 \\ 8 \end{pmatrix}, \quad b = \begin{pmatrix} 8 \\ 2 \\ 10 \\ 5 \end{pmatrix}
$$

the problem is,

$$
\max \sum_{i=1}^{5} \sum_{j=1}^{4} c_{ij} x_{ij}
$$

subject to

$$
\sum_{j=1}^{4} x_{ij} = a_i, \quad i = 1, \ldots, 5
$$

$$
\sum_{i=1}^{5} x_{ij} = b_j, \quad j = 1, \ldots, 4
$$

$$
x_{ij} \geq 0, \quad \text{all } i, j
$$

46. Verify the complementary slackness results of Theorem 3.5 where applicable for every problem in Problem 45.

47. If one of a primal–dual pair is solved using artificial variables, say with the M procedure, then an optimal solution to its dual may involve the number M. When and only when is this possible? What would it mean?

48. With the linear-programming problem
$$
\max 3x_1 - x_2 + 2x_3
$$
subject to
$$
x_1 + x_2 + x_3 = 6
$$
$$
x_1 - 2x_2 - x_3 \geq 2
$$
$$
x_1, x_2, x_3 \geq 0
$$

designated P, and the problem with surplus and artificials P_A, write the dual of P and the dual of P_A. Supposing that the M method is to be used on P_A, show that an optimal solution to dual P_A is optimal for dual P. Can you generalize?

49. Show that when the rth primal variable is unrestricted the rth dual constraint is an equality.

50. Show that when the rth primal constraint is an equality the rth dual variable is unrestricted.

51. Show that if the tableau transformation operations are also applied to the $c_j - z_j$ row, then those new values are as found by recomputing the $c_j - z_j$ at the start of the next iteration.

PERT AND NETWORK FLOWS

4.1. Introduction

It is a well-established fact that the flow of current in a circuit is a function of the various elements of the circuit. In the past decade an increasing amount of attention has been focused on network flow theory and its relevant applications to fields other than electrical engineering, such as transportation problems, minimum-cost flow problems, and shortest route problems. Consequently, network flow theory has become a powerful aid to systems analysts.

One problem that has been solved by techniques closely related to those of network flow theory is the planning and control of research and development projects. Both PERT (Program Evaluation and Review Technique) and CPM (Critical Path Method) have treated special facets of this problem. A second problem in network flow theory has been the determination of the optimum method of allocating flows to a network to maximize the total flow through it. The determination of the shortest route between two points through a connecting graph constitutes another problem area in network theory. These various problems will be covered in this chapter.

4

4.2. PERT

i. *Introduction*

The technique called PERT (Program Evaluation and Review Technique) was created in the late fifties as a management tool in the Navy Polaris Fleet Ballistic Missile Program. PERT was credited with moving up the projected completion date of Polaris by more than 2 years. The Navy's success in applying PERT led to its subsequent application in other Defense Department projects, and in industry. Examples of industry use of PERT can be found in construction programs for dams and power stations and the installation of production control and information systems.

PERT is limited in that it cannot be applied for most repetitive production, distribution, or sales activities, except perhaps for the setting up for initial production.

Several management objectives can be attained by the use of PERT when it is applied to evaluate a project. PERT enables the manager to evaluate the probability of meeting a set schedule for completion of a task. The manager can also evaluate the effect of changes in the project, as well as locate potential delays. For example, a portion of the project may be redesigned and the manager may have to determine the impact of this change on the total schedule of the project.

ii. *Formulation of the PERT Model*

a. PERT NETWORK

Given the information necessary to evaluate a project, it is then possible to construct a *PERT network*, which is a pictorial representation of the project. An example of such a network, which can also be classified as a flow diagram, is given in Figure 4.1.

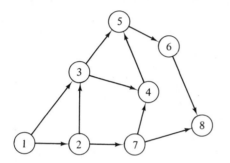

Figure 4.1. Example of a PERT network.

Each of the circled nodes in Figure 4.1 represents an *event* which in PERT is used to denote the "start" or "completion" of a task. An event is not the actual performance of a task. A PERT event must conform to three criteria: (1) It must indicate a noteworthy or significant point in the project; (2) it is the start or completion of a task; and (3) it does not consume time or resources. There is a definite sequential relationship between events in a PERT network. Events depend upon one another, such as event 1 "report started" and event 2 "report completed," and a logical step relationship is followed. Events may be represented in a PERT network by circles, ovals, squares, or other selected geometric figures.

The branches in Figure 4.1, with their directions indicated by arrowheads, represent activities. A PERT *activity* is the actual performance of a task. It is the time-consuming portion of the PERT network and requires manpower, material, space facilities, or other resources. It is assumed that the time necessary to complete one activity has no effect on the elapsed time for any other activity in the network. Events are connected by activities to form a PERT network. Consequently, an event is the culmination of all the activities leading to that event, and no activity can be initiated until the preceding event is reached. The event or events that immediately follow another event without any intervening events are called *successor events*. Similarly, *predecessor events* can be defined as the event or events that immediately come before another event without any intervening events.

Thus a PERT network can be redefined as a pictorial representation of the interrelationships of all necessary events and activities that constitute a project. In Example 4.1 it is shown how it is possible to construct a PERT network.

Example 4.1

From the list of events describing an oil change and lubrication performed on a car, construct a PERT network.

(1) Finish pouring in new oil.
(2) Engine started.
(3) Old oil drained.
(4) Car on lift.
(5) Suspension lubricated.
(6) Start pouring in new oil.
(7) Oil filter installed.

For any set of events describing a project, there is usually one event that marks the commencement of the project and one event that marks the completion of the project. Examining the list of events, it is observed that event 4 is the first event and event 2 is the last event. The remaining five events occur between events 4 and 7. The old oil cannot be drained and the suspension cannot be lubricated until the car is on a lift. Thus event 4 is a predecessor event to events 3 and 5. The new oil cannot be poured in until the oil filter is installed and the suspension is lubricated. Consequently, event 6 is a successor event to events 7 and 5. Event 1 is a predecessor event to 2 and a successor event to 6 (see Figure 4.2).

In the example just considered, events 4 and 2 are called the *source* and *sink* nodes, respectively. However, it is not necessary to restrict a PERT

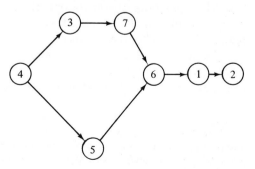

Figure 4.2. Solution to Example 4.1.

network to a single source and sink. Furthermore, PERT networks are not unique, since it is up to the manager to determine which milestones in their proper sequence should be PERT events.

b. ACTIVITY TIMES

In order to obtain estimates for the performance times for each activity, three time estimates are introduced. These time estimates are obtained for each activity from the person who is most familiar with the activity, and are labeled optimistic time, most likely time, and pessimistic time. *Optimistic time*, denoted by *a*, is the minimum possible period of time in which the activity can be accomplished. This is the time it would take to complete the activity if everything proceeded better than normally expected. The best estimate of the period of time in which the activity can be accomplished is the *most likely time*, denoted by *m*. Finally, *pessimistic time*, denoted by *b*, is the maximum possible period of time it would take to accomplish the activity (i.e., everything went wrong, excluding major catastrophes).

The estimates represent calendar times and not actual working time, since one object of PERT analysis is to find the overall time for the project. Once made, the time estimates are considered firm and should not be changed unless a corresponding change in the work scope or application of resources takes place.

To transform the three estimates into estimates of the expected value μ and the variance σ^2 of the elapsed time required for the activity we make two basic assumptions. We first assume that σ, the standard deviation, equals one sixth the difference between the pessimistic time estimate and the optimistic time estimate; that is,

$$\sigma = \frac{b-a}{6} \tag{4.1}$$

This assumption is based on the common practice used in quality control charts to separate the control limits by six standard deviations. Thus the wider the separation between the optimistic and pessimistic time estimates, the greater the uncertainty associated with the times for the activity, and, consequently, the higher the variance.

We next assume that the distribution of activity times is approximately beta,[†] where the mode is *m*, the lower and upper bounds are *a* and *b*, respectively, and $\sigma = (b-a)/6$. Then the expected value of the elapsed time μ is

---

[†]The beta distribution was chosen since it is unimodal (see Chapter 6) and has finite end points, which are restricted to be nonnegative. In addition, the beta distribution is not necessarily symmetrical (e.g., see Chapter 1 of *Analysis of Systems in Operations Research*, by the same authors).

given by

$$\mu = \frac{2m}{3} + \frac{a+b}{6} \tag{4.2}$$

Note that the mode is attached to a weight which is twice the value of the midrange $\frac{1}{2}(a+b)$. The relative positions of a, m, b, and μ are displayed in Figure 4.3 for three different distributions.

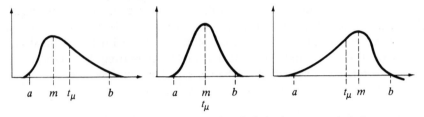

Figure 4.3. Relative positions of optimistic time a, pessimistic time b, most likely time m, and expected activity time μ.

c. Earliest, Latest, and Slack Times

Before the PERT analysis of a project can be undertaken, it is necessary to discuss the concepts of *earliest*, *latest*, and *slack times*. Consider the network in Figure 4.4, where the number above each branch represents the fixed

Figure 4.4. Sample network.

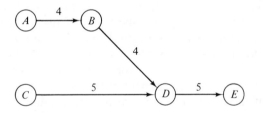

time (e.g., in weeks) to terminate each activity. It is noted that event D cannot take place sooner than 8 weeks from the initiation of the project. Though activity CD can have its start delayed by 3 weeks, that would not delay events D and E and, thus, not prevent the completion of the project on time. It is thus proper to introduce the concepts of earliest time and latest time. The *earliest time* for an event, denoted by T_E, is the earliest possible time that an event can occur if the preceding activities have begun as soon as possible. Thus, for events D and E, the earliest times are 8 and 13 weeks, respectively, given that the earliest times for the source nodes are set equal to zero. Similarly, the *latest time* for an event, denoted by T_L, is the latest allowable time that an event can take place and still not delay termination of the project past its earliest time.

Usually, the latest time for the sink node is set equal to its earliest time. Thus, for events B and E, the latest times are 4 and 13 weeks, respectively.

Once the latest and earliest times for an event have been determined, the *slack time* for an event, which is T_L minus T_E, can be evaluated. Positive slack is an indication of an ahead-of-schedule condition. Similar interpretations may be given for zero slack and negative slack. Negative slack for an event can occur when the latest time for the last event of the project differs from the earliest time for that event. If T_S is the contractual obligation date for the project, then T_S equals T_L for the last event in all cases. Consequently, negative slack occurs when the contractual obligation date for the project is less than the earliest time for the completion of the project. In Figure 4.4, the slacks for events A, B, C, D, and E are 0, 0, 3, 0, and 0, respectively. The importance of slack is related to determining whether events are on schedule, and if any delay will be tolerated in reaching the next event.

iii. *PERT Analysis: The Critical Path of a PERT Network*

a. DETERMINISTIC ACTIVITY TIMES

Once a project has been reduced to a network of events with connecting activities, and the elapsed times for activities have been estimated [e.g., by using relation (4.2)], the PERT analysis of a project commences with the determination of a critical path of the corresponding network. A *critical path of a network* is defined as a path through the network with events on the path having minimum slack. The *minimum slack* is the difference between T_S and T_E for the last event in the project. Thus, when T_S equals T_E for the last event, the minimum slack is zero. The critical path for the network in Figure 4.4 is the sequence of events A, B, D, and E. A shorthand way of expressing the critical path of this network is $A–B–D–E$. Although the network in Figure 4.4 has only one critical path, there can exist multiple critical paths in a network, if there is more than one path with minimum slack through the network. However, the possibility of multiple critical paths in a network will not affect the computational procedure for determining one.

A procedure for determining a critical path of a network consists of the following steps:

1. Evaluate all T_E's.
2. Set T_S equal to the T_L for the last event.
3. Evaluate all T_L's.
4. Determine all slack times.
5. Find a path with minimum slack through the network.

If there is an event with minimum slack that is not on a critical path already determined, then there exists an additional critical path through that event.

Consider the following example of finding a critical path of a network in which the elapsed times for activities are deterministic. In addition, note the tabular form in which the results are displayed.

Example 4.2

Determine the critical path of the PERT network in Figure 4.5. The activity times (in days) are indicated next to the respective branches. The contractual obligation date is 43 days.

From the definition of earliest time, it follows that if there is only one

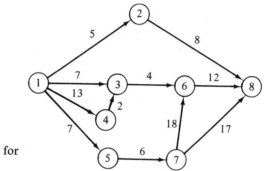

Figure 4.5. PERT network for Example 4.2.

path leading to a particular event then the earliest time for that event is the sum of elapsed times for the activities on the path leading to the event. Thus in our example the earliest time for event 7 is 13 days. However, when there exists a multiplicity of paths leading to a particular event, for example, event 6, simple addition of the activity times for any one path will not suffice for determining the earliest time for the event. The earliest time for such events will be the maximum of the total elapsed time along the distinct paths leading to the event. For event 6 there are three separate paths leading to it with elapsed times of 11, 19, and 31 days. Thus the earliest time for event 6 is $7 + 6 + 18 = 31$ days. In similar fashion, the latest time for a particular event may be determined. However, when there exists a multiplicity of paths leading from some event to the terminal event, for example, event 7, then a minimum is taken over the differences between T_S and the sum of elapsed times for the activities for each path from the event to the terminal event. Thus for event 7, with two distinct paths, the differences are $43 - (18 + 12) = 13$ and $43 - 17 = 26$, and the latest time for event 7 is 13 days.

The solution is the critical path 1–5–7–6–8. A summary of the values for T_E, T_L, and the slack time for each event in our example are found in Table 4.1. We observe that the minimum slack for the events on the critical path is zero.

Note that it was not necessary to evaluate all the paths every time in

Table 4.1. SUMMARY OF RESULTS FOR EXAMPLE 4.2

Event	T_E	T_L	Slack
1	0	0	0
2	5	35	30
3	15	27	12
4	13	25	12
5	7	7	0
6	31	31	0
7	13	13	0
8	43	43	0

order to calculate the earliest and latest times for an event. Letting $(T_E)_i$ be the earliest time for event i, then

$$(T_E)_8 = \max\{(T_E)_6 + 12; (T_E)_2 + 8; (T_E)_7 + 17\}$$
$$= \max\{31 + 12; 5 + 8; 13 + 17\} = 43$$

Similarly,

$$(T_L)_7 = \min\{(T_L)_6 - 18; (T_L)_8 - 17\} = 13$$

b. PROBABILISTIC ACTIVITY TIMES

In the examples considered thus far, the times to perform the various activities in a project have been deterministic. The introduction of expected values and variances of the elapsed times for the activities as given by relations (4.1) and (4.2), permits the computation of probabilities of meeting a schedule for all the events in a project. Recall that in the case of deterministic elapsed times for activities the earliest time for a particular event was the maximum of the total elapsed times of the activities along each of the paths leading to the event. However, with the introduction of expected value and variance of the elapsed time of an activity a simplifying assumption is introduced. It is assumed that the maximum total elapsed time takes place on the path having the maximum expected total elapsed time. Furthermore, the variance of the earliest time of an event is the variance of the path to the event with the maximum expected total elapsed time. Thus the method used to compute earliest and latest times for deterministic activity times can be applied to the situation of probabilistic activity times.

Once a project's schedule has been agreed upon and the expected value and variance of the earliest time for each event determined, the probability of meeting the schedule for a particular event can be evaluated. Since earliest time is the sum of several random variables, it is assumed that the probability distribution of earliest time is normal. A justification of this assumption is given by the central limit theorem in probability theory (e.g., see Chapter 1 of *Analysis of Systems in Operations Research*, also by B.D. Sivazlian and

L. E. Stanfel, Prentice-Hall Inc., 1975). Then the procedure to determine the probability that the earliest time will be less than the scheduled time for the event is the following.
Let for event i:

$$(T_E)_i = \text{random variable describing the earliest time}$$
$$E[(T_E)_i] = \text{expected value of the earliest time}$$
$$\text{Var}[(T_E)_i] = \text{variance of the earliest time}$$
$$(T_S)_i = \text{original scheduled time}$$

The required probability is

$$P\{(T_E)_i \leq (T_S)_i\} = P\left\{\frac{(T_E)_i - E[(T_E)_i]}{\sqrt{\text{Var}[(T_E)_i]}} \leq \frac{(T_S)_i - E[(T_E)_i]}{\sqrt{\text{Var}[(T_E)_i]}}\right\}$$

Since $(T_E)_i$ is assumed to be normally distributed, the random variable

$$\frac{(T_E)_i - E[(T_E)_i]}{\sqrt{\text{Var}[(T_E)_i]}}$$

has a normal distribution with zero mean and unit variance. Using a table of the normal distribution, the required probability may be determined.

Example 4.3

Determine the probability that the original schedule of events in Table 4.2 will be met for the PERT network in Figure 4.6. In addition, determine the critical path of the network. The expected value and variance of elapsed time for each activity are indicated next to their respective branches.

Table 4.2. ORIGINAL SCHEDULE OF EVENTS

Event	1	2	3	4	5	6
Original schedule	0	5	8	11	12	18

In Figure 4.6 the expected value and variance of the elapsed time for the activity connecting events 1 and 3 are 7 and 4, respectively. The data from the PERT network in Figure 4.6 can be placed in the tabular form displayed in Table 4.3.
The probability of meeting the schedule for each of the events in the network of Figure 4.6 is found in Table 4.4. Note from Table 4.2 that the contractual obligation date is 18.

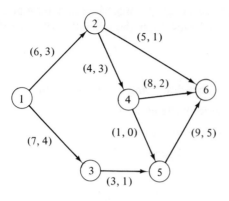

Figure 4.6. PERT network for Example 4.3.

Table 4.3. DATA FROM PERT NETWORK OF EXAMPLE 4.3

| | Predecessor Events | | | Successor Events | | |
| | | Activity Time Estimates | | | Activity Time Estimates | |
Event	Event	Expected Value	Variance	Event	Expected Value	Variance
6	2	5	1			
	4	8	2			
	5	9	5			
5	3	3	1	6	9	5
	4	1	0			
4	2	4	3	5	1	0
				6	8	2
3	1	7	4	5	3	1
2	1	6	3	4	4	3
				6	5	1
1				2	6	3
				3	7	4

Thus the probability that the project will be completed on time is 25%. Notice that all the events have negative slack. This is due to the scheduled completion time for the project being less than the expected value of earliest time for the termination event of the project. The critical path for the project is 1–2–4–5–6, which does not include event 3. If the expected value of elapsed time for the activity between events 3 and 5 were increased by 1 to 4,

Table 4.4. PROBABILITY OF MEETING SCHEDULE FOR EXAMPLE 4.3

Event	Earliest Time T_E Expected Value	Variance	Latest Time T_L Expected Value	Variance	Slack	Original Schedule T_S	Probability of Meeting Original Schedule
1	0	0	−2	11	−2	0	—
2	6	3	4	8	−2	5	.28
3	7	4	6	6	−1	8	.69
4	10	6	8	5	−2	11	.66
5	11	6	9	5	−2	12	.66
6	20	11	18	0	−2	18	.25

there would be two critical paths for the project. The two critical paths would be independent except for the activity connecting events 5 and 6.

c. DISCUSSION

The importance of a critical path is that should any event on it be delayed past its expected date of attainment the project can be expected to be delayed by the same length of time. Therefore, the activities on a critical path should be carefully monitored and efforts made to shorten the required times for the activities. However, because the analysis uses expected values, there is a probability that some path other than the critical one will be longer *and thus critical*. This is a consequence of the approximation which assumed that the maximum total elapsed time always took place on the path with the maximum expected total elapsed time.

After a project is initiated, with the appropriate PERT analysis being complete, it is possible to make the following decisions:

1. Determine whether events are on schedule.
2. Formulate a new schedule should an existing schedule become impossible to meet.
3. Identify areas of potential disturbances so that positive action can be taken to offset the identified disturbances.
4. Identify areas of potential delay and determine the significance of the delay and its impact on other events.
5. Reevaluate the project and make decisions as to upgrading the priorities of the critical path to ensure on-time completion.

It should be kept in mind that PERT aids the decision maker, but does not make decisions for him.

There exist many complex computer programs for the use of PERT with the capability of identifying a proposed change in a schedule, thus improving management control. One of the criticisms leveled at probabilistic PERT has been the assumption of the beta distribution for elapsed times of activities, and the related expressions for the expected value and variance of the elapsed times for the activities. MacCrimmon and Ryavec [8] found that these PERT assumptions could lead to errors of magnitude for the expected value and variance on the order of 30 and 15%, respectively, in the extreme cases that were examined. For the majority of cases these errors reduced to 5 or 10%, which is small in comparison to errors in time estimates, which commonly run 10 or 20%.

Another problem that has occurred has been the difficulty of obtaining three time estimates per activity, when uncertainty exists in the elapsed time for an activity. Project managers, when asking for three time estimates for an activity which has never been undertaken before, are told that even coming up with one estimate is difficult, much less three. In addition, how pessimistic is a pessimistic time estimate for an activity? As a direct consequence, many users of PERT have stopped using three time estimates for each activity. Instead, a single best guess for the elapsed time for the activity is being required. The use of single estimates for times has also been dictated by difficulty encountered in collecting data for projects involving several thousand activities.

An additional criticism has been the assumption that activities are independent; that is, the time necessary to complete one activity will have no effect on the elapsed times of the successor activities. Although this assumption seems reasonable, in practice this may not be true, as in the case of limited resources.

Despite some of the problems that have developed in the implementation and use of PERT, PERT and techniques similar to it will remain in use because of some of the notable successes and their simplicity of application.

4.3. Critical Path Method—CPM

i. *Introduction*

Since the first successful implementation of PERT, many similar techniques have been developed. Some techniques have considered the level of technical performance of the individual activities in addition to their elapsed times. Other techniques have been concerned with the cost aspects of the activities as well as their elapsed times; however, the latest government version of PERT includes cost aspects. One of the most notable of the new techniques, which was developed independently of PERT, is the *critical path method* (CPM). CPM is not to be confused with the critical path of PERT.

Originally, CPM was designed to solve scheduling problems in industry;

it is concerned with minimizing the costs associated with scheduling a project. Just as in PERT, the elapsed time for an activity in CPM is either deterministic or probabilistic. In addition, CPM allows for variation in the duration of the activity as a function of planned allocation of resources (e.g., money, machines, and men). One of the assumptions of CPM is that activity times can be reduced if additional resources are made available. Although it is possible to expedite, or crash, or shorten the time required for the activity, it is not always necessary to do so. Which activities to shorten and by what amounts are two of the problems CPM solves while minimizing the costs for the project.

ii. Cost Structure of CPM

Associated with any project are two types of costs. There are *direct costs* associated with the separate activities in the project. In addition, there exist *indirect costs*, which are the result of overhead expenses and vary with the duration of the project. Whereas direct costs increase as the individual activities are expedited, indirect costs increase with the length of time for the project. Figure 4.7 displays these ideas.

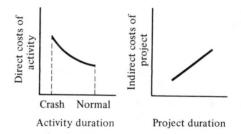

Figure 4.7. Direct and indirect costs in projects as a function of project duration.

The original CPM model assumed that the relationship between time and costs for an activity is linear. The more negative the slope of the line, the larger the cost of speeding up the activity. When an activity cannot be further expedited by the allocation of additional resources, then a vertical line represents the relationship between costs and activity duration. It may be possible to slow down the activity to a point where costs remain stable as activity duration increases. Furthermore, costs may even start to increase if activity duration exceeds a certain critical value. These relationships are illustrated in Figure 4.8. In reality, the relationship between cost and duration of an activity as displayed in Figure 4.8 might better be represented by some quadratic or higher-order curve. Subsequent versions of the original CPM model have made it possible to improve the modeling of the cost–duration relationship of an activity.

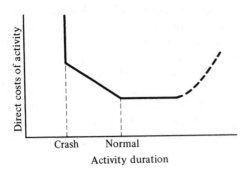

Figure 4.8. Cost-duration relationship for an activity.

iii. *Optimum Schedule of CPM*

If the tradeoff relationship of the two types of costs incurred—between rising indirect costs as the project takes longer and increasing direct costs as a price for shortening the duration of activities—is examined, it appears that there is some optimum project length. A method using the CPM model is then applied, which results in a minimum-cost or optimum schedule for the project.

The procedure for deriving the minimum-cost schedule is to obtain first an initial schedule in which all activities are assigned earliest and latest starting times, without any allocation of additional resources. This first step for deriving the minimum-cost schedule is the method for determining a critical path for the PERT model where the individual activities have deterministic elapsed times. The length of the critical path is the maximum project length, which CPM attempts to reduce through the allocation of additional resources. The reduction of the length of the project schedule is accomplished by expediting one or more of the activities at an additional cost. If the savings from fewer indirect costs, due to shortened project length, is greater than the cost of expediting an activity, then a more economical schedule can be determined. CPM proceeds in an iterative fashion to decrease project length while reducing project costs. Thus the problem is to find which activities to expedite and when to terminate the schedule-reduction method.

At each iteration only those activities along the current critical paths are examined for crashing. The minimum negative slope from among all the cost –duration slopes for activities on the critical path is determined. (Ties are broken arbitrarily.) The activity selected with this minimum negative slope is the one that can best be expedited with the smallest expenditure of additional resources. If a savings is achieved between indirect costs saved for one time period and direct costs incurred for the same length of time, then the activity is expedited until either the activity duration cannot be further shortened or because some activity has become critical along a parallel path. The procedure is repeated by examining the remaining critical activities and selecting the one with the smallest negative slope to be considered for shortening. If there exists a multiplicity of critical paths, then one activity in each path must be selected for crashing. The procedure is terminated when either it is not possible to shorten any critical activities, or the costs for shortening a critical activity are larger than the savings in indirect costs. Consider the following example, which demonstrates the application of the procedure outlined for determining a minimum-cost schedule.

Example 4.4

Determine the minimum-cost schedule for the CPM network in Figure 4.9. Associated with each activity in the network are two activity times in parentheses(–), being respectively the normal duration in days and the minimum duration for the activity in days. The number in brackets [ ] is the cost in dollars per day for crashing the activity. The overhead cost for the project is $5.25 per day.

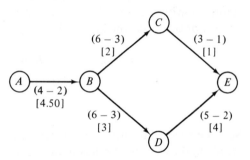

Figure 4.9. CPM network for Example 4.4.

Thus, for the activity connecting events A and B, the normal duration of the activity is 4 days. This activity cannot be accomplished in less than 2 days with a cost of $4.50 per day if it is crashed. The information from the CPM network in Figure 4.9 can be expressed in the tabular form found in Table 4.5.

Table 4.5. INPUT DATA FROM CPM NETWORK FOR EXAMPLE 4.4

Event	Successor Event	Normal Days Duration	Minimum (Crash) Days Duration	Cost in $/Day of Crashing
A	B	4	2	4.5
B	C	6	3	2
B	D	6	3	3
C	E	3	1	1
D	E	5	2	4

The first step in determining the minimum-cost schedule for Example 4.4 is to find the critical path of the network with all activities at a normal duration (i. e., no additional resources). Table 4.6 summarizes the calculations necessary to determine the critical path, which is *A–B–D–E*.

Table 4.6. FIRST STEP IN CPM ANALYSIS

Event	T_E	T_L	Slack
A	0	0	0
B	4	4	0
C	10	12	2
D	10	10	0
E	15	15	0

From Table 4.6, the maximum length of the schedule will be 15 days. Ignoring the invariant costs of the activities associated with normal duration, the cost of this first schedule is

$$\text{cost} = \text{cost of crashing} + \text{indirect costs}$$
$$= 0 + (15 \text{ days})(\$5.25/\text{day})$$
$$= \$78.75$$

Next, we proceed to shorten the schedule by determining which of the activities along the critical path has the smallest negative slope, where the slope is the cost per day of crashing. Of the three activities *AB*, *BD*, and *DE* on the critical path, activity *BD* has the smallest cost per day of crashing. Furthermore, the cost per day of crashing *BD* is less than the overhead cost incurred per day; thus expediting activity *BD* reduces the cost of the schedule. Activity *BD* can only be expedited 2 days before a parallel path (*A–B–C–E*) becomes critical. The figure of 2 days could also have been obtained from the value of the slack of one of the activities on the parallel path. The length of each critical path is now 13 days, and the cost of the new schedule is $74.25, a saving of $4.50 over the schedule of 15 days.

We observe that there are two critical paths and that the schedule can be further shortened by reducing the length of both paths. Of the five possible combinations of activities (*AB*; *BD* and *BC*; *BD* and *CE*; *DE* and *BC*; *DE* and *CE*) that can reduce project length, activities *BD* and *CE* have the minimum cost of crashing per day, and if they are expedited will reduce the cost of the schedule. Upon activities *BD* and *CE* being expedited 1 additional day, the length of each critical path becomes 12 days. The cost of this schedule is cost = cost of crashing *BD* and *CE* + indirect costs

= (3 days)($3/day) + $1 + (12 days)($5.25/day)

= $73

The procedure is repeated until the minimum-cost schedule of 9 days at a cost of $71.25 is derived. The duration of activities *AB, BC, BD, CE,* and *DE* is 2, 6, 3, 1, and 4 days, respectively, if a minimum-cost schedule is followed.

iv. *Lengthening the Duration of Activities*

In the prior discussion it was assumed that an activity was either performed at its normal pace or expedited. The question arises as to whether it would make sense for a manager to lengthen or stretch the duration of an activity by using a reduced level of resources. For most situations that would normally arise, the answer is no. There would be no point in stretching the duration of an activity if the effect were a combined increase in direct activity costs and project overhead. In addition, over the longer duration of an activity, an increase in financing costs would be incurred.

There are several situations that do arise in which lengthening of the project might be considered worthwhile. Companies engaged in multiproject activities, for which projects do not always follow each other, might lengthen the project duration to maintain a stable job level. There is also the situation in which a normal pace for an activity would not be its most efficient. Thus the minimum driving time need not be the most economical driving policy.

v. *Linear Programming Formulation of CPM Problems*

The technique illustrated in the solution of Example 4.4 is essentially an exhaustive search procedure (see Chapter 6), which means essentially, that all possibilities are evaluated, and the best is then selected. As such, it is adequate for obtaining the minimum-cost schedule for small projects with only a few paths. However, as projects become increasingly complex, the implementation of the technique becomes increasingly difficult. Consequently, mathematical optimization techniques have been applied to the problem of finding a minimum-cost schedule. Kelley [7] has formulated a linear-programming procedure for obtaining the minimum-cost schedule for networks with activities having linear cost–duration curves.

The minimum-cost schedule is obtained using a search procedure that employs linear programming at each step of the search. Recall that the costs for a schedule consist of the total direct costs for the activities and the indirect project costs. At each step of the procedure, the direct costs of the activities for a project of maximum length L are minimized using linear programming. The indirect costs for a project are added to the value found for the direct costs of the activities. The result is a cost for a schedule with a maximum length of L. A new value for L is selected to obtain a new schedule with a cost lower than the prior one. The procedure is then repeated until a minimum-cost schedule is determined.

Denote by t_{ij} the time required to perform activity ij. Let s_{ij} represent the value of the slope of the cost–duration relationship for activity ij. Thus the equation for the linear cost–duration relationship of activity ij is

$$d_{ij} = a_{ij} + s_{ij}t_{ij}$$

where a_{ij} is the intercept of the projected line on the direct-cost axis in Figure 4.10. The time t_{ij} for the activity is restricted to lie in the interval between crash and normal times, as in Figure 4.10. Letting C_{ij} and N_{ij} be the crash and normal times, respectively, for activity ij, then

$$C_{ij} \leq t_{ij} \leq N_{ij}$$

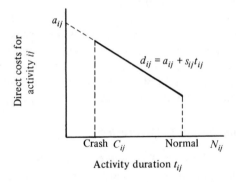

Figure 4.10. Linear cost-duration relationship for activity ij.

Finally, let x_i be the earliest time for event $i(i = 1, \ldots, m)$, where x_1 is the earliest time for the source event and x_m is the earliest time for the sink event.

The total direct costs for the activities in the project are

$$\sum_{ij}(a_{ij} + s_{ij}t_{ij})$$

Since the slope intercepts a_{ij} are constant, direct costs for a project with maximum length L can be minimized by the following linear-programming formulation:

$$\max \sum_{ij} -s_{ij}t_{ij}$$

subject to

$$x_i + t_{ij} - x_j \leq 0, \quad \text{for all } i, j$$
$$t_{ij} \leq N_{ij}, \text{for all } i, j$$
$$t_{ij} \geq C_{ij}, \text{for all } i, j$$
$$x_m - x_1 \leq L$$

The first constraint states that the difference in earliest times for two connected events must be at least as large as the length of time for the activity connecting them. The constraint involving the earliest times for the source and sink events, specifies that the project length cannot be greater than L. The time L must be feasible; that is, it cannot be smaller than the length of the critical path with all activities crashed to their minimum duration.

The search procedure using linear programming has been extended to handle nonlinear cost–duration curves. More efficient (as related to computer time) solution techniques have been devised using network flow theory.

4.4. Graph Theory and Network Flows

i. *Overview of the Theory of Graphs*

Consider a set of two or more distinct *points* (or *nodes*, or *vertices*) with certain pairs of these points joined by one or more *lines* (or *branches*, *links*, or *edges*). The resulting form is called a *graph* and is denoted by G. There are two basic types of graphs: *undirected* and *directed*. The distinction between the two types of graphs is that whereas an undirected graph has its nodes connected by edges, the nodes of a directed graph are connected by oriented lines or arcs. An *arc* is defined as an edge with direction or orientation. Figure 4.11 is an example of a directed graph. If the orientation on the

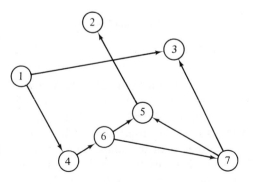

Figure 4.11. Example of a directed graph, G.

arcs of the directed graph in Figure 4.11 were removed, it would be an undirected graph.

Between two distinct nodes of a directed graph there can exist a path. Letting x_i represent a node, the sequence $x_1, (x_1, x_2), x_2, \ldots, x_i, (x_i, x_{i+1})$, $x_{i+1}, \ldots, x_{n-1}, (x_{n-1}, x_n), x_n$ is a *path* from x_1 to x_n, where either (x_i, x_{i+1}) or (x_{i+1}, x_i) is an arc of the directed graph. In Figure 4.11, one path from node 6 to 7 is 6, (6, 5), 5, (5, 7), 7. A path can exist in an undirected graph, in which case all the links (x_i, x_{i+1}) of the sequence defining the path are edges. A subset of the paths of a directed graph is the set of chains. A path is a *chain* if every link of the sequence defining the path is an arc. If a chain has $x_1 = x_n$, it is called a *cycle*. Thus the path from node 6 to 7 via node 5 is not a chain since (5, 7) is not an arc. However, the sequence 6, (6, 7), 7 is a chain from node 6 to 7. A shorthand way of expressing this chain is $6 \rightarrow 7$.

A *connected graph* is defined as a graph for which there exists a path connecting any two nodes of G. An example of a connected graph is the directed graph of Figure 4.11. Although there exists a path between every pair of nodes of a connected graph, it does not necessarily follow that there is a chain between every two nodes of a connected graph. In Figure 4.11 there is no chain from node 4 to 1.

ii. Network Flows and Capacities

A *network* is a graph in which a flow can take place in the arcs of the graph. Every network has an origin and a terminus called the *source* and *sink*, respectively. A network can be thought of as a pipeline system in which the arcs represent pipelines, the source is the inlet of water, the sink is the outlet of water, and all other nodes are connections between the lengths of pipe.

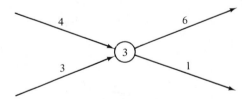

Figure 4.12. Conservation of flow for node 3.

A flow in a network may be a flow of fluids, electricity, or funds. The corresponding units of flow may be cubic feet per second, amperes, or dollars per month. In general, a flow is a rate. One of the major assumptions about networks is the conservation of flow in and out of a node. The flow into a node (other than the source and sink) is equal to the flow out of it. If the number above each arc in Figure 4.12 represents the flow in the arc, then the conservation of flow for node 3 is observed.

The upper limit on the flow in an arc is the *capacity* of the arc. The flow capacity of an arc is a positive number and may be infinite, in which case the arc is said to be *uncapacitated*. If an arc has a flow capacity of zero, removing the arc from the network does not alter the potential flow in the network. Given a chain connecting the source and sink of a network, the maximum flow that can be put across the chain is restricted by the arc of the chain with the smallest capacity. If (x, y) is an arc of G, then the capacity of (x, y) is given by $c(x, y)$. The flow in an arc (x, y), designated by $f(x, y)$, is related to its capacity by

$$c(x, y) \geq f(x, y) \geq 0$$

iii. *Maximal Flow Problem*

a. Mathematical Formulation of the Maximal Flow Problem—The Max-Flow Min-Cut Theorem

The *maximal flow* problem is to find a feasible equilibrium or steady-state flow pattern through a network such that the total flow from the source to the sink is maximized. Should there exist more than a single source (sink) node, then a supersource (supersink) node is connected to each of the sources (sinks). The arcs making the connection have infinite flow capacity.

Define *after* x, which is denoted by $A(x)$, as the set of all nodes y that succeed the node x in a network and for which there is an arc (x, y) in the network. Similarly, define *before* x, which is denoted by $B(x)$, as the set of all nodes y that precede the node x and for which there is an arc (y, x) in the network. Then the flow out of a node x is

$$\sum_{y \in A(x)} f(x, y)$$

In the same fashion, the flow into a node x is

$$\sum_{y \in B(x)} f(y, x)$$

From the conservation of flow assumption,

$$\sum_{y \in A(x)} f(x, y) - \sum_{y \in B(x)} f(y, x) = 0, \quad \text{if } x \neq s \text{ or } t$$

where s and t designate the source and sink nodes, respectively. In addition, the flow out of the source node equals the flow into the sink node; then

$$\sum_{y \in A(s)} f(s, y) = v = \sum_{y \in B(t)} f(y, t)$$

The maximal flow problem can be stated as

$$\max v$$

subject to

$$\sum_{y \in A(x)} f(x, y) - \sum_{y \in B(x)} f(y, x) = \begin{cases} v, & x = s \\ 0, & x \neq s \text{ or } t \\ -v, & x = t \end{cases}$$

$$c(x, y) \geq f(x, y) \geq 0$$

An observation can be made about the upper bound on the flow through a network. The sum of the capacities on the arcs leaving the source provides an upper bound on the flow out of the source node. Similarly, the sum of the capacities of the arcs entering the sink provides an upper bound on the flow into the sink. Since the flow out of the source equals the flow into the sink, the sum of the arc capacities that is smaller is the upper limit on the flow in the network.

Three solution techniques that solve the maximal flow problem will be discussed. They are an intuitive algorithm, a labeling algorithm, and a linear-programming formulation of the maximal flow problem.

Irrespective of which approach is used to obtain the maximal flow in a network, it would be considerably easier to recognize when optimality has been reached without either performing an exhaustive search for a nonexistent chain or relabeling all the nodes in a network. This can sometimes be accomplished by means of the central theorem of network flow theory called the max-flow min-cut theorem.

A *cut* is defined as any set of arcs containing at least one arc from every chain from source to sink. The *capacity* or *value of the cut* is the sum of the flow capacities of the arcs in the direction specified by the cut. Thus, recalling the earlier observation, a cut across the arcs leaving the source has a value that is an upper bound on the flow in the network. A cut separating the source and sink that has the smallest capacity is called a *minimum cut*.

The *max-flow min-cut theorem* states that for any network the value of the maximal flow from source to sink is equal to the capacity of the minimum cut separating sink and source. The capacity of any cut in a network specifies an upper bound on the flow in a network. The importance of the theorem lies in that if at any step of a solution procedure the capacity of a cut in the initial network equals the value of the flow presently determined, then the present flow pattern maximizes the flow through the initial network.

b. SOLUTION TO THE MAXIMAL FLOW PROBLEM USING AN INTUITIVE ALGORITHM

Consider the following intuitive approach for finding the maximal flow in a network. The network is examined for possible chains from source to sink, where each of the arcs in the chain has positive flow capacity. If such a chain is found, called a chain with positive flow capacity, then a flow equivalent to the smallest flow capacity of the arcs in the chain is imposed on the chain. The value found for the flow in the chain is subtracted from each arc of the chain, which results in a new network. The procedure is repeated with the new network until no more chains with positive flow capacity exist. Then, the maximal flow in a network is the sum of the flows in the chains found at each step of the procedure.

Example 4.5

Determine the maximal flow in the network in Figure 4.13 where the capacities of the individual arcs are indicated. Assume an initial flow of zero in the network.

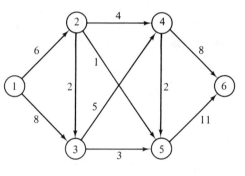

Figure 4.13. Network for Example 4.5 with zero flow.

The chain $1 \rightarrow 2 \rightarrow 4 \rightarrow 6$ in the network of Figure 4.13 has a positive flow capacity of 4. The bottleneck in the chain is the arc $(2, 4)$ with a capacity of 4. A flow of 4 is imposed on the chain, and the flow capacities of the arcs in the chain are reduced by 4. The resulting network with flow F_1 equal to 4 is shown in Figure 4.14. Next, select the chain $1 \rightarrow 3 \rightarrow 4 \rightarrow 6$ in the network of Figure 4.14, which has a minimum flow capacity of 4 on arc $(4, 6)$. A flow of 4 is imposed on this chain, and the arc capacities of the chain are decreased by

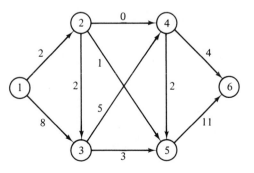

Figure 4.14. Network of Figure 4.13 with a flow of $F_1 = 4$.

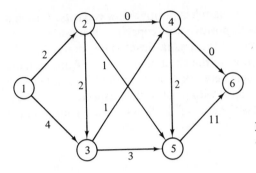

Figure 4.15. Network of Figure 4.14 with an additional flow of $F_2 = 4$.

4. The resulting network with additional flow F_2 equal to 4 is shown in Figure 4.15.

The chain $1 \longrightarrow 2 \longrightarrow 5 \longrightarrow 6$ in the network of Figure 4.15 has a positive flow capacity of 1. A flow of 1 is imposed on this chain, and the arc capacities of the chain are decreased by 1. The resulting network with additional flow F_3 equal to 1 is shown in Figure 4.16.

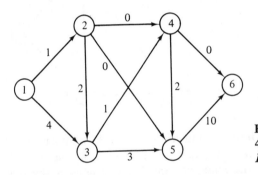

Figure 4.16. Network of Figure 4.15 with an additional flow of $F_3 = 1$.

The chain $1 \longrightarrow 3 \longrightarrow 5 \longrightarrow 6$ in the network of Figure 4.16 has a positive flow capacity of 3. A flow of 3 is imposed on this chain, and the arc capacities of the chain are decreased by 3. The resulting network with additional flow F_4 equal to 3 is shown in Figure 4.17.

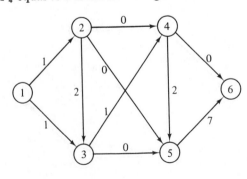

Figure 4.17. Network of Figure 4.16 with an additional flow of $F_4 = 3$.

The chain $1 \to 3 \to 4 \to 5 \to 6$ in the network of Figure 4.17 has a positive flow capacity of 1. A flow of 1 is imposed on this chain, and the arc capacities of the chain are decreased by 1. The resulting network with additional flow F_s equal to 1 is shown in Figure 4.18.

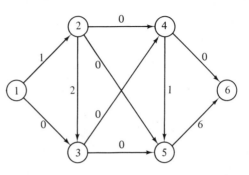

Figure 4.18. Network of Figure 4.17 with an additional flow of $F_5 = 1$.

In the network of Figure 4.18, no chains exist with positive flow capacity on each arc of the chain. The flow in the network is the sum of F_1, F_2, F_3, F_4, and F_5, or 13. The question is then posed as to whether this is the maximum flow possible through the network. For the network of Figure 4.13, the answer is yes, since the value of the minimum cut for this network, indicated in Figure 4.19, is 13. Nevertheless, the procedure used in solving Example 4.5 will not always guarantee a maximal flow's being determined, as is demonstrated in Example 4.6.

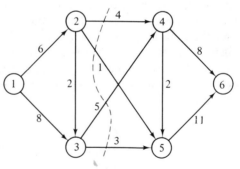

Figure 4.19. Minimum cut of the network of Figure 4.13.

Example 4.6

Determine the maximal flow in the network in Figure 4.20 where the capacities of the individual arcs are indicated. Assume an initial flow of zero in the network.

The chain $1 \to 3 \to 4 \to 6$ in the network of Figure 4.20 has a positive flow capacity of 6. A flow F_1 of 6 is imposed on the chain, and each arc capacity in the chain is decreased by 6. Figure 4.21 is the resulting network with a

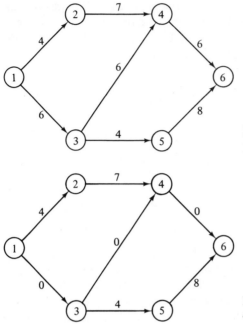

Figure 4.20. Network for Example 4.6 with zero flow.

Figure 4.21. Network of Figure 4.20 with a flow of $F_1 = 6$.

flow of 6, in which no chains remain from source to sink with positive flow capacity on each arc of the chain. Nevertheless, the maximal flow in the network has not been obtained. Consider the path 1–2–4–3–5–6. A positive flow in this path moves in the wrong direction along arc (3, 4). However, since the flow $f(3, 4)$ equals 6, it is possible to allow a flow $f'(4, 3)$ of as much as 6 in the wrong direction. This can be rationalized by equating the capacity of an arc to the net flow that can be imposed in the direction of the arc. If (x, y) is an arc in a network, the concept of net flow in an arc is expressed by

$$c(x, y) \geq f(x, y) - f'(y, x) \geq 0$$
$$f(x, y) \geq 0 \quad \text{and} \quad f'(y, x) \geq 0$$

where f' denotes a flow against the direction of the arc.

Thus the minimum flow capacity along the path 1–2–4–3–5–6 is 4. If a flow F_2 of 4 is imposed along this path, the resulting network is shown in Figure 4.22. The net flow in arc (3, 4) of Figure 4.22 is $f(3, 4)$ minus $f'(4, 3)$, or 2 units of flow. Since the initial capacity of the arc was 6 units of net flow, the remaining capacity is 4 units. Recall that an upper bound on the flow in a network is the sum of the capacities of the arcs leaving the source. For Figure 4.20, this sum is 10 which is equal to the sum of F_1 and F_2. Therefore, the

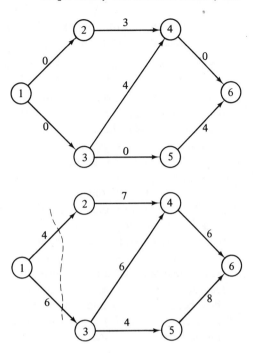

Figure 4.22. Network of Figure 4.21 with an additional flow of $F_2 = 4$.

Figure 4.23. Minimum cut of the network of Figure 4.20.

maximal flow in the network in Figure 4.20 is 10. The minimum cut for this network is shown in Figure 4.23.

From the solution to the two preceding examples of maximal flow, an algorithm for determining the maximal flow in a network can be stated as

Step 1. If in the initial network more than one source (sink) exists, append a supersource (supersink) where the corresponding arcs making the connections have infinite flow capacity. Assume an initial flow of zero unless otherwise specified.

Step 2. In the network obtained, determine a chain going from the source to the sink such that the capacity on each of the arcs of this chain is positive. If no such chain can be found, then the procedure terminates and the flow through the network is a maximum.

Step 3. Examine this chain for the arc with minimum flow capacity with value, for example, c'. To the flow of the chain obtained, and hence the resulting network, add the value c'.

Step 4. Subtract c' from the flow capacity of each arc in the chain determined in step 2. Construct an arc in the reverse direction for each arc in the chain. Assign a flow capacity c' to each of these reverse arcs. (If a reverse arc already exists, add c' to its flow capacity instead.) Go to step 2.

c. SOLUTION TO THE MAXIMAL FLOW PROBLEM BY A
LABELING ALGORITHM

For a large, complex network, the procedure just outlined for determining the maximal flow is tedious, due to the difficulty in keeping track of arcs
that form chains from source to sink. In addition, for a multiarc chain (e.g.,
10 or more) the search for the arc with the smallest flow capacity in a chain is
time consuming, as is the location of a flow-increasing chain.

These two difficulties can be eliminated if a labeling algorithm due to
Ford and Fulkerson [3] is used. A brief description of the algorithm follows.
The algorithm provides a systematic method of searching for all possible
flow-increasing chains from source to sink. This is accomplished by assigning labels to nodes to indicate directions that flows in arcs can be increased.
Once a flow-increasing chain from source to sink is determined, the flow in
the chain is increased to its maximum capacity and all the labels on the nodes
are erased. The labeling then recommences by assigning fresh labels to nodes
based on the new flow. The labeling algorithm consists of two steps: a labeling
process and a flow change. Iteration of the two steps continues until increasing
the flow becomes impossible. The more complex aspects of the labeling algorithm can be found in [3]. A particular application of this algorithm to the
assignment problem is presented in Chapter 5.

d. SOLUTION OF THE MAXIMAL FLOW PROBLEM BY
LINEAR-PROGRAMMING

The problem can also be formulated as a linear-programming problem.
Denote $f(i,j)$ by f_{ij} the flow between nodes i and j, and denote $c(i,j)$ by c_{ij},
the capacity of arc (i,j). Renumbering the nodes so that node 1 is the source
and node N is the sink, the conservation of flow equation is

$$\sum_i f_{ik} - \sum_j f_{kj} = 0, \quad \text{for } k = 2, 3, \ldots, N-1$$

The flow out of node 1 and into node N is given by

$$\sum_k f_{1k} = v = \sum_k f_{kN}$$

Then the linear-programming formulation to the maximal flow problem is

$$\max v = \sum_k f_{1k}$$

subject to

$$\sum_i f_{ik} - \sum_j f_{kj} = 0, \quad \text{for } k = 2, \ldots, N-1$$

$$f_{ij} \leq c_{ij}, \quad \text{for all } i, j$$
$$f_{ij} \geq 0, \quad \text{for all } i, j$$

The problem can then be solved by the simplex algorithm (Chapter 3) to obtain the maximum flow possible in the network.

iv. Shortest-Route Problem

a. GENERAL FORMULATION OF THE SHORTEST-ROUTE PROBLEM

The shortest-route problem considers the determination of the minimum distance from an origin to a destination through some connecting graph. An example is the determination of the shortest route to take while traveling by car between two cities, S and T. Between cities S and T are intermediate cities, some of which will have to be passed through before arriving at T (i.e., no direct route from S to T). The intermediate cities and the cities S and T can be regarded as belonging to a set of nodes V. Pairs of cities are connected by roads in such a way that it is always possible to find a path between every pair of cities. Letting the roads be members of a set of edges E with associated distances, then the sets E and V define a graph. Furthermore, the graph is connected, since there is a path between every pair of nodes of the graph. Thus there is a graph connecting S and T, which are the origin and destination for the car, respectively.

Consider the graph in Figure 4.24. The nodes are cities and the edges are

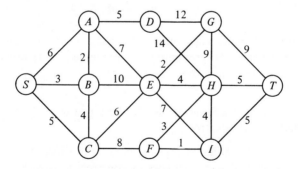

Figure 4.24. Example of a shortest route graph.

roads, with associated distances indicated near the edges. Thus the shortest route from S to T necessitates deciding which cities to visit in turn while minimizing the total distance traveled. Enumerating all the paths from S to

T with their associated distances is one method of determining the shortest route from S to T. However, as the graph becomes increasingly complex, this method becomes impractical, and we therefore consider other solution techniques for the shortest-route problem.

The shortest-route problem can be formulated as an uncapacitated maximal flow problem with costs. Consider a network where each arc has an infinite capacity for flow. Associated with each arc of the network is a cost for imposing a flow in the arc which is proportional to that flow. Thus, if a unit of flow is available at the source and a unit of flow is required at the sink, the shortest-route problem is the problem of finding the smallest cost of getting the unit of flow to the sink, where the unit of flow is not divisible.

The problem can also be solved by integer programming. Renumber the nodes such that node 1 is the source and node N is the sink. Let a_{ij} represent the cost (or distance) of putting one unit of flow on arc (i, j), and f_{ij}, the flow on arc (i, j). The integer-programming formulation of the shortest-route problem is

$$\min \sum_{i,j} a_{ij} f_{ij}$$

subject to

$$\sum_i f_{ik} - \sum_j f_{kj} = 0, \quad \text{for } k = 2, \ldots, N - 1$$

$$\sum_k f_{1k} = 1$$

$$\sum_k f_{kN} = 1$$

$$f_{ij} = 0 \text{ or } 1, \quad \text{for all } i, j$$

There are other solution techniques that solve the shortest-route problem, among them dynamic programming (see Example 7.5). However, only one will be discussed in detail in the present chapter.

b. Solution of the Shortest-Route Problem by Dijkstra's Algorithm

An algorithm by Dijkstra [1] determines the shortest chain from the source to all other nodes in a network by constructing a *tree* of arcs, within the graph, a *tree* being a connected graph with no cycles. Each arc (i, j) of the network has a positive distance d_{ij} associated with it. If there is no arc from node j to k, the distance d_{jk} is infinite. This is logical, since an infinite distance from node j to k implies that node k is never reached when a departure is made from node j, and no other nodes are visited prior to node k.

Furthermore the algorithm does not require that distances between a pair of nodes be symmetric (i.e., $d_{jk} \neq d_{kj}$ is possible). If there is an edge between nodes i and j in a network, it is replaced by two arcs (i, j) and (j, i) with distances d_{ij} and d_{ji} equal to the distance associated with the edge.

Arcs that form a tree are called *tree arcs*, and arcs not in a tree are called *nontree arcs*. The algorithm commences with a tree containing no arcs (i.e., empty), and all arcs in the network are nontree arcs. It then proceeds to add arcs to the tree one at a time until there are $n - 1$ of them, where n is the number of nodes in the network, at which time there will be exactly one shortest path from s to every other node.

Before proceeding to a statement of the algorithm, it is necessary to define some of the notation used in its description. Let L_{sk} represent the actual shortest distance from the source s to node k, and L'_{sk} represent the shortest distance from the source to node k using tree arcs and at most one nontree arc. For all chains from the source to some node k that require more than one nontree arc, L'_{sk} equals infinity. A node k is a *neighbor* of the tree if for a node i of the current tree, there exists either an arc (i, k) or an arc (k, i).

The following is Dijkstra's algorithm for finding the shortest chain from the source to all other nodes of a network.

Step 1. Let $L'_{sk} = d_{sk}$. Initially, s (source) is the only node in the tree and $L_{ss} = 0$.

Step 2. $L_{sr} = \min_k L'_{sk} = L_{sj} + d_{jr}$. The k are neighbor nodes of the current tree.

Step 3. Make the arc (j, r) a tree arc.

Step 4. If the number of tree arcs is $n - 1$, terminate the algorithm. Otherwise, proceed to step 5.

Step 5. $L'_{sk} := \min(L'_{sk}, L_{sr} + d_{rk})$, where $:=$ means to be replaced by. Go to step 2.

The algorithm may be implemented by labeling the nodes. A label of the type (L, i) will be given to each node of the network. The L in the label is the value L'_{sk} or L_{sk}, and i refers to the last node on the shortest chain from the source to node k. There are two types of labels: temporary and permanent. A label is temporary if $L = L'_{sk}$, and it is permanent if $L = L_{sk}$. Example 4.7 illustrates the application of the algorithm.

Example 4.7

Determine the shortest chain from the source to all other nodes of the network in Figure 4.25, where the distances associated with the arcs and edges are indicated.

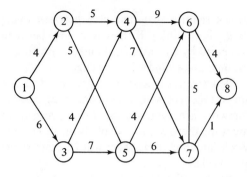

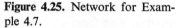
Figure 4.25. Network for Example 4.7.

Initially, there are no tree arcs and node 1 is the sole tree node. The neighbor nodes of 1 are 2 and 3. Since $L'_{1k} = d_{1k}$, we attach the temporary labels $(4, 1)$ to node 2 and $(6, 1)$ to node 3. By step 1 of the algorithm, $L_{11} = 0$; thus $L_{1r} = \min\{4, 6\} = 4$, and $(4, 1)$ is a permanent label of 2. In addition, the arc $(1, 2)$ is now a tree arc.

The neighbor nodes of the two-node tree are 3, 4, and 5:

$$L'_{13} := \min (L'_{13}, L_{12} + d_{23}) = \min (6, 4 + \infty) = 6$$
$$L'_{14} := \min (L'_{14}, L_{12} + d_{24}) = \min (\infty, 4 + 5) = 9$$
$$L'_{15} := \min (L'_{15}, L_{12} + d_{25}) = \min (\infty, 4 + 5) = 9$$

Since it is noted that the smallest of the L'_{1k} is L'_{13}, then $(6, 1)$ becomes a permanent label for node 3, and $(1, 3)$ is a tree arc. The tree now consists of two tree arcs, and there are only five more to determine. The neighbor nodes of the current tree are 4 and 5. Then

$$L'_{14} := \min (L'_{14}, L_{13} + d_{34}) = \min (9, 6 + 4) = 9$$
$$L'_{15} := \min (L'_{15}, L_{13} + d_{35}) = \min (9, 6 + 7) = 9$$

Since L'_{14} equals L'_{15}, the tie is broken arbitrarily, and $(9, 2)$ is selected as a permanent label for node 4. Also, $(2, 4)$ is a tree arc. The neighbor nodes of the current tree are 5, 6, and 7. Thus

$$L'_{15} := \min (L'_{15}, L_{14} + d_{45}) = \min (9, 9 + \infty) = 9$$
$$L'_{16} := \min (L'_{16}, L_{14} + d_{46}) = \min (\infty, 9 + 9) = 18$$
$$L'_{17} := \min (L'_{17}, L_{14} + d_{47}) = \min (\infty, 9 + 7) = 16$$

Since the smallest of the L'_{1k} is L'_{15}, then $(9, 2)$ becomes a permanent label for node 5, and $(2, 5)$ is a tree arc. The neighbor nodes of the current tree are 6 and 7. Hence

$$L'_{16} := \min(L'_{16}, L_{15} + d_{56}) = \min(18, 9 + 4) = 13$$
$$L'_{17} := \min(L'_{17}, L_{15} + d_{57}) = \min(16, 9 + 6) = 15$$

Since the smaller of the L'_{1k} is L'_{16}, then (13, 5) becomes a permanent label for node 6, and (5, 6) is a tree arc. The neighbor nodes of the current tree are 7 and 8. The tree now consists of five tree arcs, and there are only two more to determine. Continuing,

$$L'_{17} := \min(L'_{17}, L_{16} + d_{67}) = \min(15, 13 + 5) = 15$$
$$L'_{18} := \min(L'_{18}, L_{16} + d_{68}) = \min(\infty, 13 + 4) = 17$$

Since L'_{17} is the smaller, then (15, 5) becomes a permanent label for node 7, and (5, 7) is a tree arc. Finally,

$$L'_{18} := \min(L'_{18}, L_{17} + d_{78}) = \min(17, 15 + 1) = 16$$

Thus (7, 8) becomes a tree arc, and node 8 receives the permanent label (16, 7). There are now seven tree arcs, and the algorithm terminates. Figure 4.26 illustrates the complete tree where the permanent labels are next to each node.

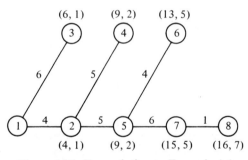

Figure 4.26. Tree solution to Example 4.6.

If in examining the neighboring nodes 4 and 5 the tie had not been broken arbitrarily, both nodes could have been incorporated in the tree simultaneously. In addition, they would have the same label. Then, nodes 6 and 7 are the neighboring nodes of the new tree and must be considered in turn with the two tree arcs just added. Therefore, the computation in step 5 of the algorithm would have been

$$L'_{16} := \min(L'_{16}, L_{14} + d_{46}, L_{15} + d_{56})$$
$$L'_{17} := \min(L'_{17}, L_{14} + d_{47}, L_{15} + d_{57})$$

REFERENCES

[1] DIJKSTRA, E. W., "A Note on Two Problems in Connection with Graphs," *Numerische Mathematik*, 1, pp. 269–271, 1959.

[2] EVARTS, H. F., *Introduction to PERT*, Allyn and Bacon, Inc., Boston, 1964.

[3] FORD, L. R., JR., and D. R. FULKERSON, *Flows in Networks*, Princeton University Press, Princeton, N.J., 1962.

[4] HADLEY, G., *Linear Programming*, Addison-Wesley Publishing Company, Inc., Reading, Mass., 1962.

[5] HILLIER, F. S., and G. J. LIEBERMAN, *Introduction to Operations Research*, Holden-Day, Inc., San Francisco, 1967.

[6] HU, T. C., *Integer Programming and Network Flows*, Addison-Wesley Publishing Company, Inc., Reading, Mass., 1969.

[7] KELLEY, J., "Critical-Path Planning and Scheduling: Mathematical Basis," *Operations Research*, 9, pp. 296–320, 1961.

[8] MACCRIMMON, K. R., and C. A. RYAVEC, "An Analytical Study of the PERT Assumptions," *Operations Research*, 12, pp. 16–37, 1964.

[9] WIEST, J. D., and F. K. LEVY, *A Management Guide to PERT/CPM*, Prentice-Hall, Inc., Englewood Cliffs, N.J., 1969.

PROBLEMS

1. Consider the following PERT network, where the deterministic activity times (in days) are indicated next to the respective branches. Determine the earliest time, latest time, and slack time for each event, when the contractual obligation date is 16 days. In addition, determine the critical path of the network.

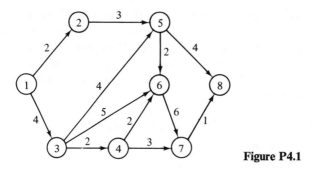

Figure P4.1

2. Consider the following PERT network, where the expected value and variance of elapsed time for each activity is indicated next to its respective branch. The original schedule of events is given in the following table:

Event number	1	2	3	4	5	6	7	8
Original schedule	0	5	7	3	13	14	20	24

Determine the probability that the original schedule of events will be met for the PERT network. Display the output information in the format of Table 4.4. In addition, determine the critical path of the network.

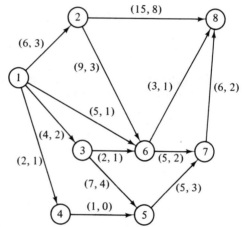

Figure P4.2

3. Given the information contained in the following table, determine the maximum length of the schedule and the minimum-cost schedule when the overhead cost is $5.50 per day.

Event	Successor Event	Normal Days Duration	Minimum (Crash) Days Duration	Cost in $/Day of Crashing
A	B	6	3	4
B	C	5	2	3
B	D	3	1	4.50
C	E	4	2	2
D	F	7	2	2.25
D	G	4	1	2
G	F	6	4	1
E	H	5	2	1
F	H	4	3	5

4. Given the information contained in the following table, where the duration times are in hours, determine the maximum length of the schedule and the minimum-cost schedule when the overhead cost is $20 per hour.

		Normal		Minimum	
Event	Successor Event	Duration	$ Cost	Duration	$ Cost
A	B	6	56	2	88
A	C	7	90	5	102
A	D	5	49	3	55
B	E	5	45	3	65
E	F	4	49	2	73
C	F	4	62	2	76
D	F	7	32	3	46

5. Consider the following network where the arc capacities are indicated. Determine the maximal flow in the network assuming an initial flow of zero.

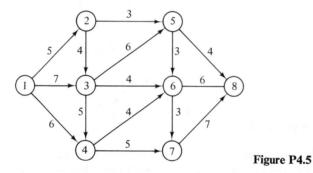

Figure P4.5

6. Consider the following graph where the distances for each edge are indicated. Determine the shortest route from node 1 to all the other nodes of the graph.

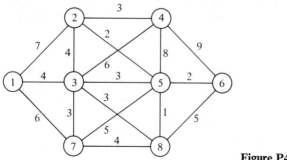

Figure P4.6

7. Consider the graph in Figure 4.24, where the distances for the edges are indicated. Determine the shortest route from S to all the other nodes of the graph.

TRANSPORTATION AND
ASSIGNMENT PROBLEMS

5.1. Introduction

As efficient as the computational methods of linear programming are, there are certain classes of linear-programming problems, some with the added complication of integer requirements on the variables, for which even better solution procedures are available. Two such classes of problems, which are of sufficient importance in applications to warrant special attention, are *transportation* and *assignment* problems.

In Chapter 3 we formulated a particular transportation problem and showed how it could be treated as a linear-programming problem. The problem may be stated as follows: Given

1. m origins O_i ($i = 1, 2, \ldots, m$), which has a total available quantity a_i,

2. n destinations $D_j(j = 1, \ldots, n)$, which has a requirement b_j, where

$$\sum_{i=1}^{m} a_i = \sum_{j=1}^{n} b_j$$

3. a cost c_{ij} for shipping one unit from origin i to destination j,

5

it is required to allocate units from origins to destinations in such a way as to exhaust all supplies, satisfy each and every requirement, and to minimize the total shipping cost.

The original transportation problem was formulated by Hitchcock [5] in 1941, who presented a method of solution to the problem similar to the simplex technique of linear programming. Kantorovich [6] also examined certain aspects of the transportation problem. In 1951, Dantzig [1] demonstrated the applicability of the simplex technique and showed how its specialized use resulted in an efficient computational procedure.

In its simplest form the assignment problem may be stated as follows: Given n tasks to be performed by n persons, each person being capable of performing any of the tasks, and given the effectiveness of each person for each task, the problem is to assign each person to one and only one task so as to maximize the total effectiveness.

Kuhn [8], in 1955, developed an efficient algorithm to solve the assignment problem and used results originally stated by the Hungarian mathematicians König and Egevary. The method has come to be known as the Hungarian method.

5.2. A Transportation Problem

i. *Mathematical Expression*

Consider the situation described in Example 3.9, where we have m locations that serve as points of origin, such as warehouses, for some commodity. Suppose that there are n points of destination, such as retail stores, to which the commodity is to be shipped from the m origins. The cost of shipping one unit of the commodity from origin i to destination j is given as c_{ij}, $i = 1, \ldots, m, j = 1, \ldots, n.$

There are a_i units of the commodity available at origin i, $i = 1, \ldots, m$; destination j requires b_j units of the commodity, $j = 1, \ldots, n.$

The problem then is to determine a shipping strategy, that is, determine how many units should be shipped from origin i to destination j, for all i, j, so as to minimize the total cost, while observing all origin limitations and destination requirements.

The number of variables whose optimal values must be determined is mn, since each origin may ship to each destination. Let

$$x_{ij} = \text{number of units to be shipped from origin } i \text{ to}$$
$$\text{destination } j$$

The formulation that followed from the description there was

$$\min \sum_{i=1}^{m} \sum_{j=1}^{n} c_{ij} x_{ij}$$

subject to

$$\sum_{j=1}^{n} x_{ij} \le a_i, \quad i = 1, \ldots, m \qquad (5.1)$$

$$\sum_{i=1}^{m} x_{ij} \ge b_j, \quad j = 1, \ldots, n$$

$$x_{ij} \ge 0$$

Clearly, the problem has no solution unless the supply is at least equal to the demand; that is,

$$\sum_{i=1}^{m} a_i \ge \sum_{j=1}^{n} b_j$$

Frequently, the commodity of interest exists in indivisible units; for example, we may be interested in sending trucks or airplanes from various origins to various destinations. Alternatively, the cargo being transported is frequently indivisible; the items may be men, pieces of machinery, or anything of which fractions of units are not feasible. But, then, it is also reasonable to expect that the a_i, b_j are integers. Furthermore, although we shall soon dispense with the assumption, it is convenient for the moment to assume that the demands must be met *exactly*, and not exceeded. We still allow, however, that we need not transport the entire supply. Thus the first problem of interest is

$$\min \sum_{i=1}^{m} \sum_{j=1}^{n} c_{ij} x_{ij}$$

subject to

$$\sum_{j=1}^{n} x_{ij} \le a_i, \quad i = 1, \ldots, m \tag{5.2}$$

$$\sum_{i=1}^{m} x_{ij} = b_j, \quad j = 1, \ldots, n$$

$$x_{ij} \text{ a nonnegative integer}$$

and a_i, b_j are given positive integers for all i, j. The costs c_{ij} are allowed to be noninteger, and are not even constrained in sign; they may be positive or negative.

Adding slack variables $s_1, s_2, \ldots, s_m$ to (5.2), we obtain

$$\min \sum_{i=1}^{m} \sum_{j=1}^{n} c_{ij} x_{ij}$$

subject to

$$\sum_{j=1}^{n} x_{ij} + s_i = a_i, \quad i = 1, \ldots, m \tag{5.3}$$

$$\sum_{i=1}^{m} x_{ij} = b_j, \quad j = 1, \ldots, n$$

$$x_{ij} \text{ a nonnegative integer,} \quad \text{all } i, j$$

$$s_i \ge 0, \quad \text{all } i$$

If the x_{ij} and a_i are integers, then the s_i are necessarily so, and it is unnecessary to impose integer requirements on them.

ii. *Special Properties*

In array form the equality constraints of (5.3) may be written as in (5.4), where the slack columns have been written last on the left side of (5.4).

$$
\begin{array}{c}
\overbrace{\qquad n \qquad}\ \ \overbrace{\qquad n \qquad}\ \ \overbrace{\qquad n \qquad}\qquad\quad \overbrace{\qquad n \qquad}\ \ \overbrace{\qquad m \qquad}
\end{array}
$$

```
        n                n               n                      n               m
     ┌───────────┬───────────┬───────────┬   ┬───────────┬───────────┐
     │1 1 1 ⋯ 1  │0 0 0 ⋯ 0  │0 0 0 ⋯ 0  │⋯  │0 0 0 ⋯ 0  │1 0 0 ⋯ 0  │
     │0 0 0 ⋯ 0  │1 1 1 ⋯ 1  │0 0 0 ⋯ 0  │   │0 0 0 ⋯ 0  │0 1 0 ⋯ 0  │
     │0 0 0 ⋯ 0  │0 0 0 ⋯ 0  │1 1 1 ⋯ 1  │   │0 0 0 ⋯ 0  │0 0 1 ⋯ 0  │
  m ⎨ │    .      │    .      │    .      │   │    .      │    .      │
     │    .      │    .      │    .      │   │    .      │    .      │
     │    .      │    .      │    .      │   │    .      │    .      │
     │0 0 0 ⋯ 0  │0 0 0 ⋯ 0  │0 0 0 ⋯ 0  │⋯  │1 1 1 ⋯ 1  │0 0 0 ⋯ 1  │
     ├───────────┼───────────┼───────────┼   ┼───────────┼───────────┤   (5.4)
     │1 0 0 ⋯ 0  │1 0 0 ⋯ 0  │1 0 0 ⋯ 0  │⋯  │1 0 0 ⋯ 0  │0 0 0 ⋯ 0  │
     │0 1 0 ⋯ 0  │0 1 0 ⋯ 0  │0 1 0 ⋯ 0  │   │0 1 0 ⋯ 0  │0 0 0 ⋯ 0  │
     │0 0 1 ⋯ 0  │0 0 1 ⋯ 0  │0 0 1 ⋯ 0  │   │0 0 1 ⋯ 0  │0 0 0 ⋯ 0  │
  n ⎨ │    .      │    .      │    .·     │   │    .      │    .      │
     │    .      │    .      │    .      │   │    .      │    .      │
     │0 0 0 ⋯ 1  │0 0 0 ⋯ 1  │0 0 0 ⋯ 1  │⋯  │0 0 0 ⋯ 1  │0 0 0 ⋯ 0  │
     └───────────┴───────────┴───────────┴   ┴───────────┴───────────┘
```

We obtain an array of coefficients with $m + n$ rows and $m(n + 1)$ columns. Aside from the slack columns, which are unit vectors, the activity vectors may be characterized as follows. There is one and only one column vector for each possible way of choosing two components of an $m + n$ vector, the first from among the first m components, the second from among the last n components. The selected components are 1's; all other components are 0.

This array of coefficients has a very special property. Cross out any number of rows and columns such that the remaining elements comprise a $k \times k$ matrix, for some k. This remaining structure, which is called a *submatrix of order k* of the given matrix A, will have determinant $0, 1$, or -1. No other values are possible. Such a matrix A is called *unimodular*.

iii. *Solution by the Simplex Method*

Suppose that we disregarded the integer requirements and set out to solve problem (5.3) by linear programming. We might add artificial variables and vectors to assist in obtaining an initial basic feasible solution, but the array of coefficients would remain unimodular.

Denoting the requirements vector in the problem by $\mathbf{b}$, a basis by B, and the corresponding basic feasible solution by $\mathbf{x}(B)$, we have

$$\mathbf{x}(B) = B^{-1}\mathbf{b}$$

Using a well-known expression for the inverse (see Section 5-ii-a of Chapter 1), we have

$$\mathbf{x}(B) = \frac{1}{\det B} \cdot (\text{adj } B)\mathbf{b}$$

where $\det B$ denotes the determinant of B, and adj B, the classical adjoint. But by the unimodularity of A and the fact that B has integer elements,

$(1/\det B) \cdot \text{adj } B$ is a square array, *all of whose elements are integers.* Since **b** is a vector of positive integers, $\mathbf{x}(B)$ a basic feasible solution implies that $\mathbf{x}(B)$ is a vector of nonnegative integers.

Thus solving (5.3) with the simplex method, which does not guarantee integer solutions in general, will, in this case, result in obtaining the desired optimal solution in integers.

Notice that there is no possibility of an unbounded solution in solving (5.3), or (5.2) for that matter. Furthermore, not only does

$$\sum_{j=1}^{n} b_j > \sum_{i=1}^{m} a_i$$

imply no feasible solution, but also

$$\sum_{i=1}^{m} a_i \geq \sum_{j=1}^{n} b_j$$

guarantees the existence of a feasible solution. The reader may wish to attempt to establish the validity of this last statement—perhaps by construction. Later we shall prove the statement.

Exercise 5.1

Solve the following transportation problem by means of the simplex method. The per unit shipping cost from origin i to destination j is the (i, j)th element of the following array:

	7	14	5	4
15	2	1	3	4
8	5	3	7	1
12	2	8	4	1

The supply at origin i is listed to the *left* of row i of the cost array, and the demand at destination j is written *above* the jth column.

Exercise 5.2

Consider the array of coefficients from the preceding exercise, where the slack columns have been removed. What is the rank of the array? How would you generalize for an arbitrary problem of type (5.3)?

Exercise 5.3

In case

$$\sum_{i=1}^{m} a_i > \sum_{j=1}^{n} b_j$$

in the problem (5.2) consider the following tactic for creating an equivalent problem where *all the constraints are equalities.* Introduce an $(n + 1)$ st destination with a demand $\sum_{i=1}^{m} a_i - \sum_{j=1}^{n} b_j$, and define

$$c_{i,n+1} = 0, \quad \text{for } i = 1, \ldots, m$$

If we require that each origin ship *all* its supply and that the demand at each destination be satisfied *exactly*, show that an optimal solution to the new problem provides an optimal solution to the given problem (5.2).

iv. *Improving upon the Simplex Method*

Actually, we may improve greatly upon the straightforward simplex method in dealing with transportation problems. Our development may appear to digress from a linear programming approach, but, as will be seen, that is not actually the case. We shall use the results of Exercise 5.3 to assume equality constraints with total supply equal to total demand. The simplest transportation problem may thus be formulated as

$$\min \sum_{i=1}^{m} \sum_{j=1}^{n} c_{ij} x_{ij}$$

subject to

$$\sum_{j=1}^{n} x_{ij} = a_i, \quad i = 1, \ldots, m \tag{5.5}$$

$$\sum_{i=1}^{m} x_{ij} = b_j, \quad j = 1, \ldots, n$$

$$x_{ij} \text{ a nonnegative integer}$$

The problem may be completely displayed by the tableau in Figure 5.1.

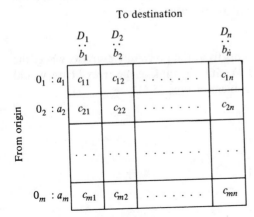

Figure 5.1. Representing a transportation problem,

$$\sum_{i=1}^{m} a_i = \sum_{j=1}^{n} b_j.$$

The element c_{ij} in the tableau corresponds to the unit shipping cost from origin O_i to destination D_j. The supply a_i at origin i is listed to the left of row i, and the demand b_j at destination j is listed above the jth column.

The method of solution is quite in the spirit of the simplex method. The first step is the finding of an initial feasible solution, and a sequence of iterations leads one to an optimal solution. The solution procedure is illustrated using the data exhibited in problem (5.6).

Destination

		D_1	D_2	D_3	D_4	D_5	D_6	
		..	..	..	..	..	..	
		8	7	12	2	3	10	
	O_1 : 15	10	4	7	2	3	4	(5.6)
	O_2 : 10	1	1	13	8	7	6	
Origin	O_3 : 5	2	8	-1†	5	1	4	
	O_4 : 9	7	9	2	7	3	2	
	O_5 : 3	3	4	15	7	-2†	5	

†Negative costs might indicate some sort of bonus, or net profit, realized by shipping from the corresponding origins to the corresponding destinations.

a. LOCATING AN INITIAL FEASIBLE SOLUTION

The following is a procedure for generating an initial feasible solution.

Begin with origin O_1. Allocate from O_1 to destination D_1 until either (1) supply at O_1 is exhausted and demand at D_1 is satisfied, (2) supply at O_1 is exhausted and demand at D_1 is not satisfied, or (3) supply at O_1 is not exhausted and demand at D_1 is satisfied. In case (1) we begin anew allocating from O_2 to D_2. In case (2), we begin allocating from O_2 to D_1. In case (3), we begin allocating from O_1 to D_2.

Again we are led to three cases. The allocation process continues, supplies at origins being exhausted in the order $O_1, O_2, \ldots, O_m$. Simultaneously, demands at destinations are satisfied in the order $D_1, D_2, \ldots, D_n$.

At every step either an origin supply is exhausted or a destination demand is satisfied, or both. Since we have $\sum_{i=1}^{m} a_i = \sum_{j=1}^{n} b_j$, then at every stage the supply unallocated and the amount of demand not satisfied are identical. Therefore, the last stage must exhaust supply m and satisfy demand n.

Omitting costs from problem (5.6) and placing the allocation x_{ij} from origin i to destination j in the block (i, j), the successive stages of the allocation procedure described for our numerical example are illustrated in the nine steps of Figure 5.2. The allocation always proceeds from left to right and top to bottom, or from west to east and north to south, thus this technique of obtaining a feasible solution has become known as the *northwest-corner method*.

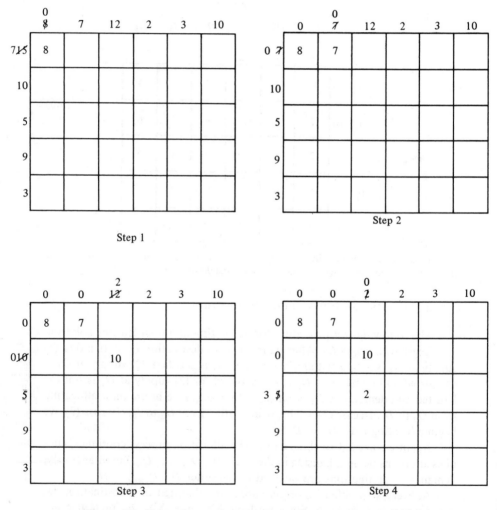

Figure 5.2. Using the northwest corner rule to generate an initial feasible solution.

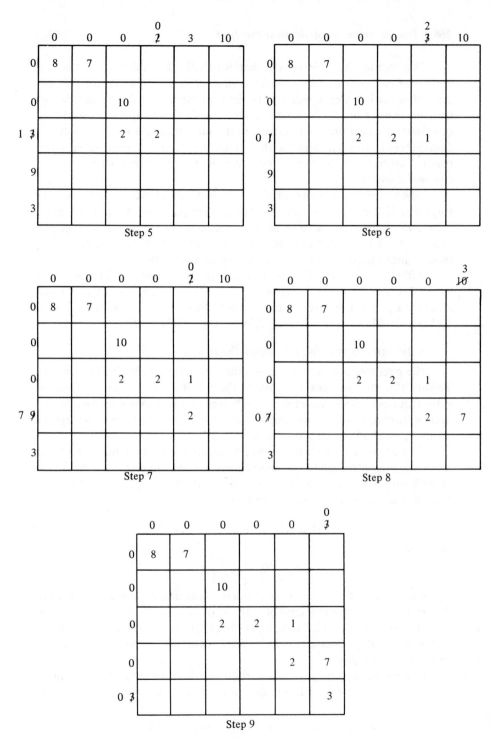

Figure 5.2. (*continued*)

This method has a property that we shall later utilize. It is that the *number of positive variables provided is at most $m + n - 1$*. To see this, observe that the last step satisfies both the mth supply constraint and the nth demand constraint. Each preceding step satisfies one supply constraint or one demand constraint, and perhaps both. Since there are $m + n$ supply and demand constraints, it must be that at most $m + n - 1$ positive variables result. Furthermore, we must always end with a positive variable in every row *and* column.

Having taken no notice of the costs in this method, it should not be surprising if the objective function value corresponding to the first feasible solution obtained proves to be far from optimal. The value of the objective function for a particular solution is easily available from our array representation, since the per unit costs will be written inside the boxes.

With respect to the feasible solution just obtained, we have $x_{11} = 8$, $x_{12} = 7, x_{23} = 10, x_{33} = 2, x_{34} = 2, x_{35} = 1, x_{45} = 2, x_{46} = 7, x_{56} = 3$, and all other $x_{ij} = 0$. The value of the objective function is $Z = 8(10) + 7(4) + 10(13) + 2(-1) + 2(5) + 1(1) + 2(3) + 7(2) + 3(5) = 282$.

b. IMPROVING THE FIRST FEASIBLE SOLUTION

The next step is to determine whether the solution obtained is optimal. Toward this end, we seek to evaluate the net change in objective function value that occurs if we allow an x_{rs} which is currently zero to become positive, *while a feasible solution is maintained.* When it is possible to improve upon a solution, we shall obtain a new one.

To evaluate cell (r, s), that is to investigate the profitability of allowing x_{rs} that is presently zero to become positive, we need to define a certain kind of path within our array. Denoting a cell by an ordered pair of integers, we must find a sequence of cells of the form

$$(r, s), (r, t), (u, t), (u, v), \ldots, (z, s)$$

such that

$$x_{rt} > 0, \ldots, x_{zs} > 0$$

and $s \neq t, r \neq u, t \neq v, \ldots, r \neq z$. Such a sequence defines a *path* through the array. To illustrate this kind of path, let $(r, s) = (2, 1)$ in an array of cells. The sequence of cells† $(2, 1), (2, 5), (1, 5), (1, 3), (5, 3), (5, 1)$, which forms a path if we connect successive cells by lines, including the last cell to the first, appears as in Figure 5.3.

In general, beginning with a particular cell to be evaluated, we find a *second cell* in the *same row* as the *first;* a *third* in the *same column* as the sec-

---

†Positivity of variables was ignored in constructing a sample path.

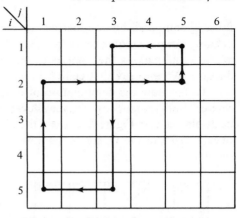

Figure 5.3. A sequence of cells forming a path.

ond; a *fourth* in the *same row* as the *third;* . . . .; a *p*th in the same column as the first. It will turn out that such a path will always be unique, if one exists; and we shall later take up the question of nonexistence.

Returning to our particular first feasible solution displayed in step 9 of Figure 5.2, we select as our (r, s) cell, cell $(2, 5)$ (note $x_{25} = 0$). The unique closed path corresponding to cell $(2, 5)$ is shown in Figure 5.4.

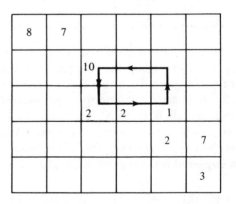

Figure 5.4. Path corresponding to sequence of cells $(2, 5)$, $(2, 3)$, $(3, 3)$, and $(3, 5)$.

The path allows the measurement of the desirability of allowing x_{25} to become positive, and, in addition, a way for constructing a new feasible solution if x_{25} is allowed to increase.

Suppose that we allow $x_{25} = 1$. Then, to satisfy the supply constraint for origin 2, we let x_{23} decrease by 1; then to satisfy the demand at destination 3, we increase x_{33} by 1; and then to simultaneously satisfy the supply constraint for origin 3 and the demand at destination 5, let x_{35} decrease by 1. The net change in objective function value, corresponding to a unit increase in the zero-valued variable x_{ij}, will be called the *evaluator* for x_{ij}. The evaluator for x_{25}, computed by considering the changes that took place in the cells defining our path sequence, is $7 - 13 - 1 - 1 = -8 < 0$; therefore, letting $x_{25} = 1$ and making the changes to maintain feasibility results in a cost reduction. We also say that cell (2, 5) has been evaluated.

To facilitate the computation of evaluators, it is customary to enter in each cell both the unit cost and the allocated quantities. Thus Figure 5.4 would be displayed as in Figure 5.5, where the c_{ij} are in the upper right-hand corners of each cell and the x_{ij} are in the lower left-hand corners, the zero x_{ij} being omitted.

10 / 8	4 / 7	7	2	3	4
1	1	13 / 10	8	7	6
2	8	−1 / 2	5 / 2	1 / 1	4
7	9	2	7	3 / 2	2 / 7
3	4	15	7	−2	5 / 3

Figure 5.5. Double entry array incorporating c_{ij} and x_{ij}.

Clearly, it would seem we could derive a greater reduction by allowing $x_{25} > 1$ and making the changes around the loop to maintain feasibility. But not in this case, since a reduction is made in cell (3, 5) and $x_{35} = 1$ initially. In general, if allowing a zero variable x_{ij} to become positive is profitable, we allow x_{ij} *to become equal to the minimum allocation associated with cells in our path where reductions take place. Thus at least one formerly positive variable becomes zero.* If we make the change just computed, we get an improved feasible solution, as shown in Figure 5.6.

10 / 8	4 / 7	7	2	3	4
1	1	13 / 9	8	7 / 1	6
2	8	−1 / 3	5 / 2	1	4
7	9	2	7	3 / 2	2 / 7
3	4	15	7	−2	5 / 3

Figure 5.6. An improved feasible solution.

The evaluator for cell $(5, 3)$ in Figure 5.6 is $3 > 0$, so x_{53} should remain zero, but a second improved feasible solution may be obtained by selecting the cell $(4, 3)$, whose path is as shown.

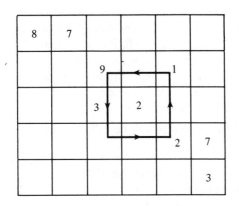

The evaluator for x_{43} is $2 - 13 + 7 - 3 < 0$, and in this case the new variable x_{43} may assume the value 2. The new solution appears as

10	4	7	2	3	4	
8	7					
1	1	13	8	7	6	
		7	3			
2	8	−1	5	1	4	(5.7)
	3	2				
7	9	2	7	3	2	
	2			7		
3	4	15	7	−2	5	
					3	

c. SELECTING A VARIABLE TO BECOME POSITIVE AND OPTIMALITY CRITERIA

Analogous to the simplex method, one chooses that variable to become positive whose evaluator indicates maximum possible improvement, rather than also considering the maximum value the new variable can assume consistent with feasibility. This is consistent with the simplex method, where we chose the variable to become nonzero on the basis of the $c_j - z_j$ alone, disregarding the new variable's value, which also influenced the improvement achieved. Also analogous to the simplex procedure, *an optimal solution for the minimization problem is indicated by all the empty cells having nonnegative evaluators.*

d. SUMMARY OF STEPS

In general, once an initial feasible solution is obtained, the procedure for obtaining an optimal solution consists of the following steps:

1. Evaluate all empty cells.
2. If no cell has an evaluator indicating that an improved solution is possible, then the present solution is optimal, and we stop.
3. If some cells have negative evaluators (positive for the maximization problem), choose a cell with the best per unit evaluator and allow the associated variable to become positive; then obtain a new feasible solution, and return to step 1.

One passage through these steps will be called an *iteration*, as with the simplex method.

e. NONEXISTENCE OF LOOPS—AN ARTIFICE, THE ϵ-PERTURBATION METHOD

Referring to (5.7) we find it *impossible* to construct a loop of the desired sort, starting with cell (1, 3); we then resort to an artifice to allow evaluation of *all* the empty cells. The method is referred to as the ϵ-*perturbation method*.

Beginning with step 9 of Figure 5.2, we first discover that no loop exists for cell (1, 3) when attempting to evaluate cells during iteration 1. (Notice that there are a *number* of cells for which loops do not exist.) We allow $x_{13} = \epsilon$, where $0 < \epsilon < 1$, so that ϵ will be the smallest variable in any loop in which it might appear. Clearly, we get an infeasible solution; but this should not be considered too serious, so long as we rectify matters later. After all, we introduced artificial variables to expedite proceedings in the simplex method and worked with infeasible solutions there. Our infeasible solution then appears as (5.8). The reader should examine (5.8) carefully and conclude that

10	4	7	2	3	4
8	7	ϵ			
1	1	13	8	7	6
		10			
2	8	−1	5	1	4
	2	2	1		
7	9	2	7	3	2
			2	7	
3	4	15	7	−2	5
				3	

(5.8)

every cell may now be evaluated. This would be true if the ϵ had been placed in *any* cell having no loop.

We emphasize four points:

1. To evaluate all the cells during one iteration may require several such operations of placing ϵ's in empty cells.
2. If placing one or more ϵ's during one iteration allows evaluation of all cells, it may occur that more need to be placed in future iterations.
3. The ϵ's are maintained *throughout the solution procedure. When the optimality criteria are satisfied, the ϵ's are set equal to zero, resulting in a feasible solution that is also optimal.*
4. ϵ is treated as any other number in an iteration. If ϵ is associated with a reduction in the loop corresponding to a variable about to become positive, that variable will assume the value ϵ, and ϵ will be added and subtracted elsewhere around the loop.

f. APPLICATION OF THE ITERATIVE TECHNIQUE TO GENERATE AN OPTIMAL SOLUTION TO PROBLEM (5.6)

Referring to problem (5.6), let us return to the initial feasible solution in step 9 of Figure 5.2 and proceed several iterations toward an optimal solution for the problem. Array (5.9) is the tableau of step 9 of Figure 5.2, after inclusion of the ϵ, incorporating the unit costs and some of the evaluators within small circles in the lower right-hand corners.

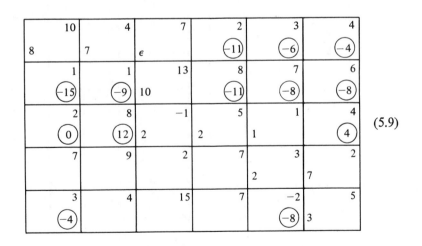

(5.9)

The most negative evaluator is -15, so the incentive to increase x_{21} from zero is greatest, and it assumes the value 8. We obtain tableau (5.10), which is an improved solution.

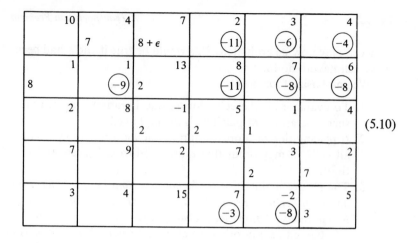

(5.10)

In tableau (5.10) the nonnegative evaluators for the next iteration have been omitted since they do not influence the selection procedure.

Exercise 5.4

To complete the presentation of the example, we illustrate a sequence of successive feasible solutions, obtained by the algorithm. The costs and cell evaluators have been omitted from these arrays, so the reader should work through the sequence shown, supplying the missing numbers and verifying that the steps shown are proper.

ε	7	8+ε			
8			2		
		4		1	
				2	7
					3

ε	7	8+ε			
8			2		
		4		1	
				2	7
					3

	7	8+ε	ε		
8+ε			2−ε		
		4		1	
				2	7
					3

		5+ε	8+ε	2	
8+ε	2−ε				
		4		1	
				2	7
					3

	$5+\epsilon$	$8+\epsilon$	2		
$8+\epsilon$	$2-\epsilon$				
		4		1	
				9	
			2	1	

		$5+\epsilon$	$7+\epsilon$	2	1
$8+\epsilon$	$2-\epsilon$				
			5		
					9
				3	

	$5+\epsilon$	$7+\epsilon$	2	ϵ	1
$8+\epsilon$	$2-\epsilon$				
		5			
				9	
			3		

	$5+\epsilon$		2	ϵ	$8+\epsilon$
$8+\epsilon$	$2-\epsilon$				
		5			
		$7+\epsilon$			$2-\epsilon$
			3		

An optimal solution to the required problem is

	D_1	D_2	D_3	D_4	D_5	D_6
	8	7	12	2	3	10
$0_1 : 15$		5		2		8
$0_2 : 10$	8	2				
$0_3 : 5$			5			
$0_4 : 9$			7			2
$0_5 : 3$					3	

The corresponding value of the objective function is

$$5 \times 4 + 2 \times 2 + 8 \times 4 + 8 \times 1 + 2 \times 1$$
$$+ 5 \times -1 + 7 \times 2 + 2 \times 2 + 3 \times -2 = 73$$

5.3. Theoretical Basis for the Transportation Algorithm

i. *Coefficient Matrix*

Let us consider the coefficient matrix in the equality-constrained problem (5.5),

$$
m\begin{cases}
\begin{pmatrix}
1 & 1 & \cdots & 1 & 0 & 0 & \cdots & 0 & & 0 & 0 & \cdots & 0 \\
0 & 0 & \cdots & 0 & 1 & 1 & \cdots & 1 & & 0 & 0 & \cdots & 0 \\
\cdot & & & & 0 & 0 & \cdots & 0 & & \cdot & & & \cdot \\
\cdot & & & & \cdot & & & & \cdots & \cdot & & & \cdot \\
\cdot & & & & \cdot & & & & & 0 & 0 & \cdots & 0 \\
0 & 0 & \cdots & 0 & 0 & 0 & \cdots & 0 & & 1 & 1 & \cdots & 1 \\
\hline
1 & 0 & \cdots & 0 & 1 & 0 & \cdots & 0 & & 1 & 0 & \cdots & 0 \\
0 & 1 & \cdots & 0 & 0 & 1 & \cdots & 0 & & 0 & 1 & \cdots & 0 \\
\cdot & & 0 & & \cdot & & 0 & & \cdots & \cdot & & 0 & \cdot \\
\cdot & & & 0 & \cdot & & & 0 & & \cdot & & & 0 \\
0 & 0 & \cdots & 1 & 0 & 0 & \cdots & 1 & & 0 & 0 & \cdots & 1
\end{pmatrix}
\end{cases} n
$$

(5.11)

This array, with $m + n$ rows and mn columns, has rank $m + n - 1$. To observe this, let us first show that the rank is not $m + n$. First, consider any linear combination of the first m rows, which gives a vector $\boldsymbol{\alpha}$ of the form

$$\boldsymbol{\alpha} = (\overbrace{\alpha_1, \alpha_1, \ldots, \alpha_1}^{n}, \overbrace{\alpha_2, \alpha_2, \ldots, \alpha_2}^{n}, \ldots, \overbrace{\alpha_m, \alpha_m, \ldots, \alpha_m}^{n})$$

Suppose that we select the components α_i $(i = 1, \ldots, m)$ of the vector to be identical and nonzero, that is,

$$\alpha_1 = \alpha_2 = \cdots = \alpha_m \neq 0,$$

so that we obtain the vector $\boldsymbol{\alpha}$ with mn identical components:

$$\boldsymbol{\alpha} = (\alpha_1, \alpha_1, \ldots, \alpha_1)$$

We now define the vector $\boldsymbol{\beta}_j$ with mn components as

$$\boldsymbol{\beta}_j = -\alpha_1 \text{ row } j, \quad j = m + 1, m + 2, \ldots, m + n$$

and form the vector

$$\boldsymbol{\beta} = \sum_{j=m+1}^{m+n} \boldsymbol{\beta}_j$$

Clearly, $\boldsymbol{\alpha} + \boldsymbol{\beta} = \mathbf{0}$, so the rows of the array are not linearly independent. By the same sort of argument, one can show that no linear combination of $m + n - 1$ or fewer rows is linearly dependent. Consequently, $m + n - 1$ is the largest number of linearly independent rows available in the array; thus the rank of the array is $m + n - 1$.

It follows that one of the constraints is redundant and could be omitted from a simplex solution procedure. It is also clear from any feasible transportation problem tableau, say (5.7), that if any $m + n - 1$ constraints are satisfied, the remaining one is automatically so. Recall also that the northwest-corner procedure always provides a feasible solution with *at most* $m + n - 1$ positive variables.

ii. *Column Vectors, Cells, and Loops*

Denoting the array (5.11) by A, we have a corresponding system of constraints $Ax = b$. In terms of the *columns* of A, this system may be written as

$$
x_{11}
\begin{bmatrix} 1 \\ 0 \\ 0 \\ \cdot \\ \cdot \\ \cdot \\ 0 \\ 1 \\ 0 \\ \cdot \\ \cdot \\ \cdot \\ 0 \end{bmatrix}
+ x_{12}
\begin{bmatrix} 1 \\ 0 \\ 0 \\ \cdot \\ \cdot \\ \cdot \\ 0 \\ 0 \\ 1 \\ \cdot \\ \cdot \\ \cdot \\ 0 \end{bmatrix}
+ \cdots + x_{ij}
\begin{bmatrix} 0 \\ \cdot \\ 1 \\ \cdot \\ \cdot \\ \cdot \\ 0 \\ 0 \\ \cdot \\ 1 \\ \cdot \\ \cdot \\ 0 \end{bmatrix}
+ x_{mn}
\begin{bmatrix} 0 \\ 0 \\ \cdot \\ \cdot \\ \cdot \\ 0 \\ 1 \\ 0 \\ \cdot \\ \cdot \\ \cdot \\ 1 \end{bmatrix}
=
\begin{bmatrix} a_1 \\ a_2 \\ \cdot \\ \cdot \\ \cdot \\ a_m \\ b_1 \\ b_2 \\ \cdot \\ \cdot \\ \cdot \\ b_n \end{bmatrix}
\tag{5.12}
$$

In (5.12) we have partitioned our column vectors so that there are m components above the dividing line, n below, and x_{ij} is the coefficient of a vector with a 1 in the ith component among the first m, a 1 in the jth component among the last n, and zeros elsewhere.

Consider the first feasible solution of the numerical example (5.6) and the loop we constructed to evaluate cell (2, 5). To do this we utilized cells (2, 3), (3, 3) and (3, 5). We had decreases associated with (2, 3) and (3, 5), and an increase associated with (3, 3).

Cell (i, j) is associated with variable x_{ij}, and x_{ij}, in terms of column vectors in (5.12), is associated with a column

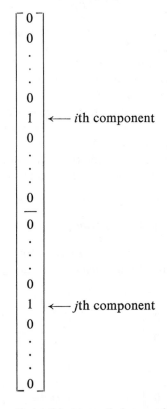

If we consider the cells in this loop, their associated columns, and how increases or decreases occurred to the variables in the loop, we find

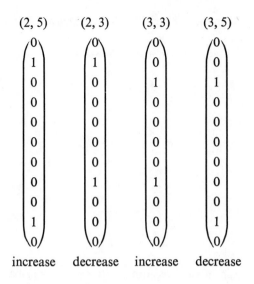

If we associate $+1$ with increase and -1 with decrease, and use these numbers as coefficients of the corresponding vectors, we get

$$
+1\begin{pmatrix}0\\1\\0\\0\\0\\0\\0\\0\\0\\1\\0\end{pmatrix}
-1\begin{pmatrix}0\\1\\0\\0\\0\\0\\0\\1\\0\\0\\0\end{pmatrix}
+1\begin{pmatrix}0\\0\\1\\0\\0\\0\\0\\1\\0\\0\\0\end{pmatrix}
-1\begin{pmatrix}0\\0\\1\\0\\0\\0\\0\\0\\0\\1\\0\end{pmatrix}
\qquad (5.13)
$$

$(2,5)\quad (2,3)\quad (3,3)\quad (3,5)$

Performing the indicated addition of vectors results in the zero vector

$$
\begin{pmatrix}
0 \\
1-1 \\
1-1 \\
0 \\
0 \\
0 \\
0 \\
-1+1 \\
0 \\
1-1 \\
0
\end{pmatrix} = \mathbf{0}
$$

In other words, each of the preceding four vectors may be represented as a linear combination of the remaining three. In particular, the vector corresponding to cell (2, 5) may then be represented as

$$
\begin{pmatrix}
0 \\ 1 \\ 0 \\ 0 \\ 0 \\ 0 \\ 0 \\ 0 \\ 0 \\ 1 \\ 0
\end{pmatrix}
=
\begin{pmatrix}
0 \\ 1 \\ 0 \\ 0 \\ 0 \\ 0 \\ 0 \\ 1 \\ 0 \\ 0 \\ 0
\end{pmatrix}
-
\begin{pmatrix}
0 \\ 0 \\ 1 \\ 0 \\ 0 \\ 0 \\ 0 \\ 1 \\ 0 \\ 0 \\ 0
\end{pmatrix}
+
\begin{pmatrix}
0 \\ 0 \\ 1 \\ 0 \\ 0 \\ 0 \\ 0 \\ 0 \\ 0 \\ 1 \\ 0
\end{pmatrix}
\tag{5.14}
$$

Thus, *the vector associated with the variable we are considering to allow to become positive may be expressed in terms of the vectors associated with the positive variables in the loop.*

This, of course, sounds very much like expressing a vector in terms of a basis in our usual linear-programming methods. Suppose for a moment that the vectors associated with the positive variables in the loop *were* part of a basis. Then as we added and subtracted costs around the loop, we would be computing what in the linear-programming problem were the $c_j - z_j$, or what might be termed $c_{ij} - z_{ij}$ in the present problem. The ± 1's we attached to these "basic" cells would be just the y_{ij} in the simplex method, that is, the unique scalars which allow the expression of a given vector in terms of a basis. Also, what would be the value of the new variable? Just

$$Q = \min \left\{ \frac{x_{ij}(B)}{y_{ij}} \,\middle|\, y_{ij} > 0 \right\} \tag{5.15}$$

where both i and j are variable. Actually, our y_{ij} would have two additional indices if we were to be consistent with the notation used in the simplex procedure. Q is exactly the minimum of the cells in the loop associated with -1's, as we determined by employing the loop.†

iii. *Basic Feasible Solutions*

But is it actually the case that the vectors associated with positive variables in this problem form a basis? A basis we know will have $m + n - 1$ elements. In Figure 5.4 we observe there are $m + n - 2 = 9$ positive variables; thus the corresponding set of vectors would not constitute a basis. But something else occurred at that point in Figure 5.4. Some of the cells, such as cell $(4, 1)$, did not possess a loop; that is, the vector associated with the variable x_{41} could not be represented in terms of the nine given vectors. The reader should satisfy himself, though, that cells $(1, 1)$, $(1, 2)$, $(2, 3)$, $(3, 3)$, $(3, 4)$, $(3, 5)$, $(4, 5)$, $(4, 6)$, $(5, 6)$, and $(4, 1)$ do have their associated vector's comprising a linearly independent set.

It seems then that the solution in Figure 5.4 if we are on the right track regarding bases is a *degenerate* basic feasible solution. However, in this case we do not even have a basis, only a portion of one. Looking back at the simplex method, we can see that if we did not have a complete basis present then for some of the activity vectors we would have no y_{ij}; consequently, we could not compute z_j or $c_j - z_j$. Is this not the difficulty in Figure 5.4? There, we cannot evaluate some cells.

It may be shown that not only does the existence of a loop imply linear dependence, but also that linear dependence implies the existence of a loop. Consider then a feasible solution provided by the northwest-corner rule. By the nature of that method, no loop can exist among the positive cells provided. (The proof of this statement is quite straightforward and is called for in the problems.) Thus the vectors associated with positive cells in our first feasible solution are linearly independent. Of course, if they number less than $m + n - 1$, we do not have a legitimate basis. But when we attached an ϵ to a cell, we selected only those vectors which were not part of a loop involving positive cells or other ϵ cells; that is, we selected only those vectors which were not expressible as linear combinations of the existing linearly independent set.

---

†The -1's in the loop become $+1$'s in the expression of the new vector in terms of the others, as seen in the transition from (5.13) to (5.14).

Thus we begin iterations with a set of $m + n - 1$ linearly independent columns, some of whose associated variables are perhaps zero—in truth a possibly degenerate basic feasible solution. It should be clear that the sole purpose of the ϵ's is to allow us to identify a basis at each stage.

The fact that succeeding iterations merely implement the usual simplex operations assures that we have a basic feasible solution, perhaps a degenerate one, at *every* stage. Again, in the presence of degeneracy we need to locate ϵ cells.

5.4. Alternative Methods for Locating Initial Feasible Solutions

As was pointed out earlier, the northwest corner rule may provide an initial value of the objective function that is far from optimal. The same, of course, was true for initial basic feasible solutions found for more general linear programming problems. There, the expected difficulty of finding a *better* starting point was sufficient to make an easily-obtainable, but farther-from-optimal, point more desirable.

As one may observe from the nature of the transportation problem constraints, generating feasible solutions is not so difficult. Therefore, methods have evolved that generate first feasible solutions with only a nominal amount of time and effort in excess of that of the northwest-corner rule and which, simultaneously, produce much better initial solutions.

We discuss three such methods and apply the technique to generating a first feasible solution to problem (5.6).

i. *Row Minimum Method*

Let c_{1k} be the minimum element in the first row of the unit cost matrix. Then, if $a_1 \leq b_k$, we allocate a_1 units to D_k; that is, we let $x_{1k} = a_1$. If, on the other hand, $a_1 > b_k$, we allocate b_k units to D_k; that is, we let $x_{1k} = b_k$ and go on to the next smallest cost element in the first row, say c_{1l}. The process is repeated with $a_1 - b_k$ units available at O_1 and b_l units required at D_l, and so forth, until all units available at O_1 are exhausted. The same procedure is applied to all successive rows until all units at the origins are allocated to the destinations. Ties for row minimum may be broken by an arbitrary selection.

Example 5.1

Referring to problem (5.6) and applying the row minimum method, we generate the following initial feasible solution:

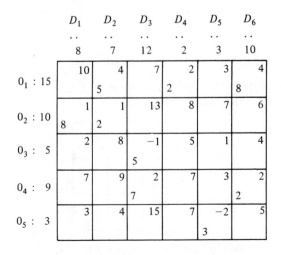

	D_1	D_2	D_3	D_4	D_5	D_6
	..	..	..	..	..	..
	8	7	12	2	3	10
$O_1 : 15$	10	4	7	2 ·2	3 ·3	4 ·10
$O_2 : 10$	1 ·8	1 ·2	13	8	7	6
$O_3 : 5$	2	8	−1 ·5	5	1	4
$O_4 : 9$	7	9 ·2	2 ·7	7	3	2
$O_5 : 3$	3	4 ·3	15	7	−2	5

The sequence of allocations to the cells is (1, 4), (1, 5), (1, 6), (2, 1), (2, 2), (3, 3), (4, 3), (4, 2), and (5, 2). The corresponding value of the objective function is $2 \times 2 + 3 \times 3 + 4 \times 10 + 1 \times 8 + 1 \times 2 - 1 \times 5 + 2 \times 7 + 9 \times 2 + 4 \times 3 = 102$.

ii. Column Minimum Method

A similar calculation involving the same steps may be used on each column in succession, until all units have been allocated.

Example 5.2

Referring again to problem (5.6), the application of the column minimum method generates the following initial feasible solution:

	D_1	D_2	D_3	D_4	D_5	D_6
	..	..	..	..	..	..
	8	7	12	2	3	10
$O_1 : 15$	10	4 ·5	7	2 ·2	3	4 ·8
$O_2 : 10$	1 ·8	1 ·2	13	8	7	6
$O_3 : 5$	2	8	−1 ·5	5	1	4
$O_4 : 9$	7	9	2 ·7	7	3	2 ·2
$O_5 : 3$	3	4	15	7	−2 ·3	5

The allocations were performed in the sequence $(2, 1), (2, 2), (1, 2), (3, 3),$ $(4, 3), (1, 4), (5, 5), (4, 6),$ and $(1, 6)$. The corresponding value of the objective function is $1 \times 8 + 1 \times 2 + 4 \times 5 - 1 \times 5 + 2 \times 7 + 2 \times 2 - 2 \times 3 + 2 \times 2 + 4 \times 8 = 73.$

iii. *Vogel's Approximation Method*

Although the column minimum procedure generated an optimal solution to problem (5.6), an indiscriminate use of low cost cells may force us into using some very high cost cells later in the process. Consider the following example.

Example 5.3

	D_1 : 7	D_2 : 7	D_3 : 7
O_1 : 5	1	5	6
O_2 : 6	4	1	7
O_3 : 10	4	7	20

The application of the row or column minimum method will yield the solution

	D_1 : 7	D_2 : 7	D_3 : 7
O_1 : 5	1 (5)	5	6
O_2 : 6	4 (6)	1	7
O_3 : 10	4 (2)	7 (1)	20 (7)

and the associated objective function value is $5 \times 1 + 6 \times 1 + 2 \times 4 + 1 \times 7 + 7 \times 20 = 166$. The feasible solution

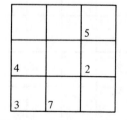

provides a value of the objective function equal to 121, which is much better and which utilizes *none* of the least-cost cells.

This example should not be construed to imply that smallest cost cells are to be avoided; rather, the lesson is that we should be interested in *differences* in cost. The penalty associated with having to utilize a cell of cost 6 rather than one of cost 1, in our Example 5.3, is not nearly so severe as having to utilize one of cost 20 rather than one of cost 4 or 7.

We describe a cost penalty method due to Vogel.

For row i compute $\Delta c_i = c_{ij'} - c_{ij''}$ where

$$c_{ij''} = \min_{j} \{c_{ij}\}$$

$$c_{ij'} = \min_{j \neq j''} \{c_{ij}\}$$

In words, for each row compute the absolute difference between the second smallest cost and the smallest cost.

Then for column $j, j = 1, \ldots, n$, compute a set of Δc_j in the same fashion; that is,

with
$$\Delta c_j = c_{i'j} - c_{i''j}$$

$$c_{i''j} = \min_{i} \{c_{ij}\}$$

$$c_{i'j} = \min_{i \neq i''} \{c_{ij}\}$$

Note that Δc_i and Δc_j may be zero.

The Δc_i, Δc_j that we compute are precisely penalties of the kind mentioned. They measure the penalty associated with eventually having to utilize the second lowest cost in a row or column if we choose to utilize the associated least cost. We shall make our allocation, then, to *the least-cost cell in the row or column whose associated penalty is largest*, thus decreasing, perhaps eliminating entirely, the number of units that may have to be assigned to the higher-cost cell later on. We assign as much as possible to the cell determined. (What is the procedure for a maximization problem?) The utilization of Vogel's approximation method is illustrated by an example.

Example 5.4

With regard to problem (5.6) again, and writing the Δc_i, Δc_j to the right and below the tableau, we get

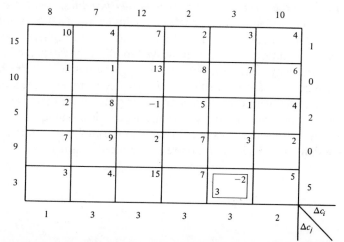

Since 5 is the largest penalty, we assign as much as possible to the least-cost cell in the fifth row, that is, 3 units to cell (5, 5). As in the northwest-corner method, every time we perform an allocation we either exhaust a supply, satisfy a demand, or, as in the present example, accomplish both. Thus, in computing penalties for the next allocation, we may shade (cross off) a row, a column, or both. In the present case, we eliminate both row 5 and column 5, and obtain a reduced cost matrix and new penalty figures as follows:

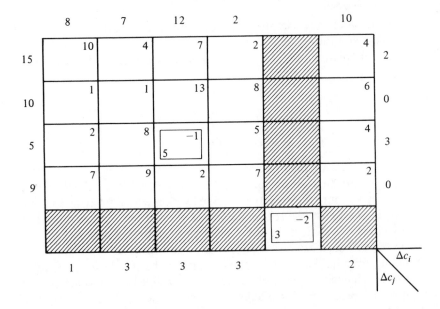

Now there is a four-way tie for the largest penalty, which is 3, and we may choose one at random, or, to attempt to be even more far-sighted, we may compute second-order penalties—differences between second smallest and third smallest costs—for those rows and columns having largest penalties, and choose the *largest of these second-order penalties.*

Ties among second order penalties could be resolved by third order penalties, and a valid question is whether pursuing these higher-order penalties produces a solution sufficiently improved to offset the expense of calculating them. One may, after all, be computing penalties with cells that will never be utilized. A reasonable rule of thumb would be to select the minimum-cost cell associated with the tied second-order penalties, if there is a tie after computing second order penalties. If that least cost is not unique, choose any of those. Doing that in the example results in the choice of column 3. Thus we next assign 5 to cell (3, 3) in the present reduced cost matrix.

Without further comment we illustrate the successive reduced arrays and the associated penalties. Ties were resolved by random choice.

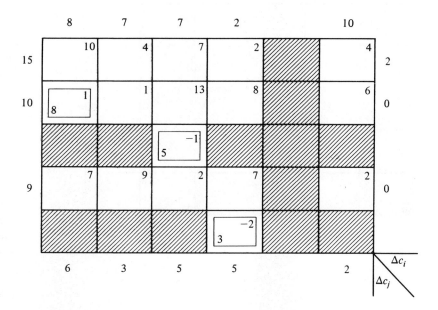

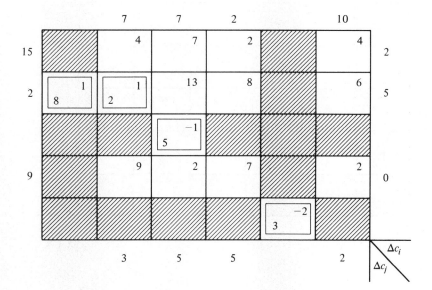

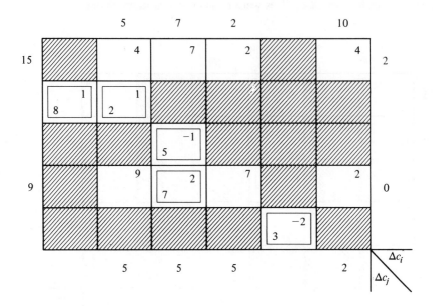

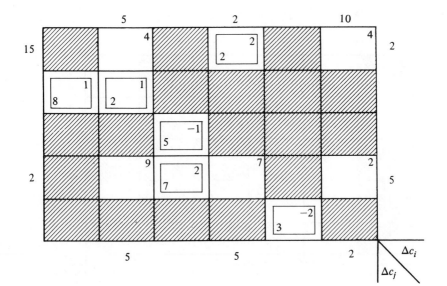

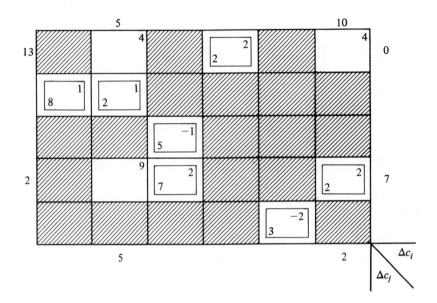

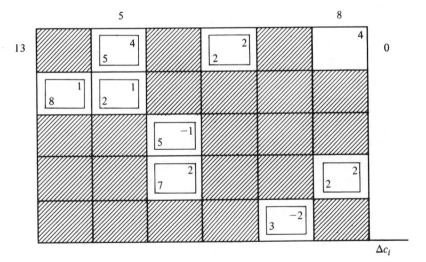

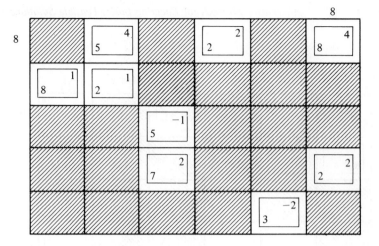

The feasible solution obtained from this tableau yields an objective function value of 73, which we may recall as being the minimum value of the objective function.

Is it the case that this alternative method of constructing a first feasible solution always produces a *basic* feasible solution? The answer, of course, is that it does, and utilizing the fact that at each stage either a supply is exhausted, a demand satisfied, or both, the reader will be able to establish that no loop can exist among the positive cells resulting. After a first (basic) feasible solution is obtained, the solution procedure continues exactly as before.

In relation to the remainder of the solution procedure, the most time-consuming operation is constructing the loops for evaluating the cells. For small problems, these operations may be carried out virtually by inspection; but for large arrays, they represent a considerable effort, and for computer solution a great deal of searching about is implicit.

We describe next a much simpler method of cell evaluation and later discover its theoretical foundation.

5.5. Duality in the Transportation Problem

For a particular solution of the transportation problem, we associate a variable u_i with row i of the array, $i = 1, \ldots, m$, and a variable v_j with column j of the array, $j = 1, \ldots, n$.

We seek to specify values for the u_i, v_j such that

$$u_i + v_j = c_{ij} \quad \text{for all } i, j, \text{ with } x_{ij} > 0 \qquad (5.16)$$

The equations (5.16) represent a system of at most $m + n - 1$ equations in $m + n$ variables. We shall always have a set of equations that is linearly independent. There will always be fewer equations than variables, since no more than $m + n - 1$ variables are ever positive. Therefore, some of the variables can be arbitrarily specified, and unique values obtained for the others (See Section 4-iii of Chapter 1). Since there will always be at least one such "free" variable, let us fix u_1 ($u_1 = 0$ is conventional) and observe that we may solve the system by inspection, working on one equation at a time, although we may need to fix some other variables.

Referring to the initial feasible solution of problem (5.6) (see Figure 5.5), we have 9 equations in 11 variables. Setting $u_1 = 0$, system (5.16) becomes

$$0 + v_1 = 10$$
$$0 + v_2 = 4$$
$$u_2 + v_3 = 13$$
$$u_3 + v_3 = -1$$
$$u_3 + v_4 = 5$$
$$u_3 + v_5 = 1$$
$$u_4 + v_5 = 3$$
$$u_4 + v_6 = 2$$
$$u_5 + v_6 = 5$$

From the first two equations, we get $v_1 = 10$ and $v_2 = 4$, respectively. No values are known for any remaining variables, so arbitrarily let $u_2 = 0$. Then we obtain successively, $v_3 = 13$, $u_3 = -14$, $v_4 = 19$, $v_5 = 15$, $u_4 = -12$, $v_6 = 14$, $u_5 = -9$, and the system is solved. Two variables have been arbitrarily specified, which could have been anticipated, since there were *two* more variables than equations.

The reader will find it more convenient and quite simple to write down the u_i and v_j in their appropriate places outside the tableau as they are discovered, by inspection, without writing down a system of equations. This is illustrated in the following array.

10	4	7	2	3	4	0
8 7	1	1	13	8	7	6	0
1	1 10	13	8	7	6	0
2	8 2	-1 2	5 1	1	4	-14 (5.17)
7	9	2	7 2	3 7	2	-12
3	4	15	7	-2	5 3	-9
10	4	13	19	15	14	u_i / v_j

The claim now is that for cells with $x_{ij} = 0$, the evaluator is simply $c_{ij} - (u_i + v_j)$, in other words we are claiming that $z_{ij} = u_i + v_j$. To see this let the cell being evaluated be the cell (i_1, j_1). Let the sequence of basis cells involved in the loop be (i_2, j_1), (i_2, j_2), (i_3, j_2), (i_3, j_3), ..., (i_k, j_k), and (i_1, j_k). The appropriate evaluator for cell (i_1, j_1) is then

$$c_{i_1 j_1} - c_{i_2 j_1} + c_{i_2 j_2} - c_{i_3 j_2} + c_{i_3 j_3} + \cdots + c_{i_k j_k} - c_{i_1 j_k}$$
$$= c_{i_1 j_1} - (u_{i_2} + v_{j_1}) + (u_{i_2} + v_{j_2}) - (u_{i_3} + v_{j_2})$$
$$\quad + (u_{i_3} + v_{j_3}) + \cdots + (u_{i_k} + v_{j_k}) - (u_{i_1} + v_{j_k})$$
$$= c_{i_1 j_1} - u_{i_1} - v_{j_1}$$
$$= c_{i_1 j_1} - (u_{i_1} + v_{j_1})$$

the desired result.

The reader will thus observe that one does obtain the same number as if he constructed a loop and added and subtracted costs around it. After locating the cell to become positive in this fashion, one constructs its loop and modifies the solution as before, but only one loop needs to be constructed during each iteration.

i. *Dual of the Transportation Problem*

Multiplying each constraint of our equality-constrained transportation problem (5.5) by -1, dropping the superfluous integer constraints, and choosing to treat a maximization problem, we obtain the equivalent problem:

$$\max -\sum_{i=1}^{m} \sum_{j=1}^{n} c_{ij} x_{ij}$$

subject to

$$
\begin{array}{llll}
-x_{11} - x_{12} - \cdots - x_{1n} & & & = -a_1 \\
 & -x_{21} - x_{22} - \cdots - x_{2n} & & = -a_2 \\
 & & \vdots & \vdots \\
 & & -x_{m1} - x_{m2} - \cdots - x_{mn} & = -a_m \\
-x_{11} & -x_{21} & \cdots \; -x_{m1} & = -b_1 \\
\quad -x_{12} & \quad -x_{22} & \cdots \quad -x_{m2} & = -b_2 \\
\quad\quad \vdots & \quad\quad \vdots & & \vdots \\
\quad\quad -x_{1n} & \quad\quad -x_{2n} & \cdots \quad -x_{mn} & = -b_n \\
& & & x_{ij} \geq 0
\end{array}
$$

Let us write the *dual* of this problem, where the dual variables are in order, $\hat{u}_1, \hat{u}_2, \ldots, \hat{u}_m, \hat{v}_1, \hat{v}_2, \ldots, \hat{v}_n$. The dual problem is

$$\max \sum_{i=1}^{m} a_i \hat{u}_i + \sum_{j=1}^{n} b_j \hat{v}_j$$

subject to

$$\hat{u}_1 + \hat{v}_1 \leq c_{11}$$
$$\hat{u}_1 + \hat{v}_2 \leq c_{12}$$
$$\vdots \qquad \vdots$$
$$\hat{u}_1 + \hat{v}_n \leq c_{1n}$$
$$\hat{u}_2 + \hat{v}_1 \leq c_{21}$$
$$\vdots \qquad \vdots$$
$$\hat{u}_i + \hat{v}_j \leq c_{ij}$$
$$\vdots \qquad \vdots$$
$$\hat{u}_m + \hat{v}_n \leq c_{mn}$$

$\hat{u}_i, \hat{v}_j$ unrestricted; $1 \leq i \leq m$, $1 \leq j \leq n$.

Clearly, the u_i, v_j we computed to evaluate $z_{ij} = u_i + v_j$ agree in number and in their unrestricted nature with the dual variables, $\hat{u}_i$ and $\hat{v}_j$. It is not, however, until we obtain an optimal solution to the primal that we realize a *feasible* solution to the dual; the *primal optimality conditions are exactly the dual inequality constraints.*

Now, recall we chose our u_i, v_j so as to satisfy the system of equations (5.16). How does that relate to the dual problem? It was indicated in Chapter 3 that when the (i_1)th, (i_2)th, (i_3)th, $\ldots$, (i_m)th primal variables are basic, then in the corresponding dual solution the (i_1)th, (i_2)th, $\ldots$, (i_m)th dual constraints are satisfied as strict equalities. It is this *complementary slackness* condition that was invoked to obtain the system (5.16).

In the case of the transportation problem, corresponding to our primal feasible solution will be an infinite number of solutions to (5.16).

Example 5.5

We illustrate the procedure using the initial solution generated by the northwest-corner rule in problem (5.6).

The u_i and v_j have already been computed in (5.17). The $z_{ij} = u_i + v_j$ are given by the following table, where the initial allocations to the cells are also entered:

10	4	13	19	15	14	u_i
8 / 7	4	13	19	15	14	0
10	4 (10)	13	19	15	14	0
−4	−10 (2)	−1 (2)	5 (1)	1	0	−14
−2	−8	1	7 (2)	3 (7)	2	−12
1	−5	4	10	6	5 (3)	−9
10	4	13	19	15	14	v_j

We now compute a table for the differences $c_{ij} - z_{ij}$:

0	0	−6	−17	−12	−10
8 / 7	0	−6	−17	−12	−10
−9	−3	0 (10)	−11	−8	−8
6	18	0 (2)	0 (2)	0 (1)	4
9	17	1	0 (2)	0 (7)	0
2	9	11	−3	−8	0 (3)

The least value of $c_{ij} - z_{ij}$ corresponds to cell $(1, 4)$, which identifies the best cell; however, since the solution is degenerate, there is no loop associated with cell $(1, 4)$. (Degeneracy does not preclude the existence of all loops.) Thus we may simply allow x_{14} to become ϵ, and we obtain a non-degenerate solution. For this new solution, the $z_{ij} = u_i + v_j$ are computed as before to yield the following table:

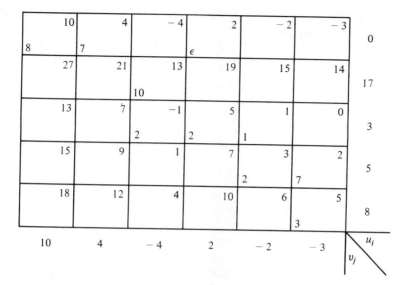

Again we compute a table for the difference $c_{ij} - z_{ij}$ and enter the allocations in the appropriate cells:

0	0	11	0	5	7
8	7		ε		
−26	−20 [10]	0	−11	−8	−8
−11	1 [2]	0 [2]	0 [1]	0	4
−8	0	1	0 [2]	0 [7]	0
−15	−8	11	−3	−8 [3]	0

The least value of $c_{ij} - z_{ij}$ corresponds to cell $(2, 1)$. The associated loop is $(2, 1)$, $(2, 3)$, $(3, 3)$, $(3, 4)$, $(1, 4)$, and $(1, 1)$. Variable x_{21} is introduced at a level of 2.

In practice, it is customary to generate a single tableau by computing for each cell directly the difference $c_{ij} - z_{ij}$, which are the evaluators. These are entered in the lower right-hand corner and circled; the evaluators are normally omitted for cells having allocations. The allocations are as usual entered in the lower left-hand corners and the unit costs are entered in the upper right-hand corners. This is illustrated for the previous computational step by the following tableau:

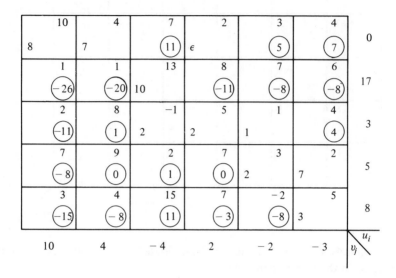

    The reader is encouraged to proceed in the same fashion and generate the optimal allocation for the problem.

5.6. Other Transportation Problems

i. *Transportation Problem with Excess Supply over Demand*

    In some realistic transportation problems it may not be necessary to exhaust all the supplies; that is, we need not ship all the items available in order to satisfy demand.

    Such a problem has a simple mathematical statement:

$$\min \sum_{i=1}^{m} \sum_{j=1}^{n} c_{ij} x_{ij}$$

subject to

$$\sum_{j=1}^{n} x_{ij} \leq a_i, \quad i = 1, \ldots, m \tag{5.18}$$

$$\sum_{i=1}^{m} x_{ij} = b_j, \quad j = 1, \ldots, n$$

$$x_{ij} \geq 0, \quad \text{all } i, j$$

Problem (5.18) has a feasible and hence optimal solution so long as

$$\sum_{i=1}^{m} a_i \geq \sum_{j=1}^{n} b_j$$

and it may be solved in the same fashion as the equality-constrained problem (see Exercise 5.3) by introducing an $(n + 1)$st destination with demand

$$\sum_{i=1}^{m} a_i - \sum_{j=1}^{n} b_j$$

and defining $c_{i,n+1} = 0$ for all i and

$$b_{n+1} = \sum_{i=1}^{m} a_i - \sum_{j=1}^{n} b_j \tag{5.19}$$

The problem may then be stated as

$$\min \sum_{i=1}^{m} \sum_{j=1}^{n+1} c_{ij} x_{ij}$$

subject to

$$\sum_{j=1}^{n+1} x_{ij} = a_i, \quad i = 1, \ldots, m \tag{5.20}$$

$$\sum_{i=1}^{m} x_{ij} = b_j, \quad j = 1, \ldots, n + 1$$

$$x_{ij} \geq 0, \quad \text{all } i, j$$

If in an optimal solution to (5.20) we find $x_{i,n+1} > 0$ for one or more i, then $x_{i,n+1}$ units will be retained at origin i and not shipped. While the strategy of defining the new destination seems intuitively valid, the formulation (5.20) assures it—we have again the problem structure of our simplest version.

ii. *Transportation Problem with Lower Bound Requirements at Destinations*

Another meaningful generalization occurs if it is not required that demands be met exactly; that is, the requirements may be phrased so that *at least* b_j units are required at destination j. The corresponding linear-programming problem is

$$\min \sum_{i=1}^{m} \sum_{j=1}^{n} c_{ij} x_{ij}$$

subject to

$$\sum_{j=1}^{n} x_{ij} = a_i, \quad i = 1, \ldots, m \qquad (5.21)$$

$$\sum_{i=1}^{m} x_{ij} \geq b_j, \quad j = 1, \ldots, n$$

$$x_{ij} \geq 0, \quad \text{all } i, j$$

Again, of course, it is necessary that

$$\sum_{i=1}^{m} a_i \geq \sum_{j=1}^{n} b_j$$

for a solution to exist to problem (5.21).

Suppose that we begin treating problem (5.21) as we would any linear-programming problem. Define surplus variables $x_{m+1,j}$ with $c_{m+1,j} = 0, j = 1, \ldots, n$, and obtain the problem

$$\min \sum_{i=1}^{m} \sum_{j=1}^{n} c_{ij} x_{ij}$$

subject to

$$\sum_{j=1}^{n} x_{ij} = a_i, \quad i = 1, \ldots, m \qquad (5.22)$$

$$\sum_{i=1}^{m} x_{ij} - x_{m+1,j} = b_j, \quad j = 1, \ldots, n$$

$$x_{ij} \geq 0, \quad i = 1, \ldots, m+1, j = 1, \ldots, n$$

Although (5.22) is an accurate representation of our problem, it lacks the essential structure of a transportation problem. However, if we annex the *redundant* constraint

$$-\sum_{j=1}^{n} x_{m+1,j} = a_{m+1} = \sum_{j=1}^{n} b_j - \sum_{i=1}^{m} a_i \leq 0 \qquad (5.23)$$

to problem (5.22), we obtain the proper structure and a transportation problem interpretation. This is illustrated by the following numerical problem.

	D_1	D_2	D_3	D_4	
	4	3	4	2	
$0_1 : 10$	5	2	1	3	(5.24)
$0_2 : 8$	4	-1	2	7	

Let us consider the transportation problem in which the least amounts required at the destinations $D_1, D_2, D_3,$ and D_4 are 4, 3, 4, and 2, respectively.

In this problem $m = 2$ and $n = 4$, and the matrix of coefficients corresponding to problem (5.22) is

$$
\begin{array}{cccccccccccc}
1 & 1 & 1 & 1 & 0 & 0 & 0 & 0 & 0 & 0 & 0 & 0 \\
0 & 0 & 0 & 0 & 1 & 1 & 1 & 1 & 0 & 0 & 0 & 0 \\
1 & 0 & 0 & 0 & 1 & 0 & 0 & 0 & -1 & 0 & 0 & 0 \\
0 & 1 & 0 & 0 & 0 & 1 & 0 & 0 & 0 & -1 & 0 & 0 \\
0 & 0 & 1 & 0 & 0 & 0 & 1 & 0 & 0 & 0 & -1 & 0 \\
0 & 0 & 0 & 1 & 0 & 0 & 0 & 1 & 0 & 0 & 0 & -1 \\
\end{array}
\qquad (5.25)
$$

Annexing the constraint (5.23) as a new third constraint to the problem (5.22) yields the matrix of coefficients

$$
\begin{array}{cccccccccccc}
1 & 1 & 1 & 1 & 0 & 0 & 0 & 0 & 0 & 0 & 0 & 0 \\
0 & 0 & 0 & 0 & 1 & 1 & 1 & 1 & 0 & 0 & 0 & 0 \\
0 & 0 & 0 & 0 & 0 & 0 & 0 & 0 & -1 & -1 & -1 & -1 \\
1 & 0 & 0 & 0 & 1 & 0 & 0 & 0 & -1 & 0 & 0 & 0 \\
0 & 1 & 0 & 0 & 0 & 1 & 0 & 0 & 0 & -1 & 0 & 0 \\
0 & 0 & 1 & 0 & 0 & 0 & 1 & 0 & 0 & 0 & -1 & 0 \\
0 & 0 & 0 & 1 & 0 & 0 & 0 & 1 & 0 & 0 & 0 & -1 \\
\end{array}
\qquad (5.26)
$$

which, disregarding the negatives, is precisely the array obtained for a simple transportation problem with $m = 3, n = 4$.

It occurs that (5.26) is unimodular, and in fact, it may be shown that the coefficient array for the general problem, (5.22) and (5.23), is unimodular. Thus the linear-programming problem (5.22) and (5.23) may be solved by the simplex method and an optimal solution in *integers* obtained. This also implies that the usual transportation algorithm can be successfully applied if one defines a new $(m + 1)$st origin such that its supply is the quantity

$$
\sum_{j=1}^{n} b_j - \sum_{i=1}^{m} a_i \le 0
$$

The unit costs $c_{m+1,j}$ are taken as zero for $j = 1, \ldots, n$, and, as in the former problems, we exhaust all supplies and satisfy all demands exactly.

Referring to problem (5.24), the annexation of a third origin O_3 will result in the following problem:

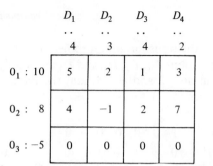

$$(5.27)$$

The application of the northwest-corner rule results in a first feasible solution:

5	2	1	3
4	3	3	
4	−1	2	7
		1	7
0	0	0	0
			5†

At this point one proceeds to evaluate the zero cells, and therein lies the *first* computational difference necessitated by the surplus variables. To understand the difference, let us write the primal and dual problems for the example.

Primal

$$\min 5x_{11} + 2x_{12} + x_{13} + 3x_{14} + 4x_{21} - x_{22} + 2x_{23} + 7x_{24}$$

subject to

$$x_{11} + x_{12} + x_{13} + x_{14} = 10$$
$$x_{21} + x_{22} + x_{23} + x_{24} = 8$$
$$-x_{31} - x_{32} - x_{33} - x_{34} = -5$$
$$x_{11} + x_{21} - x_{31} = 4$$
$$x_{12} + x_{22} - x_{32} = 3$$
$$x_{13} + x_{23} - x_{33} = 4$$
$$x_{14} + x_{24} - x_{34} = 2$$

$$x_{ij} \geq 0, \quad i = 1, 2, 3,$$
$$j = 1, 2, 3, 4$$

---

†The third constraint reads $-x_{31} - x_{32} - x_{33} - x_{34} = -5$, so if the cells are to hold the variable values, those in row 3 must remain positive.

Dual

$$\max 10u_1 + 8u_2 - 5u_3 + 4v_1 + 3v_2 + 4v_3 + 2v_4$$

subject to

$$u_1 + v_1 \geq 5$$
$$u_1 + v_2 \geq 2$$
$$u_1 + v_3 \geq 1$$
$$u_1 + v_4 \geq 3$$
$$u_2 + v_1 \geq 4$$
$$u_2 + v_2 \geq -1$$
$$u_2 + v_3 \geq 2$$
$$u_2 + v_4 \geq 7$$
$$-u_3 - v_1 \geq 0$$
$$-u_3 - v_2 \geq 0$$
$$-u_3 - v_3 \geq 0$$
$$-u_3 - v_4 \geq 0$$

$$u_i, v_j \text{ unrestricted}, \quad i = 1, 2, 3$$
$$j = 1, 2, 3, 4$$

The computational difference then is that whereas we previously had $u_i + v_j = z_{ij}$ in our computation of $c_{ij} - z_{ij}$, now, for any dummy origin k with negative supply, we have $-u_k - v_j = z_{kj}$ for all j.

A set of (u_i, v_j) for our first feasible solution, along with the evaluators, is then obtained as follows:

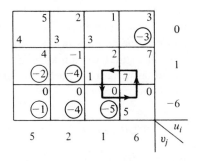

Thus, x_{33} is to become positive, and we next obtain

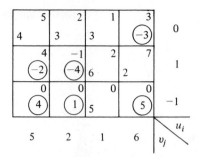

Our second computational difference is in obtaining a new solution, as just observed.

Denoting our corresponding activity vectors $\mathbf{a}_{ij}$, we would previously have had

$$\mathbf{a}_{33} - \mathbf{a}_{23} + \mathbf{a}_{24} - \mathbf{a}_{34} = 0$$

But in the present problem, all vectors $\mathbf{a}_{3j}$ are the *negatives* of what they previously were. Thus we presently have, indicating vectors in the present problem with primes,

$$-\mathbf{a}'_{33} - \mathbf{a}'_{23} + \mathbf{a}'_{24} + \mathbf{a}'_{34} = 0$$

or

$$\mathbf{a}'_{33} + \mathbf{a}'_{23} - \mathbf{a}'_{24} - \mathbf{a}'_{34} = 0$$

Thus, if we add to cell (3, 3), we also add to (2, 3), subtract from (2, 4), aud subtract from (3, 4). This is in agreement with what would be expected, since if something new becomes positive in row 3, one must also add the same amount into column 3, since the new positive row 3 number actually represents a subtraction. But then, having added into row 2 we must subtract from elsewhere in row 2, since it is a normal origin. Finally, having subtracted from column 4, it was necessary also to *subtract* from (3, 4) since the number in (3, 4) again, represented a negative amount. The rule for determining the initial value of the variable becoming positive is the same as before —it is the smallest value associated with −1's in the loop—in this case, the 5 in (3, 4).

We leave the completion of the example as an exercise. Clearly, we can combine both the variations—inequalities in the origin and destination constraints—and obtain a problem that we can solve by combining the tactics used for treating the individual variations. Such a problem is also included in the problems at the end of this chapter.

5.7. Transportation Problems with Transhipment

Frequently, the physical locations in a transportation problem cannot be dichotomized into two disjoint sets—one of origins, the other of destinations. Some, or all, of the locations may act simultaneously as origins and destinations. For example, in a distribution system it may occur that warehouse *i* ships some items to warehouse *j*, which in turn ships items to retail stores. Here, warehouse *j* acts as an origin *and* a destination. As another example, for reasons of efficiency in handling volumes of items, it may be cheaper, perhaps, for warehouse *i* to ship items to warehouse *j*, which then ships to retail store *k*, rather than for warehouse *i* to ship to store *k* directly. In both sample cases, warehouse *j* would be called a point of *transhipment*. Clearly, for similar kinds of reasons, store *k* could become a transhipment point, rather than a pure destination.

Thus, whereas we might have illustrated the transportation problem as the network of Figure 5.7, where directed lines indicate possible shipping routes, a problem in which transhipment is allowed might be better described in terms of the network of Figure 5.8, where directed segments indicate possible shipping routes, each point has associated a supply or a demand, and, in general, any point can both ship and receive goods.

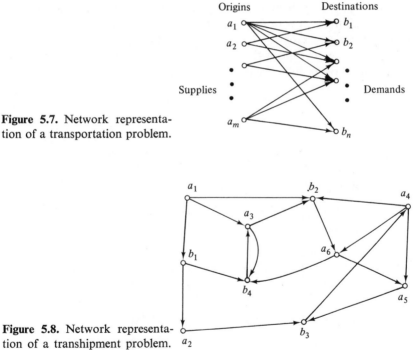

Figure 5.7. Network representation of a transportation problem.

Figure 5.8. Network representation of a transhipment problem.

Transhipment problems may be formulated and solved in terms of networks; however, we shall dwell here on the method we used for solving transportation problems.

i. *Model Involving a Single Destination as a Transhipment Point*

Let us suppose that we have a transportation problem in which exactly *one destination* acts as a point of transhipment; that is, it can ship to other destinations. We arbitrarily choose the nth destination as the transhipment point. We are given shipping costs per unit transhipped from this point, that is, costs for shipping from destination n to all other destinations.

Recall that although the nth destination is actually a transhipment point, that point retains its demand b_n; that is, there must be b_n units that reach and remain at the nth destination.

Define a new origin, the $(m+1)$st, with a supply $\sum_{i=1}^{m} a_i$, and redefine the demand b_n at destination n to be $b_n + \sum_{i=1}^{m} a_i$. $\sum_{i=1}^{m} a_i$ is chosen as the supply amount at the new origin and the amount by which b_n is increased *in order that the possible transhipment amounts be completely unconstrained.* That is, if it is most profitable to tranship everything, then we must allow that possibility. The transhipment cost from destination n to destination j is $c_{m+1,j}, j = 1, \ldots, n$, where we define $c_{m+1,n} = 0$, so that the cost of transhipping from destination n to itself is zero. The variable $x_{m+1,j}$ represents the amount *transhipped* from destination n to destination j, $j = 1, \ldots, n-1$, and $x_{m+1,n}$ is the amount of the commodity *not transhipped*. The equivalent transportation problem is displayed in Figure 5.9.

	D_1 b_1	D_2 b_2		D_n $b_n + \sum_{i=1}^{m} a_i$
$0_1 : a_1$	c_{11}	c_{12}	$\cdots$	c_{1n}
$0_2 : a_2$	c_{21}	c_{22}	$\cdots$	c_{2n}
$\cdot$	$\cdot$	$\cdot$	$\cdots$	$\cdot$
$0_m : a_m$	c_{m1}	c_{m2}	$\cdots$	c_{mn}
$0_{m+1} : \sum_{i=1}^{m} a_i$	$c_{m+1,1}$	$c_{m+1,2}$	$\cdots$	0

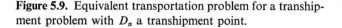

Figure 5.9. Equivalent transportation problem for a transhipment problem with D_n a transhipment point.

To illustrate, suppose that we are given the following supply and demand requirements array:

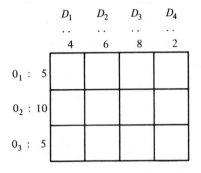

	D_1	D_2	D_3	D_4
	4	6	8	2
$0_1 : 5$				
$0_2 : 10$				
$0_3 : 5$				

Destination D_4 is assumed to be the transhipment point. By introducing a new origin O_4, we first obtain

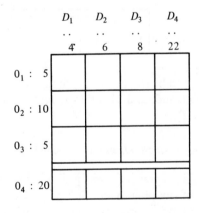

	D_1	D_2	D_3	D_4
	4	6	8	22
$0_1 : 5$				
$0_2 : 10$				
$0_3 : 5$				
$0_4 : 20$				

and total supply equals total demand. A particular feasible solution is then

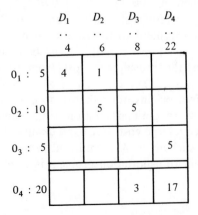

	D_1	D_2	D_3	D_4
	4	6	8	22
$0_1 : 5$	4	1		
$0_2 : 10$		5	5	
$0_3 : 5$				5
$0_4 : 20$			3	17

The interpretation of this solution, with respect to the new origin and the increased demand at destination 4, is that 5 units are being shipped to destination 4 and that 3 of these are *transhipped*, then, to destination 3. The fact that $x_{44} = 17$ implies, as stated previously, that there are 17 units not transhipped.

Assume now that the costs c_{4j}, $j = 1, 2, 3$, are such that an optimal solution is of the form

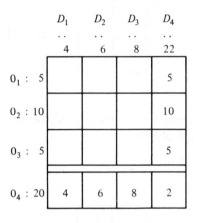

in which all 20 units are to be shipped to D_4 first, two of these retained at D_4 to satisfy D_4's own demand, and all other demands satisfied by transhipment of 18 units from D_4.

ii. *Model Involving a Single Origin as a Transhipment Point*

In the event we face a transhipment problem by virtue of one origin, say O_1, being capable of shipping to all other origins, an equivalent problem is obtained by defining a transportation problem in which the supply available at O_i is $a_i + a_1$, $i = 2, \ldots, m$, with O_1's supply remaining a_1. In addition, one defines $m - 1$ new destinations D_{n+k}, $k = 1, \ldots, m - 1$, each with demand a_1; thus D_{n+k} represents shipments from O_1 to O_{k+1}. The costs $c_{1,n+1}$, $c_{1,n+2}, \ldots, c_{1,n+m-1}$ correspond to the transhipment costs from O_1 to O_2, $O_3, \ldots, O_m$, respectively. For $i = 2, 3, \ldots, m$, the costs $c_{i,n+i-1}$ are defined to be zero, these being the transhipment costs from origins other than O_1 to themselves. Finally, $c_{i,n+j-1}$, $j = 2, \ldots, m$, $j \neq i$, $i = 2, \ldots, m$, are made *prohibitively large*, since origins other than 1 do not tranship; thus, for the minimization problem an optimal solution cannot find a positive variable corresponding to such a large cost. For hand computation a symbol can be used for these costs, such as the M from linear programming, or ∞, for example. The equivalent transportation problem is displayed in Figure 5.10.

	D_1 $\,b_1$	D_2 $\,b_2$		D_n $\,b_n$	D_{n+1} $\,a_1$	D_{n+2} $\,a_1$		D_{n+m-1} $\,a_1$
a_1	c_{11}	c_{12}	$\cdots$	c_{1n}	$c_{1,n+1}$	$c_{1,n+2}$	$\cdots$	$c_{1,n+m-1}$
$O_2 : a_2 + a_1$	c_{21}	c_{22}	$\cdots$	c_{2n}	0	∞	$\cdots$	∞
$O_3 : a_3 + a_1$	c_{31}	c_{32}	$\cdot\,\cdot$	c_{3n}	∞	0	$\cdots$	∞
$\vdots$	$\vdots$	$\vdots$	$\cdots$	$\vdots$	$\vdots$	$\vdots$	$\cdots$	$\vdots$
$O_m : a_m + a_1$	c_{m1}	c_{m2}	$\cdots$	c_{mn}	∞	∞	$\cdots$	0

Figure 5.10. Equivalent transportation problem for a transhipment problem with O_1 a transhipment point.

As an illustration, consider the following supply and demand requirements array

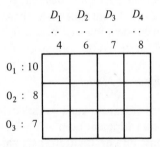

and imagine that O_1 may tranship to one or both of O_2, O_3. By introducing D_5 and D_6 as two destinations, each with a total requirement of 10 units, the equivalent transportation problem becomes

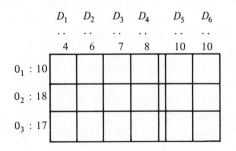

A feasible solution may be verified to be

	D_1	D_2	D_3	D_4	D_5	D_6
	..	..	..	..	..	..
	4	6	7	8	10	10
$O_1 : 10$	4					6
$O_2 : 18$			7	1	10	
$O_3 : 17$		6		7		4

The interpretation with regard to transhipment from O_1 is as follows: O_1 ships 4 of its 10 units to D_1 and tranships the remaining 6 to O_3 (D_6). O_2, receiving no units in addition to its own 8, distributes those among true destinations, D_3 and D_4. The 10 units entered as allocated from O_2 to D_5 were not actually received by O_2, so $x_{25} = 10$. O_3 received 6 additional units from O_1, and thus was able to allocate $6 + 7 = 13$ units among D_2 and D_4. The additional 4 units which it could have received were not, so $x_{36} = 4$. Notice that feasibility requires $x_{26} = x_{35} = 0$.

If we wish to choose a particular destination and allow it to tranship back to origins at perhaps different unit costs, then one obtains an equivalent transportation problem by defining one new origin and m new destinations with appropriate supplies, demands, and costs. This is left for the problems.

iii. *General Transhipment Model*

In the most general instance, each point, origin or destination, is allowed to be a possible transhipment point. That is, origins may ship to other origins or to destinations; and destinations may ship to other destinations or back to origins. A point retains its identity, origin or destination, only in that it is, respectively, a point at which a supply is originally located or a point at which some demand must ultimately be satisfied.

We have, effectively, then, $m + n$ origins each of which may ship items to any of $m + n - 1$ destinations. If some of these potential routes do not exist in a particular problem, the costs corresponding to such routes are made sufficiently undesirable that they will not be utilized. With the appropriate costs available, the supplies and demands in the new problem become as follows:

Each of the original m origins has its supply a_k increased by $\sum_{i=1}^{m} a_i$. Each of the n new origins is assigned a supply in the amount $\sum_{i=1}^{m} a_i$. Each of the m new destinations is assigned a demand $\sum_{i=1}^{m} a_i$. Each of the original n destinations now has demand $b_j + \sum_{i=1}^{m} a_i$. We have assumed that total supply equals total demand, since that may be achieved by using the appropriate artifices previously discussed in advance of the transhipment modifications.

5.8. Assignment Problem—Mathematical Formulation

The simplest assignment problem is one of the following form:

Suppose that there are n tasks which must be performed and n persons available to perform them. We assume that each person is capable of performing each of the tasks, but not necessarily equally well. It is supposed that a measure of the ith person's ($i = 1, 2, \ldots, n$) effectiveness in accomplishing task $j(j = 1, 2, \ldots, n)$ is given as some real number c_{ij}. The problem is to assign each person to one task such that each person is occupied, every task is accomplished, and such that the total effectiveness measure, *assumed* to be the sum of the individual measures, is maximized.

Suppose that we define

$$x_{ij} = \begin{array}{l} 1, \quad \text{if person } i \text{ is assigned to task } j \\ 0, \quad \text{otherwise} \end{array}$$

Then the objective is to

$$\max \sum_{i=1}^{n} \sum_{j=1}^{n} c_{ij} x_{ij}$$

That each person be assigned one task may be expressed

$$\sum_{j=1}^{n} x_{ij} = 1, \qquad i = 1, \ldots, n$$

That every task be performed may be expressed as

$$\sum_{i=1}^{n} x_{ij} = 1, \qquad j = 1, \ldots, n$$

Our assignment problem becomes

$$\max z = \sum_{i=1}^{n} \sum_{j=1}^{n} c_{ij} x_{ij}$$

subject to

$$\sum_{j=1}^{n} x_{ij} = 1, \quad i = 1, \ldots, n \tag{5.28}$$

$$\sum_{i=1}^{n} x_{ij} = 1, \quad j = 1, \ldots, n$$

$$x_{ij} = 0 \text{ or } 1, \quad \text{all } i, j$$

Similarly to the transportation problem, one may express the assignment problem in the form of a tableau consisting of an n-square cost array with unit availabilities and requirements. A feasible solution to the problem may then be expressed in the form of an n-square matrix $\mathbf{x}$ whose elements are the x_{ij}. Thus the matrix $\mathbf{x}$ has exactly one unit element corresponding to each row of the cost array and one unit element corresponding to each column of the cost array; all other elements are zero. An example of a feasible solution for a 4×4 assignment problem may appear as

$$\begin{pmatrix} 0 & 0 & 0 & 1 \\ 1 & 0 & 0 & 0 \\ 0 & 1 & 0 & 0 \\ 0 & 0 & 1 & 0 \end{pmatrix}$$

Although we have formulated a problem in which the number of tasks and the number of persons to perform them is identical, the reader will understand that meaningful assignment problems exist in which this condition is not satisfied. Suppose that there are m persons and n tasks, and $m \neq n$; one may still obtain a new problem, equivalent to the one given, in which there are exactly as many persons as tasks.

If $m > n$, we may require only that all tasks be performed. We introduce $m - n$ dummy tasks each of which may be performed by each person with zero effectiveness.

If $m < n$, we may require only that all persons be occupied. We then introduce $n - m$ dummy persons (in the sense of the preceding paragraph), each capable of performing every task with zero effectiveness.

In both cases, no possible additional contribution to the objective function was made. One obtains a square cost array and, aside from nonnegativity requirements, a set of equality constraints. Thus our presentation of the problem with $m = n$ does not actually exclude the more general cases.

i. *Conceptual Approach*

Each variable in the assignment problem may assume only the values zero or one, which results in a *zero–one programming problem*. This problem in addition, has linear constraints and linear objective function.

Without exception, zero–one programming problems are conceptually trivial to solve, nonlinearities included. If such a problem has n variables, there are at most 2^n feasible solutions, so all one need do is to generate 2^n binary vectors and substitute each in the constraints to check feasibility, discarding infeasible ones, while retaining always the one feasible vector which has provided the best objective function value up to that time. Regardless of the size of n and the number of constraints, an optimal solution will *always* be found, and in a finite number of steps. In the assignment problem it is further possible to combine the generation *and* feasibility checking of the 2^n possibilities by evaluating each of $n!$ possible ways of assigning persons to tasks.

The difficulty with this approach, however, is that "finite" can, at the same time, be quite large, so solution by enumeration of all possibilities becomes impractical even for moderate values of n. Efficient algorithms for solving zero–one programming problems do not as yet exist although a great deal of research effort is being given them. Therefore, one becomes interested in other solution procedures. (It should be pointed out however that as our computing machines become faster and faster, some brute force tactics that may have once been unthinkable become practical. Today's 100-hour problem may require 1 hour a year or two hence.)

ii. *Solution as a Transportation Problem*

We observe that the assignment problem is a transportation problem with $m = n$ and every supply and demand equal to 1. It is evident then that the transportation-problem algorithm could be used to obtain an optimal solution to the assignment problem.

But just as the transportation problem could be solved much more efficiently than as a normal linear-programming problem, so also is the assignment problem solvable in much more efficient fashion than as a transportation problem.

We should observe, of course, that the matrix of constraint coefficients in the assignment problem is unimodular. Thus, the simplex method would provide an optimal solution in zero-one variables, but we pursue more expeditious solution procedures.

iii. *Dual Solution for the Assignment Problem*

It is convenient to consider the objective to be maximization, in which case we can multiply each of our efficiencies, c_{ij}, by -1 and minimize the resulting linear form, obtaining the problem

$$\min z = \sum_{i=1}^{n} \sum_{j=1}^{n} -c_{ij}x_{ij}$$

subject to the constraints of (5.28)

If in the resulting array of costs we add a constant α_i to row i, $i = 1, \ldots, n$, and add a constant β_j to column j of the array, $j = 1, \ldots, n$, we obtain an array C' where $c'_{ij} = -c_{ij} + \alpha_i + \beta_j$.

Consider the assignment problem

$$\min z = \sum_{i=1}^{n} \sum_{j=1}^{n} c'_{ij} x_{ij}$$

subject to the constraints of (5.28)
or, in terms of c_{ij}, α_i, β_j

$$\min z = \left[-\sum_{i=1}^{n} \sum_{j=1}^{n} c_{ij} x_{ij} + \sum_{i=1}^{n} \sum_{j=1}^{n} \alpha_i x_{ij} + \sum_{i=1}^{n} \sum_{j=1}^{n} \beta_j x_{ij} \right]$$

subject to the constraints of (5.28)
But this last objective may be written

$$\min z = \left[-\sum_{i=1}^{n} \sum_{j=1}^{n} c_{ij} x_{ij} + \sum_{i=1}^{n} \alpha_i + \sum_{j=1}^{n} \beta_j \right] \qquad (5.29)$$

so if $\mathbf{x}^*$ is optimal for the problem with costs c'_{ij} then it is optimal for the identical statement (5.29) and consequently also for the given problem since the portion of (5.29) and the initial problem which differ is a constant.

We choose to specify α_i, β_j in a particular fashion,

$$\alpha_i = -\min_j \{-c_{ij}\}$$

We add the appropriate α_i and obtain an array $\bar{C}$ to which the β_j are added. The number $\beta_j = -\min_i \{\bar{c}_{ij}\}$. By definition, the resulting array C' has every element nonnegative, and has at least one zero in every row and at least one zero in every column.

To illustrate, let C be the given array of efficiencies (we allow negative efficiencies—it may be that man i always breaks the machine on task j, which is deemed worse than zero efficiency).

$$C = \begin{pmatrix} 1 & 3 & -1 & 4 \\ 5 & 7 & 9 & 1 \\ 3 & 8 & 1 & 0 \\ -1 & 5 & 8 & 3 \end{pmatrix}; \qquad -C = \begin{pmatrix} -1 & -3 & 1 & -4 \\ -5 & -7 & -9 & -1 \\ -3 & -8 & -1 & 0 \\ 1 & -5 & -8 & -3 \end{pmatrix}$$

$$\bar{C} = \begin{pmatrix} 3 & 1 & 5 & 0 \\ 4 & 2 & 0 & 8 \\ 5 & 0 & 7 & 8 \\ 9 & 3 & 0 & 5 \end{pmatrix}, \qquad C' = \begin{pmatrix} 0 & 1 & 5 & 0 \\ 1 & 2 & 0 & 8 \\ 2 & 0 & 7 & 8 \\ 6 & 3 & 0 & 5 \end{pmatrix}$$

In the problem with costs c'_{ij}, it is clear that whatever the minimum objective function value, it will never be less than zero, although the figure with costs c'_{ij} will not, in general, be the same as that with costs $-c_{ij}$, although an $\mathbf{x}^*$ which minimizes one will minimize the other.

With the primal problem (cost transformations having been performed, and the costs obtained denoted simply c_{ij})

$$\min z = \sum_{i=1}^{n} \sum_{j=1}^{n} c_{ij} x_{ij}$$

subject to

$$\sum_{i=1}^{n} x_{ij} = 1, \quad j = 1, \ldots, n \tag{5.30}$$

$$\sum_{j=1}^{n} x_{ij} = 1, \quad i = 1, \ldots, n$$

$$x_{ij} \geq 0$$

one obtains as the dual, where the dual variables corresponding to the first n constraints of (5.30) are labeled $u_1, \ldots, u_n$ and those corresponding to the next n are denoted $v_1 \ldots, v_n$

$$\max y = \sum_{i=1}^{n} u_i + \sum_{j=1}^{n} v_j$$

subject to $\hspace{6cm}$ (5.31)

$$u_i + v_j \leq c_{ij}, \quad i, j = 1, \ldots, n$$
$$u_i, v_j \text{ unrestricted}, \quad i, j = 1, \ldots, n$$

Suppose we find a feasible $\mathbf{x}$ for the primal problem and choose any vector, perhaps dual *infeasible*, $(\mathbf{u}, \mathbf{v})$. Then

$$z - y = \sum_{i=1}^{n} \sum_{j=1}^{n} c_{ij} x_{ij} - \sum_{i=1}^{n} u_i - \sum_{j=1}^{n} v_j$$

But if $\mathbf{x}$ is feasible then $\sum_{i=1}^{n} x_{ij} = 1$ for all $j = 1, \ldots, n$ and $\sum_{j=1}^{n} x_{ij} = 1$ for all $i = 1, \ldots, n$.

Consequently,

$$z - y = \sum_{i=1}^{n} \sum_{j=1}^{n} c_{ij} x_{ij} - \sum_{i=1}^{n} u_i \sum_{j=1}^{n} x_{ij} - \sum_{j=1}^{n} v_j \sum_{i=1}^{n} x_{ij}$$

$$= \sum_{i=1}^{n} \sum_{j=1}^{n} (c_{ij} - u_i - v_j) x_{ij}$$

$$= \sum_{i=1}^{n} \sum_{j=1}^{n} c'_{ij} x_{ij}$$

Owing to the unrestricted nature of u_i, v_j our arbitrary vector $(\mathbf{u}, \mathbf{v})$ is feasible if and only if the constraints of (5.31) are satisfied; that is iff

$$c'_{ij} \geq 0 \quad \text{all } i, j = 1, \ldots, n$$

Should it occur that $(\mathbf{u}, \mathbf{v})$ is feasible, then, we know

$$\sum_{i=1}^{n} \sum_{j=1}^{n} c'_{ij} x_{ij} \geq 0$$

for any feasible $\mathbf{x}$.

Suppose next that we find $(\mathbf{u}, \mathbf{v})$ and $\mathbf{x}$ such that

$$c'_{ij} x_{ij} = 0 \quad \text{all } i, j = 1, \ldots, n \tag{5.32}$$

If both $(\mathbf{u}, \mathbf{v})$ and $\mathbf{x}$ are, in addition, feasible, then *both are optimal* since (5.32) represents the difference $z - y$, which, if zero, with feasible vectors for both primal and dual, implies $z = y$, which in turn implies, from Theorem 3.3 again, that the vectors are optimal for their respective problems.

Thus a possible approach to the assignment problem is to attempt to locate feasible vectors $(\mathbf{u}, \mathbf{v})$, $\mathbf{x}$ such that (5.32) is satisfied.

Suppose we have a feasible $\mathbf{x}$ and *any* $(\mathbf{u}, \mathbf{v})$. Then

$$
\begin{aligned}
z &= \sum_{i=1}^{n} \sum_{j=1}^{n} c_{ij} x_{ij} = \sum_{i=1}^{n} \sum_{j=1}^{n} (c'_{ij} + u_i + v_j) x_{ij} \\
&= \sum_{i=1}^{n} \sum_{j=1}^{n} c'_{ij} x_{ij} + \sum_{i=1}^{n} u_i + \sum_{j=1}^{n} v_j
\end{aligned}
\tag{5.33}
$$

Now the last two terms of (5.33) are constant, since $(\mathbf{u}, \mathbf{v})$ is a particular vector. Thus, $\mathbf{x}^*$ minimizes the problem with costs c_{ij} if and only if it minimizes the problem with costs c'_{ij}.

If, in addition, we have a feasible $(\mathbf{u}^*, \mathbf{v}^*)$ that gives $c'_{ij} x_{ij} = 0$ all $i, j = 1, \ldots, n$ then equation (5.32) says

$$\min z = \sum_{i=1}^{n} \sum_{j=1}^{n} c'_{ij} x_{ij} = 0;$$

That is, the optimal objective function value for the problem with costs c'_{ij} is zero.

We also have, then,

$$\sum_{i=1}^{n} \sum_{j=1}^{n} c_{ij} x_{ij}^* = \sum_{i=1}^{n} u_i^* + \sum_{j=1}^{n} v_j^*$$

Recalling, then, that $\mathbf{x}^*$ has exactly one 1 corresponding to each row of the cost array and exactly one 1 corresponding to each column, with all other components zero, if we subtract u_i^* from every element in row $i, i = 1, \ldots, n$ and *then* subtract v_j^* from every element in column $j, j = 1, \ldots, n$, a set of costs c'_{ij} is obtained with

$$\sum_{i=1}^{n} \sum_{j=1}^{n} c'_{ij} x_{ij}^* = 0$$

iv. *Combinatorial Result*

Thus, as a starting point one obtains a set of pairs $\{(i, j)\}$ such that the costs c'_{ij} are zero. If one selects a subset N of these pairs, N is called *independent* if no two of the pairs correspond to the same row or to the same column of the array C'. As for the transportation problem, the terminology *independent* stems from the fact that the activity vectors corresponding to (i, j) in the assignment problem form a linearly independent set.

Consider the array

$$C' = \begin{pmatrix} 1 & 0 & 0 & 2 \\ 0 & 3 & 2 & 1 \\ 5 & 8 & 7 & 0 \\ 4 & 0 & 2 & 0 \end{pmatrix}$$

The set of all zero positions is

$$P = \{(1, 2), (1, 3), (2, 1), (3, 4,), (4, 2), (4, 4)\}$$

Thus

$$N_1 = \{(1, 2)\}$$
$$N_2 = \{(2, 1), (4, 2), (3, 4)\}$$
$$N_3 = \{(1, 3), (4, 4), (2, 1)\}$$

are all *independent* sets, whereas none of

$$N_4 = \{(1, 2), (4, 2), (4, 4)\}$$
$$N_5 = \{(3, 4), (4, 4), (1, 2)\}$$
$$N_6 = \{(4, 2), (4, 4), (2, 1), (1, 3)\}$$

are independent.

For the set P of *all* zero positions, consider the drawing of lines through rows and columns of the cost array, such that every element of P is covered by a line.

For example, with the present array one might accomplish the task in a variety of ways, as shown:

$$\begin{pmatrix} 1 & 0 & 0 & 2 \\ 0 & 3 & 2 & 1 \\ 5 & 8 & 7 & 0 \\ 4 & 0 & 2 & 0 \end{pmatrix} \qquad \begin{pmatrix} 1 & 0 & 0 & 2 \\ 0 & 3 & 2 & 1 \\ 5 & 8 & 7 & 0 \\ 4 & 0 & 2 & 0 \end{pmatrix}$$

$$\begin{pmatrix} 1 & 0 & 0 & 2 \\ 0 & 3 & 2 & 1 \\ 5 & 8 & 7 & 0 \\ 4 & 0 & 2 & 0 \end{pmatrix} \qquad \begin{pmatrix} 1 & 0 & 0 & 2 \\ 0 & 3 & 2 & 1 \\ 5 & 8 & 7 & 0 \\ 4 & 0 & 2 & 0 \end{pmatrix}$$

$$\begin{pmatrix} 1 & 0 & 0 & 2 \\ 0 & 3 & 2 & 1 \\ 5 & 8 & 7 & 0 \\ 4 & 0 & 2 & 0 \end{pmatrix} \qquad \begin{pmatrix} 1 & 0 & 0 & 2 \\ 0 & 3 & 2 & 1 \\ 5 & 8 & 7 & 0 \\ 4 & 0 & 2 & 0 \end{pmatrix}$$

Clearly there are many ways to accomplish that objective, just as there are many ways to select an independent subset of P.

Consider the two problems, then.

1. To find an independent subset of *maximum* size, that is, an independent subset such that no other independent subset has more elements.
2. To cover all the elements of P with the *smallest* possible number of lines; that is, no set of lines which covers all of P can have fewer elements.

For the present array, the solution to problem 1 is 4. The set $N = \{(1, 3), (2, 1), (3, 4), (4, 2)\}$ is independent, since all first elements of the pairs are distinct *and* all second elements of the pairs are distinct. Furthermore, N is maximal, since it has 4 elements and since the array is 4×4.

Solving problem 2 is also easy in this case, and we see that covering all elements of P requires at least four lines. The reason, of course, is that one line can cover one row *or* one column, and there are zeros in every row *and* in every column. Less than four lines, then, cannot possibly do the job. The result here is an example of a theorem [7] that states:

Theorem 5.1

The number of elements in a maximal independent set is equal to the number of elements in a minimal covering set.

v. *Iterative Solution Procedure for the Assignment Problem*

The solution strategy is an iterative one. Given a nonoptimal feasible solution, the existing costs c_{ij} are varied in such a way as to obtain an *equivalent problem* which makes possible an improved feasible solution to the original problem. The new costs generated will be such that if a feasible solution produces a total cost of zero with respect to those costs, that solution will be optimal.

The algorithm we present is probably the best-known algorithm for the assignment problem, and utilizes the dual of that problem. Given an array C', suppose that we solve the corresponding covering problem and find that a minimal cover has k elements.

If $k = n$, there are n independent positions available, and these determine a set of x_{ij}, which when set to unity, with all other x's zero, produce a feasible solution with objective function value zero–hence an optimal solution.

If $k < n$, we know we cannot find n independent positions, *and it is here that the duality results are invoked.*

First, construct any minimal cover. Select an element

$$c_0 = \min_{\substack{1 \le i \le n \\ 1 \le j \le n}} \{c'_{ij} \mid c'_{ij} \text{ is uncovered}\}$$

In other words, find the smallest element in C' that is not covered by some line in the minimal cover. Now, let c_0 be subtracted from every element in every *uncovered* row and then added to every element in every *covered* column, where a row or column is considered *covered* if and only if a line is drawn through it, so an uncovered row or column may contain covered elements. We obtain an array C''.

The modification of C' *looks* as that performed with the dual vectors $(\mathbf{u}, \mathbf{v})$ previously described, and may, in fact, be described in precisely those terms. *Define* a dual vector as follows

$$u_i = c_0 \quad \text{if row } i \text{ is uncovered}$$
$$u_i = 0 \quad \text{if row } i \text{ is covered}$$
$$v_j = -c_0 \quad \text{if column } j \text{ is covered}$$
$$v_j = 0 \quad \text{if column } j \text{ is uncovered}$$

(Is it feasible?)

The new array obtained, C'', is then described exactly as

$$c''_{ij} = c'_{ij} - u_i - v_j \tag{5.34}$$

As noted in Section 5.8-iii, the solution $\mathbf{x}^*$ is optimal for C'' iff it is optimal for C'.

It is also the case that C'' has every element greater than or equal to zero (so that a next iteration may be performed), which may be seen in light of the definition of $(\mathbf{u}, \mathbf{v})$ together with (5.34). Examining the *total* difference in the costs c'_{ij}, c''_{ij}, the following is true:

$$\sum_{i=1}^{n} \sum_{j=1}^{n} c'_{ij} - \sum_{i=1}^{n} \sum_{j=1}^{n} c''_{ij} = nc_0(n - k) \tag{5.35}$$

To see the validity of equation (5.35), let

$$r = \text{number of uncovered rows}$$
$$c = \text{number of covered columns}$$

Now the total decrease in array costs is nrc_0 and the total increase is ncc_0, so the difference, which is the left-hand side of equation (5.35), is

$$nrc_0 - ncc_0 = nc_0(r - c)$$

On the other hand, we have

$$k = c + (n - r)$$

or

$$r - c = n - k$$

which establishes equation (5.35).

With the new array, one can perform another iteration, that is, solve the covering problem, and if $k < n$, proceed with the selection of c_0, modify the costs, and so on.

However, we notice that the total array cost difference (5.35) is positive during every iteration. Since new arrays have every element nonnegative, there must be but a finite number of iterations during which $nc_0(n - k)$ is positive.

Thus, after a finite number of iterations, we must realize $n = k$, that is, be able to locate n independent positions each of current cost zero, which implies that the corresponding solution $\mathbf{x}^*$ is optimal for the current cost array and indeed for the original costs. The optimal objective function value may then be calculated from the array of initial costs.

This is the method of Egevary [3], or the Hungarian method.

Example 5.6

Let us illustrate the technique on a sample problem, being careful to realize that *for large cost arrays the solution of the covering problem and the location of a maximal independent set may not be trivial problems.*

Let the initial costs be

$$
\begin{array}{rrrrrr}
-1 & -2 & 3 & 4 & 5 & 1 \\
2 & -4 & 3 & 5 & 10 & -1 \\
4 & 3 & -6 & 7 & 8 & -1 \\
3 & 2 & 5 & 7 & 9 & 4 \\
2 & 4 & 0 & 3 & 7 & -1 \\
4 & -3 & 0 & 2 & 7 & 8
\end{array}
$$

with the objective to maximize

$$\sum_{i=1}^{6}\sum_{j=1}^{6} c_{ij}x_{ij}$$

Transforming the cost array as previously described an equivalent minimization problem is obtained, and the beginning array then is

2	4	0	0	0	4
4	11	5	4	0	11
0	2	12	0	0	9
2	4	2	1	0	5
1	0	5	3	0	8
0	8	6	5	1	0

where we have indicated a minimal cover and a maximal independent subset.

The minimal uncovered element is 1, so we subtract it from every uncovered row and add it to every covered column, obtaining

2	5	0	0	1	4
3	11	4	3	0	10
0	3	12	0	1	9
1	4	1	0	0	4
0	0	4	2	0	7
0	9	6	5	2	0

(The reader may prefer to think of subtracting from every uncovered element and adding to every twice-covered element.)

For this array, a minimal cover, one of which we have shown, has six elements. We also show a maximal independent set of zero positions. An optimal solution is

$$x_{13}^* = x_{25}^* = x_{31}^* = x_{44}^* = x_{52}^* = x_{66}^* = 1$$
$$x_{ij}^* = 0, \quad \text{all other } i, j$$

and the optimal objective function value is $3 + 10 + 4 + 7 + 4 + 8 = 36$.

Example 5.7

For another example, we imagine that work begins, after the usual sequence of cost transformations, on the array

$$
\begin{array}{cccccc}
2 & 3 & 0 & 0 & 3 & 2 \\
4 & 3 & 0 & 4 & 5 & 4 \\
5 & 1 & 0 & 2 & 6 & 2 \\
2 & 7 & 1 & 0 & 5 & 5 \\
0 & 6 & 4 & 8 & 0 & 0 \\
1 & 0 & 2 & 1 & 0 & 0
\end{array}
$$

whose zeros may be covered by a minimum of four lines, one such cover being shown.

The smallest uncovered element is 1, which is subtracted from each un-covered element, and added to each twice-covered element to obtain the array

$$
\begin{array}{cccccc}
1 & 2 & 0 & 0 & 2 & 1 \\
3 & 2 & 0 & 4 & 4 & 3 \\
4 & 0 & 0 & 2 & 5 & 1 \\
1 & 6 & 1 & 0 & 4 & 4 \\
0 & 6 & 5 & 9 & 0 & 0 \\
1 & 0 & 3 & 2 & 0 & 0
\end{array}
$$

This illustrates a minimal cover for the zeros in the array. A maximal independent set now has five elements, so another iteration is required. The minimum uncovered element is again 1, and the next array is

$$
\begin{array}{cccccc}
0 & 2 & 0 & 0 & 1 & 0 \\
2 & 2 & 0 & 4 & 3 & 2 \\
3 & 0 & 0 & 2 & 4 & 0 \\
0 & 6 & 1 & 0 & 3 & 3 \\
0 & 7 & 6 & 10 & 0 & 0 \\
1 & 1 & 4 & 3 & 0 & 0
\end{array}
$$

The zeros of this last array cannot be covered by five or fewer lines, so a maximal independent set has six elements and an optimal assignment problem solution is available. One such is

$$
x_{11}^* = x_{23}^* = x_{32}^* = x_{44}^* = x_{55}^* = x_{66}^* = 1
$$

and there are several others. The reader should determine four other optimal solutions and verify that all provide the same total cost.

vi. *Graph-Theoretic Approach to the Assignment Problem*

a. BIPARTITE GRAPHS

It is now our desire to view the assignment problem more in terms of graphical representations and to present an algorithm that treats the problem in terms of network flow problems.

The assignment problem, as well as the transportation problem, has a convenient representation in terms of graphs. The assignment problem may be illustrated in terms of a *bipartite graph,* that is, a graph G whose vertices V may be separated into two nonempty subsets V_1, V_2 ($V_1 \cap V_2 = \varnothing$, $V_1 \cup V_2 = V$) such that every element of the set E of G's edges is incident upon *exactly one* vertex $p_i \in V_1$ and *exactly one* vertex $q_j \in V_2$. For example the graph in Figure 5.11 is bipartite, whereas the graph in Figure 5.12 is not.

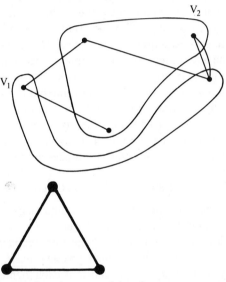

Figure 5.11. A bipartite graph.

Figure 5.12. A non-bipartite graph.

In particular, let there be m vertices in V_1 and let these correspond to persons. Let the set V_2 correspond to the n tasks. Here we require $m \leq n$, and if we choose to disallow unfilled tasks and/or unoccupied persons, we will have $m = n$. In case a given problem had $n < m$, we could simply invert the identities of V_1, V_2. Next construct one edge incident on (p_i, q_j) for each $(p_i, q_j) \in V_1 \times V_2$. Denote this set of edges by E.

Specifying, then, a set of edges $E_1 \subseteq E$ is equivalent to assigning persons to tasks. Unless some care is exercised in selecting E_1, one might assign more than one person to the same task, one person to several tasks, and so on. To generate a feasible assignment, each element of E_1 must be incident upon a different element of V_1. For three persons and five tasks, when each person

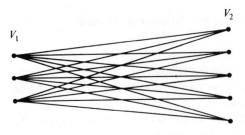

V_1

V_2

Figure 5.13. Bipartite graph representation of a 3 person 5 task assignment problem.

is capable of performing any task, the bipartite graph would be represented as in Figure 5.13.

b. GRAPH-THEORETIC STATEMENT OF THE PROBLEM

With regard to the general graph G of the variety in Figure 5.13 with m elements in V_1 and n in V_2, the problem may be characterized as the selection of a set M of edges of G such that no vertex lies on more than one element of M. In graph-theoretic terms the set M is called a *matching* of G. Furthermore, in the assignment problem context, the matching must be maximal; that is, with the property that *if any other edge is added to M to form a set M', M' is not a matching*. For if M is not maximal, then it must be possible to add an edge to M, obtaining M' that is a matching, and this is equivalent to locating a person and a task not already involved. Furthermore, the "cost" of the matching is to be minimized. Thus the assignment problem consists of finding a maximal matching that minimizes cost in a bipartite graph.

c. STATEMENT AS A NETWORK FLOW PROBLEM

The transition to a capacitated network flow problem (Chapter 4) is then straightforward. Consider each element of V_1 as a source with one unit available. Let each edge of E have unit capacity and the cost per unit flow the c_{ij} as given in the assignment problem itself. Adjoin a vertex q_0 and edges E' $= (q_j, q_0)$, all $q_j \in V_2$, such that each element of E' has unit capacity. The terminal vertex q_0 is a sink with demand m. As usual, to obtain a network with a single source, we also add a source vertex p_0 with m units available and edges $E'' = (p_0, p_i)$, all $p_i \in V_1$, each element of which has unit capacity, and E', E'' have zero costs per unit flow.

Operating on the graph of Figure 5.13 in this fashion, the modified graph appears as in Figure 5.14. Clearly, finding a feasible flow that satisfies the demand at q_0 is equivalent to finding a feasible solution to the assignment problem.

Now a matching problem can be solved by any algorithm that maximizes flow in a graph. We shall utilize the labeling algorithm, first mentioned in Chapter 4, in its tabular form, which is an efficient procedure for solving matching problems. The assignment problem itself will be solved by solving a sequence of matching problems.

Figure 5.14. Capacitated network flow representation of a 3 person 5 task assignment problem with source and sink added.

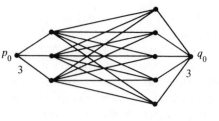

d. Use of the Labeling Algorithm

For the general graph with source x_s, sink x_t, other vertices $x_1, x_2, \ldots,$ x_l, and α'_{ik} the capacity remaining in edge (x_i, x_k) at any time, the labeling algorithm proceeds as follows.

1. Label x_s with $(\ , \infty)$.
2. Examine any vertex x_i that is already labeled (x_j, f_i). Label every vertex x_k not already labeled, and such that an edge lies on (x_i, x_k) and $\alpha'_{ik} > 0$ with the label $(x_i, \min(f_i, \alpha'_{ik}))$. If x_k could have several different labels, select one.
3. Continue labeling vertices until either x_t is labeled or until no other vertex can be labeled.
4. If labeling must stop and x_t is unlabeled, the existing flow is optimal.
5. If x_t is labeled, start at x_s and determine a sequence $x_s, x_a, x_b, \ldots,$ x_t *in which each vertex is the label of the following vertex*. In every edge of the path increase the algebraic flow by f_t and reduce the remaining capacities in each edge by the amount f_t. Erase labels and go to 1.

In words, this algorithm simply provides a systematic means for discovering new paths from source to sink over which flow may be increased. When no new paths can be found (x_t cannot be labeled), the flow is optimal. An edge whose capacity has been exhausted is called *saturated*.

Now referring to the assignment problem, we observe that the corresponding bipartite graph will have no loops and no parallel edges. Also, excluding p_0, q_0, our labels will be such that only q's can label p vertices and only p's can label q vertices. When applying the labeling algorithm in its tabular form to networks of this kind, one deals with arrays of *modified* capacities of the form

		V_1	V_2
V_1	p_1 p_2 $\vdots$ p_m	O	A_1
V_2	q_1 q_2 $\vdots$ q_n	A_2	O

(5.36)

where $V_1 = \{p_1, p_2, \ldots, p_m\}$ is, again, the set of vertices corresponding to persons, and $V_2 = \{q_1, q_2, \ldots, q_n\}$ the set of vertices corresponding to tasks.

Although our edges are not oriented, the portion of any of our tableaux for flows or capacities associated with the A_2 block of (5.36) can be neglected, for any capacities or flows taking place from V_2 to V_1 can be taken into account, and only *net flows and capacities* in the V_1 to V_2 direction maintained. Thus both current flow and remaining capacity can be kept in a single $m \times n$ array.

If we let for $i = 1, 2, \ldots, m$ and $j = 1, 2, \ldots, n$

α'_{ij} = remaining capacities at the moment

a'_i = remaining availabilities ($a'_i = 0$ or 1)

b'_j = remaining demands ($b'_j = 0$ or 1)

x_{ij} = units of flow from p_i to q_j ($x_{ij} = 0$ or 1) the $m \times n$ array will

appear as

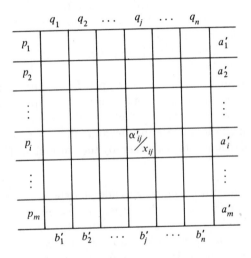

In the tabular form for the general network, labeling a vertex corresponds to labeling a row of a matrix. Owing to the nature of bipartite networks, labeling a vertex corresponds to labeling a particular row or column of a matrix. Furthermore, in our situation one may dispense with the numbers f_i, since the capacity of every edge is simply 1 in the p_0 to q_0 direction.

e. SOLVING A MATCHING PROBLEM

Let us first examine the application of the labeling algorithm to a matching problem. Let our bipartite graph with single source and sink be as in Figure 5.15, in which the capacity of each edge is 1, and a flow of $+1$ from left to right is allowed, or a flow of -1 from right to left.

Suppose that we apply the algorithm and consider the situation when we

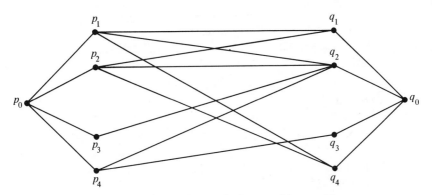

Figure 5.15. Graph of a particular matching problem.

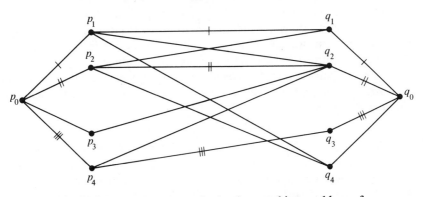

Figure 5.16. A flow of 3 units in the matching problem of Figure 5.15.

have realized a flow of 3 units from p_0 to q_0 through the sequences of vertices $p_0 p_1 q_1 q_0$, $p_0 p_2 q_2 q_0$, and $p_0 p_4 q_3 q_0$. This is illustrated in Figure 5.16.

Let us recall that if an edge is saturated in the positive direction, as for example (p_1, q_1) is in Figure 5.16, then it has a *positive* capacity in the negative direction; thus, assuming q_1 has already a positive second component in its label, p_1 *could be labeled from* q_1. Having a positive capacity in the negative direction is to say that the flow in the positive direction could be reduced. Similarly, we note in Figure 5.16 that p_2 could *not* be labeled from q_1.

Proceeding from the situation in Figure 5.16, we can label p_3 with $(p_0, 1)$, q_2 with $(p_3, 1)$, p_2 with $(q_2, 1)$, q_4 with $(p_2, 1)$, and q_0 with $(q_4, 1)$. Thus the flow has been increased by 1 unit, and we know immediately that we have a maximal matching, since 4 is the maximum possible flow. In terms of flow, the sequence of vertices labeled indicates an increase in flow by 1 in (p_0, p_3), an increase in flow by 1 in (p_3, q_2), a decrease in flow by 1 in (p_2, q_2), an increase in flow by 1 in (p_2, q_4), and an increase in flow by 1 in (q_4, q_0).

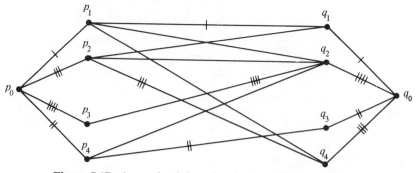

Figure 5.17. An optimal flow for the matching problem of Figure 5.15.

In terms of the network, the optimal situation is displayed in Figure 5.17, and a maximal matching is (p_1, q_1), (p_2, q_4), (p_3, q_2), (p_4, q_3).

If we carefully consider the application of this labeling algorithm to a matching problem, we can observe certain properties of the procedure that allow a quite simple *tabular* solution of the same problem.

Suppose that we have discovered a vertex sequence which allows a unit increase in flow (no greater increase will be possible, of course). The sequence will have the form

$$p_0, p_i, q_j, p_k, \ldots, q_z, q_0$$

The edges (p_0, p_i), (q_z, q_0) will be unsaturated, and remaining successive edges will always alternate between being unsaturated and being saturated. Thus (p_i, q_j) will be unsaturated, (q_j, p_k) will be saturated, and so on. Furthermore, any sequence with these properties allows a unit increase in flow.

Let us then represent the network in tabular form—a row for each p_i, a column for each q_j, and an X in any square (i, j) corresponding to *the absence of an edge* between p_i and q_j; empty squares denote the existence of an edge incident on the corresponding p_i and q_j. With respect to the graph of Figure 5.15, we have

	q_1	q_2	q_3	q_4
p_1			X	
p_2			X	
p_3	X		X	X
p_4	X			X

Unit flows in edges are indicated by a 1 in the appropriate square (i, j); for example, the flow of Figure 5.16 would be represented as

	q_1	q_2	q_3	q_4
P_1	1		X	
P_2		1	X	
P_3	X		X	X
P_4	X		1	X

$$(5.37)$$

Locating an appropriate sequence of vertices in the graph has, then, a corresponding meaning in terms of the table; one seeks to find a sequence of squares beginning in an "unsaturated" row, ending in an "unsaturated" column, and otherwise alternating as follows—a saturated square in the same column as the first square—then an unsaturated square in the same row as the second square—then a saturated square in the same column as the third square—then an unsaturated square in the same row as the fourth square—and so on. The reader should convince himself that this is equivalent to finding a vertex and, consequently, edge sequence of the desired kind in the graph.

Returning to table (5.37), the final step in the given matching problem of Figure 5.15 is where we have numbered the squares in the order of their occurrence in the sequence. To obtain the new flow, of course, one adds 1 to the square $(3, 2)$, subtracts 1 from square $(2, 2)$, and adds 1 to square $(2, 4)$.

	q_1	q_2	q_3	q_4
P_1	1		X	
P_2		②₁	X	③
P_3	X	①	X	X
P_4	X		1	X

Finally, the optimal flow in tabular form becomes

	q_1	q_2	q_3	q_4
p_1	1		X	
p_2			X	1
p_3	X	1	X	X
p_4	X		1	X

The finding of a flow-increasing vertex (edge) sequence or the discovery that none exists *can be systematized by a labeling approach in the tableau.* (Remember that squares entered with an X are considered removed from any consideration.) We proceed by labeling all unsaturated rows with the label p_0. Suppose that row i is labeled. Then label with p_i every column j which is *not* already labeled and such that square (i, j) has *no* flow. If column j is labeled, attach the label q_j to every row i not already labeled such that square (i, j) *does* have flow.

Labeling stops when an unsaturated column is labeled, in which case a flow-increasing sequence can be found, or when the former has not occurred, but no further labeling is possible, in which case the flow is maximal.

For an example, returning to (5.37), we first label row 3 with p_0. Next, column 2 is labeled p_3; next, row 2 is labeled q_2; next, column 1 is labeled p_2, and column 4 is labeled p_2. At this point the process stops, since an unsaturated column has been labeled. The result is

	q_1	q_2	q_3	q_4	
p_1	1		X		
p_2		1	X		(q_2)
p_3	X		X	X	(p_0)
p_4	X		1	X	
	(p_2)	(p_3)		(p_2)	

(5.38)

A sequence of the desired kind may then be discovered from this tableau. For example, in (5.38) we begin with a row labeled p_0 and we find the only one to be row 3. We then seek a *column* labeled p_3, and note the squares (there may be several) at the intersection of row 3 and columns thus labeled. Having identified such a square in column 2, that is, (3, 2), we ask if there is a row labeled q_2, which there is, row 2. The intersection of row 2 and column 2 thus gives another square, (2, 2). Next, we search for columns labeled p_2, of which there are two, and one of these, column 4, is unsaturated, which defines a third square (2, 4). The tableau (5.39) illustrates the squares discovered and the order of their discovery.

	q_1	q_2	q_3	q_4	
p_1	1		X		
p_2		②₁	X	③	(q_2)
p_3	X	①	X	X	(p_0)
p_4	X		1	X	
	(p_2)	(p_3)		(p_2)	

$$(5.39)$$

The new (optimal) flow is obtained by the addition of ± 1 to the squares, the first being plus, the second minus, the third plus, and so on. The new flow obtained is exhibited in (5.40) and is seen to be the same as previously found:

	q_1	q_2	q_3	q_4
p_1	1		X	
p_2			X	1
p_3	X	1	X	X
p_4	X		1	X

$$(5.40)$$

Having now a flow of value 4, it is known that an optimal flow is at hand.

f. ANOTHER ALGORITHM FOR THE ASSIGNMENT PROBLEM

We now present another algorithm for the assignment problem. The present algorithm will provide an optimal solution by solving a sequence of matching problems. (Our first algorithm accomplished the same—each maximal matching was obtained by finding a minimal cover. The covering problems may be nontrivial to solve, and no systematic procedure for their solution was given.) It differs from the first algorithm only in that the matching problems are solved via the network flow algorithm, which *is* systematic and applicable regardless of the size of the array. The present algorithm is otherwise the same: Duality results are invoked to modify costs, and the sequence of matching problems solved ends with a matching of n elements and a modified total cost of zero, at which time the assignment problem has been solved. We shall deal entirely in terms of tabular forms, which will be $n \times n$ arrays, since, as we have seen, the interjection of dummy tasks and/or dummy persons does not alter the problem.

Let us assume that the c_{ij} are measures of profit such that the objective is to *maximize* the total profit. As shown before, we can obtain an equivalent cost array with at least one zero in every row and at least one zero in every column. Considering the squares in the tableau, a solution to the assignment problem is feasible if and only if it utilizes exactly one square in each row and exactly one square in each column [a square (i, j) being utilized meaning that we assign person i to task j].

Consider then the following approach. From the array of revised but equivalent c_{ij} *solve the matching problem corresponding to the graph represented by only those squares with zero cost.* If the maximal matching has n edges, then the assignment problem has been solved, since it represents an assignment of cost zero, which is known in advance to be the best possible. If there are less than n edges in a maximal matching, a different matching problem has to be solved, and so on until one with n edges is found. We illustrate the procedure by an example.

Example 5.8

For a five-person five-task problem, let a set of initial profits be as follows

$$
\begin{array}{ccccc}
2 & 3 & 1 & 4 & 6 \\
1 & 2 & 6 & 5 & 7 \\
8 & 4 & 1 & 3 & 9 \\
2 & 7 & 8 & 1 & 2 \\
5 & 5 & 4 & 5 & 6 \\
\end{array}
$$

The usual cost modifications yield

$$
\begin{array}{ccccc}
3 & 2 & 5 & 1 & 0 \\
5 & 4 & 1 & 1 & 0 \\
0 & 4 & 8 & 5 & 0 \\
5 & 0 & 0 & 6 & 6 \\
0 & 0 & 2 & 0 & 0
\end{array}
$$

The zeros then correspond to edges in the bipartite graph in Figure 5.18, there being an edge on (p_i, q_j) if square (i, j) has a zero.

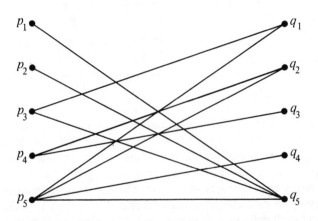

Figure 5.18. Graph of initial matching problem of Example 5.8.

We may infer immediately that there will not be five edges in a maximal matching for the graph in Figure 5.18. This is possible, though not essential to our solving the problem, because it may be shown that if a set of m vertices V_1 in a bipartite graph can be matched in the set of n vertices V_2 it is necessary that the total number of vertices of V_2 which are adjacent† to vertices in any subset V'_1 of V_1 must at least equal the number of vertices in V'_1.

In Figure 5.18 consider $V'_1 = (p_1, p_2) \subset V_1$. There is only one vertex of V_2, that is, q_5, which is adjacent to vertices of V'_1. Hence V_1 cannot be matched in V_2, which is to say that a maximal matching there cannot contain five edges.

---

†Two vertices are *adjacent* if they lie on the same edge.

To find a maximal matching for the graph in Figure 5.18, we apply the labeling algorithm, having already found the flow represented by

	q_1	q_2	q_3	q_4	q_5	
p_1	X	X	X	X	1	(q_5)
p_2	X	X	X	X		(p_0)
p_3	1	X	X	X		
p_4	X	1		X	X	
p_5			X	1		

(p_2)

From the previous remarks on maximal matchings, we know the above flow must be optimal, since it has the value 4 and there can be no flow of value 5. At any rate, the labeling procedure ends as shown, no unsaturated column is labeled, and the flow is optimal.

The next step in the algorithm is to consult the tableau again and observe which rows and columns have been labeled. Mark each row that was *not* labeled and each column that *was* labeled. Cost elements in marked rows or marked columns are considered to be marked; an element in both a marked row *and* a marked column is *doubly marked*; all other elements are *unmarked*. Then locate the smallest unmarked element, *subtract* it from each *unmarked* element, and *add* it to every *doubly marked* element.

g. Correspondence of Cost Modification to the First Algorithm

The preceding cost modification is *exactly* that employed in the first algorithm. This may be seen if it is established that a *covered* row (column) in the first algorithm is the same as a *marked* row (column) in the second algorithm. Alternatively, from the method of marking in the second algorithm, we have to show that a covered row in the first algorithm corresponds to an *unlabeled* row in the second algorithm and that a covered column corresponds to a labeled column. In other words, the set of unlabeled rows and labeled columns represents a minimal cover.

Let us show that they do comprise a cover. If we suppose that they do not, then, there must be a zero cost cell at the intersection of a *labeled row* and an *unlabeled column*. Our hypothetical cell must have flow in it (a 1), for otherwise the column would be labeled. Thus the row in question is saturated, and was not labeled as a result of being unsaturated. Consequently, it must have been labeled as a result of a labeled column intersecting it at a cell *with* flow. However, this implies that the row has more than a single unit flow, which it does not. Therefore, the set of unlabeled rows and labeled columns constitutes a cover. The proof that the cover is minimal is left as an exercise.

Example 5.9

Consider again the problem of Example 5.6. The initial array of non-negative costs with zeros in every row and column is

$$
\begin{array}{cccccc}
2 & 4 & 0 & 0 & 0 & 4 \\
4 & 11 & 5 & 4 & 0 & 11 \\
0 & 2 & 12 & 0 & 0 & 9 \\
2 & 4 & 2 & 1 & 0 & 5 \\
1 & 0 & 5 & 3 & 0 & 8 \\
0 & 8 & 6 & 5 & 1 & 0
\end{array}
$$

We shall use the labeling algorithm to find an initial maximal matching.

An initial labeling appears in (5.41), where all rows are unsaturated, and where blank squares are those eligible for flows (1's).

	q_1	q_2	q_3	q_4	q_5	q_6	
p_1	X	X				X	(p_0)
p_2	X	X	X	X		X	(p_0)
p_3		X	X			X	(p_0)
p_4	X	X	X	X		X	(p_0)
p_5	X		X	X		X	(p_0)
p_6		X	X	X	X		(p_0)

(p_1)

(5.41)

An unsaturated column is labeled, so the first labeling halts and a second labeling, without any cost modification, is

	q_1	q_2	q_3	q_4	q_5	q_6	
p_1	X	X	1			X	
p_2	X	X	X	X		X	(p_0)
p_3		X	X			X	(p_0)
p_4	X	X	X	X		X	(p_0)
p_5	X		X	X		X	(p_0)
p_6		X	X	X	X		(p_0)

(p_2)

The third, fourth, and fifth labelings are

	q_1	q_2	q_3	q_4	q_5	q_6	
p_1	X	X	1			X	
p_2	X	X	X	X	1	X	
p_3		X	X			X	(p_0)
p_4	X	X	X	X		X	(p_0)
p_5	X		X	X		X	(p_0)
p_6		X	X	X	X		(p_0)

(p_6)

	q_1	q_2	q_3	q_4	q_5	q_6	
p_1	X	X	1			X	
p_2	X	X	X	X	1	X	
p_3		X	X			X	(p_0)
p_4	X	X	X	X		X	(p_0)
p_5	X		X	X		X	(p_0)
p_6		X	X	X	X	1	

(p_3)

	q_1	q_2	q_3	q_4	q_5	q_6	
p_1	X	X	1			X	
p_2	X	X	X	X	1	X	
p_3	1	X	X			X	
p_4	X	X	X	X		X	(p_0)
p_5	X		X	X		X	(p_0)
p_6		X	X	X	X	1	

(p_5)

In the final labeling

	q_1	q_2	q_3	q_4	q_5	q_6	
p_1	X	X	1			X	
p_2	X	X	X	X	1	X	(q_5)
p_3	1	X	X			X	
p_4	X	X	X	X		X	(p_0)
p_5	X	1	X	X		X	
p_6		X	X	X	X	1	

(p_4)

(5.42)

the flow is maximal, since no further labeling is possible, and has value 5.

No solution to the assignment problem has, as yet, been obtained, and costs must be modified. We point out that getting this far by labeling is not the most rapid procedure. The flow of (5.42) or another flow of value 5 could have been easily obtained by inspection. Furthermore, in (5.42) it may be observed that the set of unlabeled rows together with the set of labeled columns does correspond to a minimal cover of the blank (zero-cost) squares.

Inserting the costs along with the existing flows, we obtain

	q_1	q_2	q_3	q_4	q_5	q_6	
p_1	2	4	0/1	0	0	4	
p_2	4	11	5	4	0/1	11	(q_5)
p_3	0/1	2	12	0	0	9	
p_4	2	4	2	1	0	5	(p_0)
p_5	1	0/1	5	3	0	8	
p_6	0	8	6	5	1	0/1	

(p_4)

The smallest cost not in an unlabeled row or a labeled column is 1. Applying the rule for adding and subtracting this element, a new array with the corresponding labels is obtained:

	q_1	q_2	q_3	q_4	q_5	q_6	
p_1	2	4	0/1	0	1	4	
p_2	3	10	4	3	0/1	10	
p_3	0/1	2	12	0	1	9	
p_4	1	3	1	0	0	4	(p_0)
p_5	1	0/1	5	3	1	8	
p_6	0	8	6	5	2	0/1	

(p_4) (p_4)

An unsaturated column is labeled; a new flow of value 6 is obtained, and an optimal solution to the assignment problem is

	q_1	q_2	q_3	q_4	q_5	q_6
p_1			1			
p_2					1	
p_3	1					
p_4				1		
p_5		1				
p_6						1

That is, $x_{13}^* = x_{25}^* = x_{31}^* = x_{52}^* = x_{66}^* = x_{44}^* = 1$, $x_{ij}^* = 0$, all other i, j. This is the same solution that was obtained using the method of Egevary.

REFERENCES

[1] DANTZIG, G. B., "Application of the Simplex Method to a Transportation Problem," *Activity Analysis of Production and Allocation*, Cowles Commission Monograph 13, John Wiley & Sons, Inc., New York, 1951, pp. 359–373.

[2] DANTZIG, G. B., *Linear Programming and Extensions*, Princeton University Press, Princeton, N.J., 1963.

[3] EGEVARY, J., *"On Combinatorial Properties of Matrices"* (a translation), *Logistics Papers*, No. 11, App. I to Quarterly Progress Report 21, Nov. 1954–Feb. 1955.

[4] HADLEY, G., *Linear Programming*, Addison-Wesley Publishing Company, Inc., Reading, Mass., 1962.

[5] HITCHCOCK, F. L., "The Distribution of a Product from Several Sources to Numerous Localities," *Journal of Mathematical Physics*, Vol. 20, pp. 224–230, 1941.

[6] KANTOROVICH, L., "On the Translocation of Masses," *Comptes. Rendus. (Doklady) Acad. Sci.* 37, pp. 199–201, 1942.

[7] KÖNIG, D., *Theorie der Endlichen und Unendlichen Graphen*, Akademie-Verlag, M.B.H., Leipzig, 1936.

[8] KUHN, H. W., "The Hungarian Method for the Assignment Problem," *Naval Research Logistics Quarterly*, Vol. 2, pp. 83–97, 1955.

[9] SIMMONARD, M., *Linear Programming*, Prentice-Hall, Inc., Englewood Cliffs, N.J., 1966.

PROBLEMS

1. Prove that the matrix of constraint coefficients for the m, n transportation problem has rank $m + n - 1$.

2. For $m = 2$, $n = 3$, prove (without enumerating all the possibilities) that the constraint coefficient array is unimodular.

3. Describe in words three or four optimization problems that have nothing to do with the transportation or distribution of products, but that may be mathematically stated as transportation problems.

4. If $\mathbf{x}^*$ is optimal for a transportation problem with c_{ij} and a is a constant, show whether $\mathbf{x}^*$ is optimal for a problem with costs $c_{ij}' = c_{ij} + a$.

5. In using the northwest-corner rule to obtain a first feasible solution, how would you describe the conditions that guarantee a degenerate solution?

6. Given a feasible solution in the transportation problem tableau, how could one proceed to obtain a basic feasible solution? Apply your method to the feasible solution $x_{11} = x_{13} = x_{21} = 4$; $x_{33} = x_{45} = 2$; $x_{46} = x_{23} = 6$; $x_{12} = 7$; $x_{34} = x_{35} = x_{36} = x_{44} = 1$; $x_{56} = 3$ for problem (5.6) in the text.

7. (a) Without reference to the constraints, show that a transportation problem will never have
   (1) An unbounded solution.
   (2) Infinite-valued variables in an optimal solution.
  (b) Could there be cycling in a transportation problem using the algorithms discussed in the text?

8. Consider the following problem: Gene, Harvey, and Percy are partners in the distribution of certain illicit beverages, which they transport in clandestine fashion to four different outlets. Their production facilities are located at different secluded spots, and their net profits per gallon, supply capability, and demand at the outlets appear in the table. The objective is to maximize profits, satisfy demand with at least the demand quantities shown, and to exhaust all supplies during each monthly operation.

Outlets $\longrightarrow$		A	B	C	D
Gene	300	.50	.28	.17	.47
Harvey	1000	.84	.35	.40	.70
Percy	800	.27	.44	.52	.40
Requires at least		200	400	700	400

Find an optimal solution to this transportation problem (whose supplies, demands, and costs are as given and where the objective is maximization) in the following ways.
  (a) Using the simplex algorithm.
  (b) Using the transportation algorithm with the northwest-corner rule and no u_i, v_j.
  (c) Same as (b) but using the u_i, v_j.
  (d) Using Vogel's method for first feasible solutions and computing the $c_{ij} - z_{ij}$ in whatever manner you select.
    Then rank the procedures in ascending order of computational effort, and, if possible, approximate the amount of effort required for each.
  (e) Repeat (b), (c), (d) with the demands satisfied exactly.

9. (a) Suppose that in Problem 8 an increased hazard owing to unfriendly competitors from Harvey's location to outlet A decreases the profit per unit to .69 for that run. Is the solution to Problem 8 still optimal for this modified problem? If not, find the optimal solution without completely resolving it.
  (b) Later, certain federal agencies make shipping from Gene's location to

outlet D and Percy's location to outlet B highly unprofitable operations. Determine the optimal routing scheme where those routes are excluded.

10. Prove that the set of cells involved in an initial feasible solution by the north-west-corner rule always corresponds to a linearly independent set of vectors. Prove that the cost-penalty method also has this property.

11. To apply the simplex algorithm to the simplest equality-constrained transportation problem, we would first add artificial variables to each constraint and attach prohibitive costs to these. Suppose that this were done to the problem. How would you interpret the resulting problem in terms of transportation problem-type concepts and language?

12. Let the following grid represent the floor plan of a particular factory. The locations marked S indicate the places at which nuts, bolts, rivets, and so forth, are stored, while the locations marked D are the machines where rivets are used. Each day personnel at S locations load barrels of rivets on carts and hand pull them down the aisles to the various machine locations. There is one such person at each S location and he makes one round trip to service each location that he visits. Thus far, there has been no order to this delivery operation, the result being that it takes too long to supply the machine operators for the day and, further, absentee rates among delivery personnel are outrageous. Thus it is decided that a program of delivery assignments is needed to minimize the total daily distance the rivets travel. The cost of sending one rivet from a particular storage place to a particular machine is the shortest distance along aisles in the factory (no diagonal movement); for example, one shortest distance shown is two units; the other shown is six units. (This distance measure is the "taxicab" metric.) The numbers indicate demands (to be satisfied "at least" at each machine location). Each storage location stores 1000 rivets. Find an optimal delivery assignment.

S		D 800	S				
	D 500						
S			S				D 1000
						D 600	
				D 500			
D 400		S			D 500		

Assuming each S location can store enough rivets to satisfy all machines, would a smaller objective function value be possible if all rivets were sent from a single location? Which location is the best for single location supply?

13. With the following cost array, availabilities, and demands, find a solution that minimizes costs.

	≥ 15	10	≥ 5	15	8	≥ 6
20	1	3	3	2	−1	−1
≤ 20	2	5	−2	7	8	2
≤ 10	8	6	2	6	4	−2
≤ 20	−1	4	1	1	2	3

14. With per unit profits, supplies, and demands as follows, suppose that origin 1 may tranship to origins 2 and 3 with profits of 1 and 2, respectively, and that destination 2 may tranship up to 5 units at profits of 2, 1, 1, and 3 to destinations 1, 3, 4, and 5, respectively. Find a shipping strategy that maximizes profit.

	15	25	5	10	10
20	4	7	3	4	6
10	2	4	3	7	2
40	3	2	8	4	6

15. Set up the following problem for solution as a transportation problem. There are six cities with routes and per unit costs between them given by the following table:

	1	2	3	4	5	6
1	0	2	∞	3	4	∞
2	8	0	5	1	∞	7
3	∞	4	0	3	1	2
4	6	8	10	0	4	2
5	8	∞	7	5	0	2
6	∞	4	1	3	1	0

Cities 1 and 5 are points of supply with 100 and 150 items of a commodity, respectively. Demands are as follows: city 2, 60; city 3, 60; city 4, 80; city 6, 20; and city 1, 20. Find a minimum-cost solution that satisfies all demands as equations.

16. Perform three iterations beyond a first feasible solution for the following problem. Origins 1 and 2 have available 80 and 75 units of a commodity, respectively. Each may tranship up to 50 units to other origins, transhipment costs being

	1	2	3	4
1	0	0.5	0.75	0.9
2	0.6	0	0.75	0.5

Other availabilities, costs, and demands are

	30	25	40	30	45	40
80	4	2	5	4	6	9
75	5	6	2	7	3	8
40	1	3	4	5	2	6
30	2	4	3	1	6	4

and not all units must be shipped to destinations.

17. Show that the cost of a feasible solution obtained by Vogel's method is at least as good as that of one obtained by the northwest-corner rule.

18. Show that the new variable values obtained by adding and subtracting around the loop are exactly as the equations from the simplex method predict.

19. Suppose that an assignment problem finds $m < n$. Without altering the problem, when is it possible to cross off a job? By the same token, if $m > n$, when is it possible to cross off a person?

20. It is easy to envision assignment-type problems in higher dimensions, those of this chapter being considered two dimensional. For example, suppose that a factory has m electricians, n apprentices, and k jobs, each job requiring exactly one apprentice and exactly one electrician, with c_{ijk} the measure of effectiveness of electrician i and apprentice j performing job k. Formulate the problem of maximizing the total effectiveness, and comment on the implications of the relative magnitudes of m, n, and k. For specific, small values of m, n, and k, determine whether or not the constraint coefficient matrix is unimodular.

21. Given the following assignment problem where $m = n = 4$ and the objective function to be maximized given by

$$\begin{pmatrix} 3 & 2 & 4 & 2 \\ 5 & 2 & 4 & 8 \\ 6 & 1 & 3 & 5 \\ 1 & 2 & 8 & 9 \end{pmatrix}$$

give a rough quantitative comparison of the computational effort required to solve the problem using the following methods:

(a) The simplex method.

(b) The transportation-problem algorithm with a first feasible solution obtained by the northwest-corner rule.

(c) The method of Egevary.

(d) The labeling algorithm.

22. Solve each of the following assignment problems (maximization) by the method of Egevary.

(a)

$$
\begin{pmatrix}
5 & 4 & 4 & 4 & 3 & 4 & 1 \\
6 & 5 & 1 & 2 & 8 & 5 & 7 \\
3 & 7 & 4 & 3 & 3 & 4 & 1 \\
2 & 8 & 2 & 3 & 5 & 5 & 3 \\
5 & 6 & 3 & 3 & 3 & 3 & 5 \\
2 & 4 & 1 & 1 & -2 & -5 & 0 \\
-1 & 6 & 2 & 1 & 7 & 1 & 2
\end{pmatrix}
$$

(b)

$$
\begin{pmatrix}
3 & 5 & 2 & 6 & 5 & 3 & 2 & -2 \\
0 & 4 & 5 & 8 & 7 & -1 & 0 & 4 \\
3 & 7 & 2 & 10 & 6 & 4 & 1 & 3 \\
8 & 9 & 6 & -3 & 6 & 9 & 6 & 9 \\
1 & 7 & 5 & 2 & 7 & 10 & 7 & 4 \\
3 & 10 & 8 & 10 & 2 & 11 & 6 & 9 \\
10 & 1 & 9 & 0 & 4 & 6 & 6 & 9 \\
8 & 3 & 4 & 7 & 8 & 7 & 7 & 1
\end{pmatrix}
$$

(c)

$$
\begin{pmatrix}
1 & 1 & 2 & 3 & 4 \\
3 & 1 & 2 & 3 & 5 \\
4 & 2 & 3 & 6 & 4 \\
5 & 1 & 3 & 7 & 6 \\
7 & 5 & 6 & 2 & 3
\end{pmatrix}
$$

(d)

$$
\begin{pmatrix}
2 & 2 & 5 & 7 & 4 & 2 \\
1 & 2 & 5 & 7 & 3 & 2 \\
6 & 4 & 4 & 2 & 1 & 3 \\
3 & 5 & 8 & 7 & 1 & 2 \\
4 & 5 & 3 & 2 & 6 & 4 \\
3 & 2 & 2 & 7 & 4 & 3
\end{pmatrix}
$$

(e)
$$\begin{pmatrix} 1 & 3 & 4 & 2 & 6 & 3 \\ 1 & 4 & 5 & 3 & 6 & 3 \\ 2 & 3 & 4 & 2 & 1 & 2 \\ 3 & 8 & 4 & 5 & 6 & 1 \\ 5 & 8 & 4 & 5 & 3 & 3 \\ 6 & 6 & 3 & 4 & 4 & 2 \end{pmatrix}$$

23. In our discussion, a covering or cover was a set of lines that covered all the zeros in a particular matrix, which had at least one zero in every row and every column. If a zero in position (i, j) indicates the existence of a certain edge,
    (a) In terms of the corresponding graph, what is the interpretation of a cover? Answer the question for both bipartite graphs and arbitrary graphs.
    (b) What is the significance, then, in terms of the graph, of a minimal cover?
    (c) For bipartite graphs, how would you interpret maximal matching = minimal cover?
    (d) Illustrate the theorem referenced in part (c).

24. Solve each part of Problem 22 by means of the labeling algorithm.

25. Does one iteration of the Hungarian method always produce a new array with more zeros than its predecessor? Does the size of a minimal cover always increase?

SEARCH TECHNIQUES

6.1. Introduction

Once again we address ourselves to optimization problems, but in the present chapter we adopt approaches quite different from those previously utilized. We shall be considering problems of the form

$$\max f = f(x_1, \ldots, x_n)$$

subject to

$$a_j \leq x_j \leq b_j, \quad j = 1, \ldots, n$$

We have had ample opportunity to observe and experience the difficulties attending the solution of such a problem for arbitrary f. Our classical methods, for example, and their immediate offspring demand the differentiability of f to provide even necessary conditions, and these provide sufficient conditions in but very special situations. Everett's method made no demands on f, but neither could it guarantee an optimal solution; and so on. Let us further confound ourselves by admitting for consideration the situation wherein *f does not exist as a mathematical statement or set of statements that tells how the function value is to be computed for particular values* $x_1^0, \ldots, x_n^0$

6

of its arguments. While this may at first appear an unusual situation to face the problem solver, typical examples abound in the real world.

Consider the medical researcher who has discovered a potential cure for cancer in rats and is attempting to discover the optimal rat dosage. On the basis of his knowledge of the drug, the animal's physiology, and the disease, he has a rough idea, but must experiment. He tries a weak dose, the disease is unchecked, and the animal dies. He increases the dose, the cancer's growth is retarded temporarily, but a month later the malignancy begins spreading anew. He increases the dosage, but the animal dies immediately from the drug's effects; so he decreases the dosage and tries again, and so on. Here, an optimal solution might be that dosage which destroyed the cancer and left the rat healthy with no side effects. Of course, there may be no dosage which provides that end result. It is desirable to find that dosage which provides the *best* end result. But in the absence of a well-defined relationship between dosage and the ultimate future of the rat, the researcher must experiment. And it must be recognized that experiments are costly—in terms of time, rats, production of the experimental drug, instruments, and so forth.

The experiments certainly may be performed in a fashion different from the preceding. First, the sequence of dosages could be permuted, and, furthermore, an entirely different philosophy of experimentation could be used. That is, rather than experimenting sequentially, one could experiment simultaneously; one could first select an entire set of dosages, administer each to a rat, observe the results, and choose that dosage as best which provides the best result.

Examples of this general kind can be found in engineering, economics, urban systems, and virtually any area.

i. *The Problem*

But now we have set the stage for the optimization technique to be used. We assume, for the moment, only that f can be evaluated for any chosen value(s) of its decision variable(s). It will be the case, alas, that to obtain powerful methods we shall again be forced into making certain restrictive assumptions about f. The problem is to observe some finite set of function values and draw (hopefully accurate) inferences about the location of the global maximum, where, as before, we choose to consider only maximization problems—without any loss of generality. In other words, on the basis of the several *observations* or *functions evaluations*, we hope to be able to specify a region, in some sense small, in which the global maximum is *certain* to lie.

We shall want to use the *smallest* such region. If, for example, it is known only that the global max lies in [0, 1], [0, $\frac{3}{4}$], and [$\frac{1}{2}$, $\frac{3}{4}$], then this last interval is the desired one. The region discussed is frequently called the "region of uncertainty," but let us here use the terminology *extremal region (interval)*.

a. BASIC CONCEPTS

Search techniques are customarily dichotomized into two broad categories, simultaneous search strategies and sequential search strategies, although "simultaneous" in this context implies nothing about time. To be precise, let us define a *simultaneous search* as one in which no outcomes of observations are used to locate other decision variable values to be examined. On the other hand, in a *sequential search* the results of some of the observations *are* used to decide where to place new experiments.

In the course of our investigations we shall encounter search techniques that appear to share properties of both sequential and simultaneous searches, but this is a minor point. In the simple medical research setting we observed examples of both kinds of search strategies, and now we proceed to examine in detail specific examples of each.

Exercise 6.1

Give four or five examples of realistic problems whose solutions might be pursued via sequential search. Do the same for simultaneous search.

6.2. Functions of a Single Variable

i. *Simultaneous Search*

The *exhaustive search* is easily the most straightforward of all search techniques and is a simultaneous search. First restricting ourselves to functions of a single variable, the typical problem will be

$$\max f = f(x)$$

subject to

$$a \leq x \leq b$$

where f is defined everywhere upon $[a, b]$.

Suppose that $[a, b]$ is partitioned via a finite set of equally spaced points, $S = \{a + \delta/2, a + \delta, a + 3\delta/2, \ldots, b - \delta, b - \delta/2\}$ and then f is evaluated at each member of this set. Plotting the results, we might find, for example, results as in Figure 6.1.

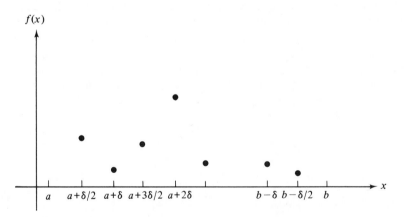

Figure 6.1. Several values of some function $f(x)$ at equally spaced points.

Suppose that we next find $\hat{x} \in S$ such that $f(\hat{x}) \geq f(p) \ \forall p \in S$, and state that the true global max $x^* \in [\hat{x} - \delta/2, \hat{x} + \delta/2]$. With regard to Figure 6.1, $\hat{x} = a + 2\delta$, and the x value selected to approximate x^* would be $\hat{x}$. Quite clearly, this deduction may be patently false, and Figure 6.2 exhibits a function f corresponding to the data of Figure 6.1 for which this is the case. But then, in general, how could this method of approximating x^* hope to

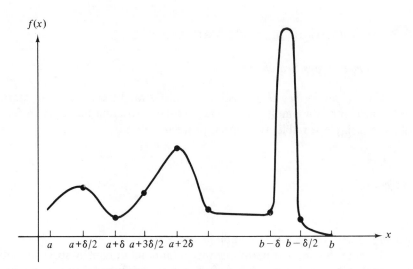

Figure 6.2. The actual function $f(x)$.

succeed? It cannot, of course, and the loose assumption regarding f is the reason. Therefore, we make an assumption about f, and it is that enough is known about f's behavior to be certain that the foregoing technique will work. In other words, enough is known that δ may be selected so that salient characteristics of the function, such as the spike in Figure 6.2, are not over-looked. Thus, as a result of this assumption, success of the exhaustive search is guaranteed, and the length of the final extremal interval is δ.

The number N of function evaluations required to yield this size interval is then easily calculated, for we must have

$$(N+1)\frac{\delta}{2} = b - a$$

or $$N = \frac{2}{\delta}(b - a) - 1 \tag{6.1}$$

and the price one pays in function evaluations for achieving small δ is obvious.

A valid and practical question, though, is "What kinds of typical real-world information guide the choice of δ?" That is, what could provide "enough" information to specify the necessary δ?

Suppose that we consider f to be continuous and to have a derivative everywhere except perhaps at a finite set of points. Also, assume that $|f'| \leq A$, that is, a bound on f' is known. Then over an interval of length β the maximum increase in f is $A\beta$. Suppose that the next best sample point value is $\bar{x}$ and that $f(\hat{x}) - f(\bar{x}) \geq \epsilon > 0$. The maximum value f could achieve on the interval $[\bar{x} - \delta/2, \bar{x} + \delta/2]$ is $f(x) = f(\bar{x}) + A\delta/2$.

If δ had been chosen so that $A\delta/2 \leq \epsilon$, then no point outside the interval $[\hat{x} - \delta/2, \hat{x} + \delta/2]$ can provide a better function value than $\hat{x}$ itself. Thus, if ϵ is first specified, δ chosen to satisfy $A\delta/2 \leq \epsilon$, and there is no point $\bar{x}$ with $f(\hat{x}) - f(\bar{x}) < \epsilon$, then the interval $[\hat{x} - \delta/2, \hat{x} + \delta/2]$ may be safely specified as containing x^*. If we discovered, with this choice of δ, $\bar{x}$ such that $f(\hat{x}) - f(\bar{x}) < \epsilon$, then there is no guarantee that $\hat{x}$ is the proper choice. One might consider reducing ϵ, and consequently δ, and resolve the problem.

With this example we have attempted to illustrate how a piece of real-world information—bounds on derivatives are frequently assumed in the analysis of the accuracy and errors of numerical techniques—might be translated into guides for executing one procedure.

ii. *Exhaustive Search in Stages*

In the absence of knowledge about f's behavior, what does one do? Small δ implies a costly search, whereas large δ would lead one to question the results. By letting δ vary, in a certain way, over the course of the search, one can hopefully strike a balance between economy and accuracy.

One first selects δ_1, which is not "too small," partitions $[a, b]$ accordingly, evaluates f at the partition points, and chooses several of the "more promising" subintervals. Each selected subinterval is then partitioned with $\delta_2 < \delta_1$ and searched exhaustively. The hierarchy of searches can be continued, of course, with "finer" searches being performed in the neighborhoods of several better values from the previous search. The searcher is thus enabled to concentrate the fine, costly search in those neighborhoods that appear promising, using cheaper searches to locate these neighborhoods, and eliminate most of the original interval.

In addition to the cost of evaluating the function, there is another reason for not crowding the sample points quite close together, and that is that one may be unable to distinguish any difference in function values for two distinct but close sample points. For example, if evaluating the function consists of executing a sequence of arithmetical operations, roundoff error could produce the result mentioned. Or, if evaluating the function consists of observing some physical system, the measuring devices might not be sufficiently sensitive. Now, information of these kinds *would* be known in advance, and could, therefore, be of assistance.

The reader will have observed that the two preceding examples which provided the indistinguishability of distinct function values actually constitute erroneous data, and if observation errors are admissible, then there must also be instances of specifying the wrong extremal interval.

Pursuing the problems associated with noisy observations, though, is not essential in an introduction to the topic, and we shall assume the ability to evaluate our functions without error, so long as we do not place our points of evaluation too near one another.

Exercise 6.2

Assuming that the function is evaluated in some other program segment, write a computer program that will implement the exhaustive search, successively reducing the partition size until some input criterion for "closeness" is met. Run the program for several functions of different behavior and for different closeness criteria, and compare the results with the true optima. Count the number of function evaluations required.

iii. *Extremal Interval Magnitude Versus Search Length*

Let L_N denote the length of the extremal interval after N observations. The equations we shall be obtaining expressing L_N in terms of N may be used in two ways:

1. To compute L_N when N is given. Thus the accuracy obtainable may be computed in advance when the number of evaluations one can afford is given.
2. To determine N, given a desirable value of the extremal interval, say L^0. In this case, the cost of the search may be computed in advance, given the desired accuracy. It should be realized, of course, that if N, L_N are related via $L_N = t(N)$, where t is some function, and we fix $L_N = L^0$, then there may not exist a positive integer $N \ni L^0 = t(N)$. In such a case one would likely be interested in finding N', the smallest positive integer with the property $t(N') < L^0$.

iv. *Sequential Search*

a. A CLASS OF FUNCTIONS—AN ASSUMPTION

Now suppose that we apply a sequential search technique to the function f, defined everywhere upon $[a, b]$. In fact, let the function already have been evaluated at $x_1, x_2, \ldots, x_k$. What information is available to help locate x_{k+1} in meaningful and efficient fashion? Stated alternatively, what do the data already obtained, that is, $\{(x_j, f(x_j)) | j = 1, \ldots, k\}$, imply about the location of the global maximum? As before, it is obvious that one could construct an infinite number of continuous functions on $[a, b]$, every one of which has its global maximum in a most "unlikely" (on the basis of the other observations) part of $[a, b]$, in which case the pursuit of sequential techniques seems unjustifiable.

But by making one assumption on the nature of f, extremely efficient search techniques have been devised. Let us assume f is *unimodal* on $[a, b]$; that is, there exists α, $a \le \alpha \le b$, such that either

1. f is strictly increasing on $[a, \alpha]$ and strictly decreasing on $(\alpha, b]$ (Figure 6.3a) or
2. f is strictly decreasing on $[a, \alpha]$ and strictly increasing on $(\alpha, b]$ (Figure 6.3b).

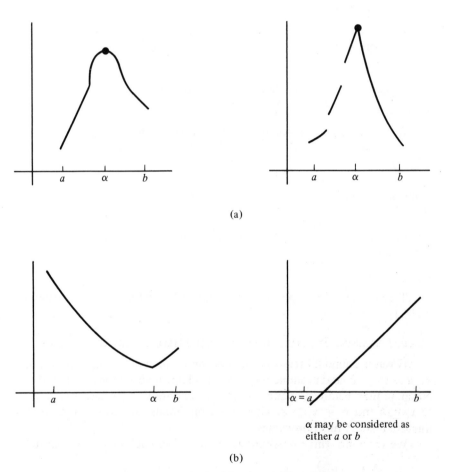

(a)

(b)

α may be considered as either a or b

Figure 6.3. Examples of unimodal functions.

By definition, in case 1, α is the unique global max, whereas in case 2, α is the unique global min. There being always exactly one extreme value, or mode, results in the naming of this class of functions. Another property of f is that if x^* is a relative extremum then x^* is a global extremum.

The real-world problem solver is not impressed by highly efficient methods if his encountering an instance where the methods are applicable is a rare event, but the "function" to be optimized in many real problems *is* unimodal.

Exercise 6.3

Give three examples of functions that are concave or convex on [a, b] but not unimodal there.

Exercise 6.4

Prove that every function which is strictly convex or concave on [a, b] is unimodal on [a, b].

Exercise 6.5

Is the converse of the statement of Exercise 6.4 true?

Exercise 6.6

If $f_1, \ldots, f_k$ are unimodal (of the same kind) on [a, b], is

$$\sum_{i=1}^{k} f_i$$

unimodal on [a, b]?

Exercise 6.7

Prove that if f is unimodal on [a, b] and $f(x) = 0$ nowhere on [a, b] then $1/f$ is unimodal on [a, b].

Exercise 6.8

True or false. $f > 0$ on [a, b] and f unimodal on [a, b] $\Rightarrow \ln f$ unimodal on [a, b].

b. Obtaining Information from Sequential Observations

Given a unimodal (case 1) function on an interval [a, b], select a point $x_1, a \leq x_1 \leq b$, and examine $f(x_1)$. What is known about the location of x^*? No more than before. But suppose that we choose x_1, x_2 where, say, $a \leq x_1 < x_2 \leq b$ and $x_2 - x_1 \geq \epsilon$, where ϵ is a minimum separation needed to distinguish distinct function values.

We recognize three possibilities and consider each of them separately:

1. $f(x_1) < f(x_2)$
2. $f(x_1) = f(x_2)$
3. $f(x_1) > f(x_2)$

1. $f(x_1) < f(x_2)$ (Figure 6.4). We know, in this case, that $x_1 < x^* \leq b$, since if $a \leq x^* \leq x_1$, f could not be unimodal. Thus two function evaluations reduce the extremal interval by $x_1 - a$, to one of length $b - x_1$.

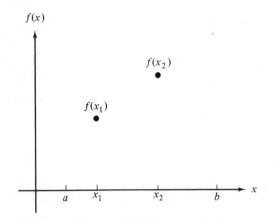

Figure 6.4. $f(x)$ unimodal and $f(x_1) < f(x_2)$.

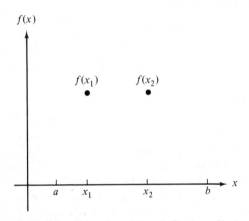

Figure 6.5. $f(x)$ unimodal and $f(x_1) = f(x_2)$.

2. $f(x_1) = f(x_2)$ (Figure 6.5). Now, it must be that $x_1 < x^* < x_2$, for otherwise f would not be unimodal. Notice also that $f(x_1) = f(x_2)$ is most unlikely for any two points x_1, x_2 of $[a, b]$, for no one function value may be assumed more than two times. In our discussions, therefore, we shall generally assume that (2) does not occur, but will comment on its consequences from time to time.

3. $f(x_1) > f(x_2)$ (Figure 6.6). Here it is immediate that $a \le x^* < x_2$. Otherwise, f would not be unimodal.

The simple principle just demonstrated may be utilized in a variety of different search techniques. Let us first investigate the *dichotomous search*.

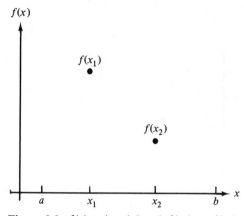

Figure 6.6. $f(x)$ unimodal and $f(x_1) > f(x_2)$.

c. Dichotomous Search

The function is first evaluated twice and

$$f\left(\frac{b+a}{2} - \frac{\epsilon}{2}\right), \quad f\left(\frac{b+a}{2} + \frac{\epsilon}{2}\right)$$

obtained; that is, the first observations are made symmetric about the mid-point of $[a, b]$ and separated by the "distinguishability" constant ϵ, which, of course, does not have to be explicitly known, so long as it is known that the points are not "too close" for the particular means of evaluating f.

The result is, depending upon whether $f(x_1) > f(x_2)$ or $f(x_1) < f(x_2)$, a new extremal interval I_1 of length $L_1 = [(b - a)/2] + (\epsilon/2)$ (Figure 6.7). The process is entirely repetitive; one next evaluates f at two points symmetric about the midpoint of I_1 and ϵ distance apart, after which we have an extremal interval I_2 of length $L_2 = \frac{1}{2}L_1 + (\epsilon/2) = [(b - a)/4] + (\epsilon/4) + (\epsilon/2)$. Defining an iteration as the taking of two new function values, after k iterations we have an extremal interval I_k of length

$$L_k = \frac{1}{2}L_{k-1} + \frac{\epsilon}{2}, \quad k = 2, 3, \dots$$

or

$$L_k = \frac{b - a}{2^k} + \sum_{j=1}^{k} \frac{\epsilon}{2^j}$$

or

$$L_k = \frac{b - a}{2^k} + \epsilon\frac{1/2 - 1/2^{k+1}}{1/2}$$

$$= \frac{b - a}{2^k} + \epsilon(1 - 1/2^k)$$

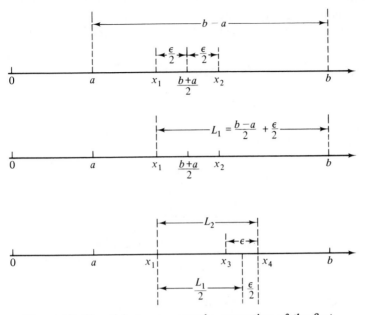

Figure 6.7. The dichotomous search: generation of the first two extremal intervals in a sample case.

If L_N is used to denote the length of the extremal interval after N function evaluations (N even), then we may write

$$L_N = \frac{b - a}{2^{N/2}} + \epsilon(1 - 1/2^{N/2}) \tag{6.2}$$

d. COMPARING SEARCH EFFICIENCIES

It will be convenient, both for purposes of comparing search techniques and for simplifying some expressions, to assume that the initial interval $[a, b]$ has unit length. If this is not the case, a simple linear transformation will map the given interval $[a, b]$ onto $[0, 1]$ in $1:1$ fashion.

Let $y = (x - a)/(b - a)$, and as x varies over $[a, b]$, y varies over $[0, 1]$, and, clearly, the transformation is invertible; which is to say, given y, a unique x is determined from

$$x = (b - a)y + a$$

Thus a final extremal interval of length $b' - a'$ defined by $[a', b']$ in the original problem is equivalent to an interval defined by $[(a' - a)/(b - a), (b' - a)/(b - a)]$ with length $(b' - a')/(b - a)$. Thus, since we can always determine how we would have done had we been working on [0, 1], we may as well assume [0, 1] *is* the given interval.

With $a = 0$, $b = 1$, equation (6.2) may be simplified to yield

$$L_N = \frac{1}{2^{N/2}} + \epsilon(1 - 1/2^{N/2})$$

or

$$L_N - \epsilon = \frac{1}{2^{N/2}}(1 - \epsilon)$$

Solving for N, we obtain

$$2^{N/2} = \frac{1 - \epsilon}{L_N - \epsilon}$$

$$\frac{N}{2}\ln 2 = \ln\left(\frac{1 - \epsilon}{L_N - \epsilon}\right)$$

$$N = 2\frac{\ln\left(\dfrac{1 - \epsilon}{L_N - \epsilon}\right)}{\ln 2} \tag{6.3}$$

e. EQUAL-INTERVAL SEARCHES

Let us next examine a family of search techniques, the *Z-point equal-interval searches*, for $Z = 2, 3, \ldots$, concentrating first upon the two-point equal-interval search.

Here each iteration consists of evaluating the function at *two* different points placed so as to partition the current extremal interval into three equal subintervals. Thus the first iteration consists of evaluating f at $x = \frac{1}{3}$, $x = \frac{2}{3}$. If $f(\frac{1}{3}) < f(\frac{2}{3})$, we get a new interval of length $\frac{2}{3}$, that is, $(\frac{1}{3}, 1]$, and f is evaluated at $x = \frac{5}{9}$, $x = \frac{7}{9}$ for iteration 2, and so on. If $f(\frac{1}{3}) > f(\frac{2}{3})$, the new extremal interval is $[0, \frac{2}{3})$, and iteration 2 evaluates f at $x = \frac{2}{9}$, $x = \frac{4}{9}$, and so on. [Should $f(\frac{1}{3}) = f(\frac{2}{3})$, we get a new extremal interval of length but $\frac{1}{3}$ and next evaluate f at $x = \frac{4}{9}$, $x = \frac{5}{9}$. We neglect this unlikely, but fortuitous, contingency in measuring the effectiveness of the search method.] Each iteration provides an extremal interval two thirds the size of the previous. After k iterations, we have an extremal interval of length

$$L_k = (\tfrac{2}{3})^k$$

and since $N = 2k$

$$L_N = (\tfrac{2}{3})^{N/2}$$

given that the process began on the interval [0, 1].

Next, consider the three-point equal-interval search. One first evaluates

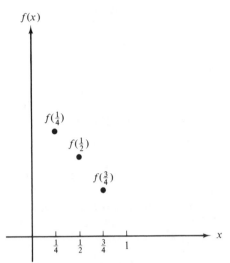

Figure 6.8. Example outcome of one iteration of a three point equal interval search.

f at $x = \frac{1}{4}$, $x = \frac{1}{2}$, $x = \frac{3}{4}$, and let us suppose that $f(\frac{1}{4}) > f(\frac{1}{2})$, $f(\frac{1}{4}) > f(\frac{3}{4})$, as in Figure 6.8. We know, then, that the global maximum lies in $[0, \frac{1}{2})$—we shall always halve the extremal interval using this method (excluding again, the case where the largest function value is not unique, whereupon the extremal interval is one fourth what it was). When we perform iteration 2, we notice an interesting and attractive property of the three-point equal-interval search. We should next partition $[0, \frac{1}{2})$ into four equal subintervals using three points, which must be $x = \frac{1}{8}$, $x = \frac{1}{4}$, $x = \frac{3}{8}$. But the function has *already* been evaluated at $x = \frac{1}{4}$; thus only *two* new evaluations are required for iteration 2, and, indeed, for all subsequent iterations. Furthermore, as can be seen, this property of requiring one less evaluation after iteration 1 is enjoyed by all those Z-point equal-interval searches for which Z is odd, and only by them.

Let us write expressions for the efficiency of the general search from this family. For the Z-point equal-interval search the reduction factor for the extremal interval is $[2/(Z + 1)]$, since x^* is known to lie in a subinterval of length equal to the length of two of the smaller intervals produced by the partition, and which is centered at the sample point giving the largest function value. Thus after k iterations we have

$$L_k = \left(\frac{2}{Z+1}\right)^k$$

For Z even, k iterations require $kZ = N$ function evaluations, but for Z odd, the first iteration requires Z function evaluations, the remainder requiring $Z - 1$. So for Z odd, k iterations require $Z + (k - 1)(Z - 1) = N$

function evaluations. We then have

$$L_k = \left(\frac{2}{Z+1}\right)^{N/Z}, \quad Z \text{ even} \tag{6.4}$$

$$L_k = \left(\frac{2}{Z+1}\right)^{(N-1)/(Z-1)}, \quad Z \text{ odd} \tag{6.5}$$

f. Finding a Best Equal-Interval Search

We observe that as Z increases the reduction factor $[2/(Z+1)]$ decreases, but so also does the exponent, and we are led to ask, what is the best value of Z? "Best" is here taken to mean that value of Z which minimizes L_k, given N.

It is fairly simple to verify that $Z = 3$ is the best value of Z for Z-point equal-interval searches, for all values of function evaluations $N \geq 3$. Assuming N fixed, consider first the case when Z is even. Define the function $\tilde{L}(Z)$ of the continuous variable Z:

$$\tilde{L}(Z) = \left(\frac{2}{Z+1}\right)^{N/Z} \tag{6.6}$$

Taking logarithms of both sides of (6.6),

$$\ln \tilde{L}(Z) = \frac{N}{Z} \ln \frac{2}{Z+1}$$

Differentiating both sides with respect to Z, we obtain

$$\frac{1}{\tilde{L}(Z)} \cdot \tilde{L}'(Z) = \frac{N}{Z} \cdot \frac{Z+1}{2} \cdot \frac{-2}{(Z+1)^2} - \frac{N}{Z^2} \ln \frac{2}{Z+1} \tag{6.7}$$

If $\tilde{L}'(Z) = 0$, then the right-hand side of equation (6.7) must be zero; that is

$$\frac{N}{Z} \left[\frac{-2(Z+1)}{2(Z+1)^2} - \frac{1}{Z} \ln \frac{2}{Z+1} \right] = 0$$

so that

$$\ln \frac{2}{Z+1} = \frac{-Z}{Z+1}$$

or

$$\ln \frac{Z+1}{2} = \frac{Z}{Z+1} \tag{6.8}$$

Equation (6.8) is independent of N, and it is easily verified that $\tilde{L}'(Z) = 0$ possesses a root lying between $Z = 3$ and $Z = 4$. If the reader will then establish that the continuous function $\tilde{L}(Z)$ possesses certain properties, it will be clear that the root is a global minimum of the function $\tilde{L}(Z)$. Thus the minimizing even integer Z for L_k in equation (6.4) is either 2 or 4, and it may easily be verified that $\tilde{L}(4) < \tilde{L}(2)$. Consequently, if Z is even, then, for all N, $Z = 4$ minimizes L_k in equation (6.4).

The same procedure may be applied for the case when Z is odd, and it may be shown that, for all N, $Z = 3$ minimizes L_k in equation (6.5). The reader is next asked to show that, for all $N \geq 3$, $Z = 3$ is the minimizing integer for L_k. Of course some interpretation is in order. For example, $Z = 3$ does not give a search requiring exactly $N = 8$ function evaluations.

Exercise 6.9

Write a computer program to implement the dichotomous search. One should be able to input a desired interval and a desired ϵ. Furthermore, the program should calculate the number of evaluations required and should be usable with minimum change on any given function.

Exercise 6.10

Implement the two- and three-point equal-interval searches in computer programs.

g. Concept of an Optimal Search Strategy

Sooner or later one must ask what is the best search technique for unimodal functions of a single variable and, more fundamentally, how does one decide when one search technique is better than another. The question about *how* to measure the performance of a search technique is also implicitly involved.

To point out the difficulties in answering these questions, let us consider a one-iteration search, in which we evaluate the function at two points x_1 and x_2. Suppose that we place the two points in two different ways (a) and (b), as in Figure 6.9. Now, if the unknown function were f_1 as in Figure 6.10, (a) would provide a small extremal interval, of length $1 - x_1$. Strategy (b), on the other hand, would give an interval of length $\frac{1}{2} + (\epsilon/2)$, and in the case of

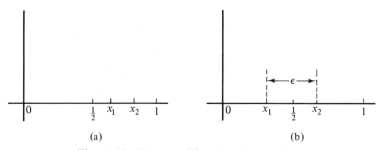

Figure 6.9. Two possible two-point searches.

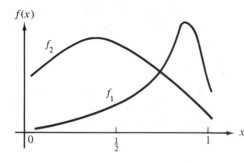

Figure 6.10. Interval location and effect on extremal interval length.

$f = f_1$, (a) enjoys a clear-cut advantage over (b). On the other hand, if we had $f = f_2$, as in Figure 6.10, (a) would yield a large extremal interval of length x_1, while (b) would provide an interval of length $\frac{1}{2} + (\epsilon/2)$. In this case, (b) is clearly superior to (a).

Since the relative merits of (a) and (b) apparently depend upon f, how can we decide which of (a) or (b) is better? Our measure of performance is selected as the length of the extremal interval remaining after the two evaluations.

The reader familiar with game theory will recall this is precisely the kind of situation encountered there. In the present context, the problem solver is one of the players—his available strategies are (a) and (b). We can imagine Nature is the other player, and her available strategies include all unimodal functions (of kind 1) defined on [0, 1]. Suppose that the problem solver plays strategy (a). As in game-theoretic studies, (a) may prove to be a wise choice for some of Nature's plays and a very unfortunate choice for other of Nature's possible choices.

We can construct a more realistic problem by allowing the problem solver to select any two distinct points, $|x_1 - x_2| \geq \epsilon$, in [0, 1]. Then, he is interested in choosing the "best" search strategy from among all possible placements of x_1, x_2.

Given the similarities to a game-theory problem, the reader should find appealing the fact that in dealing with search techniques an optimal search strategy is defined in minimax terms. Therefore, letting S_n be the set of all search strategies involving n function evaluations, we say $S^* \in S_n$ is *optimal if it minimizes the maximum extremal interval that can exist after n function evaluations.*

From this definition it is clear that strategy (b) in our recent example is not only superior to (a) but is, in fact, optimal among all search strategies in S_2. *Any other placement* of the two points admits the possibility that the final extremal interval is greater than $\frac{1}{2} + (\epsilon/2)$, which is the maximum length of the extremal interval when (b) is used; therefore, (b) minimizes this maximum length.

We observe that we guarantee ourselves a certain performance, regardless of how pathological a function Nature may choose to present, and if she proves significantly benign in some instances, we shall do better. Recall, also, that certain nice conditions were eliminated from consideration in measuring the performance of the dichotomous and equal-interval searches. The motivation was the same as here, that is to measure performance in the least favorable circumstances, thus knowing the strategy would do *at least* that well.

h. A NEAR-OPTIMAL SEARCH—THE GOLDEN-SECTION SEARCH

Let L_k = length of extremal interval at the end of the kth iteration, $k = 1, 2, \ldots$, and L_0 = length of the initial interval; then we describe a search in terms of two assumptions:

$$L_{k-1} = L_k + L_{k+1}, \quad k = 1, 2, \ldots \tag{6.9}$$

$$\frac{L_k}{L_{k-1}} = \frac{L_{k+1}}{L_k}, \qquad k = 1, 2, \ldots \tag{6.10}$$

Assumption (6.10) states that the ratio of successive extremal intervals is a constant, and our first step is to determine that constant, which turns out to be unique.

Substituting (6.9) into (6.10), we obtain

$$\frac{L_k}{L_k + L_{k+1}} = \frac{L_{k+1}}{L_k}$$

or, taking reciprocals,

$$1 + \frac{L_{k+1}}{L_k} = \frac{L_k}{L_{k+1}} \tag{6.11}$$

Writing $u = L_{k+1}/L_k$, equation (6.11) becomes

$$1 + u = \frac{1}{u}$$

or

$$u^2 + u - 1 = 0$$

which has the roots

$$u = \frac{-1 \pm \sqrt{5}}{2}$$

or

$$u \simeq .618, \qquad u \simeq -1.618$$

Our lengths being positive, only the root $u = .618$ has meaning for the problem. At any rate, the search technique is now completely specified, for

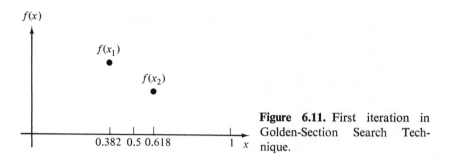

Figure 6.11. First iteration in Golden-Section Search Technique.

we must first evaluate the function at $x_1 = .618$, $x_2 = 1 - .618 = .382$ in order to assure $L_1/L_0 = .618$ (L_0 assumed 1).

Let us imagine $f(x_1) > f(x_2)$ (Figure 6.11). The new extremal interval is $[0, .618)$, and we must next evaluate f such that $L_2/L_1 = .618$ or $L_2 = .618L_1$. By assumption, we shall operate so as to maintain $L_{k+1} = .618L_k$, which would appear to provide about the same reduction power as the two-point equal-interval search, where $L_{k+1} = \frac{2}{3}L_k$; but let us proceed with the next iteration of our new search.

The function must next be evaluated at x_3, x_4 such that

$$\frac{.618 - x_3}{.618} = .618 \quad \text{and} \quad \frac{x_4}{.618} = .618$$

(where we have assumed $x_3 < x_4$), since one of $[0, x_4)$ and $(x_3, .618)$ will be the extremal interval, and these must have equal length. Solving for x_3, x_4, we find $x_3 \simeq .236$ and $x_4 \simeq .382$.

But f has *already* been evaluated at x_4, so only *one* new function evaluation is required to achieve the prescribed reduction of the extremal interval. Therein lies the power of this search technique; it *always* occurs that one of the two evaluations needed for the next iteration has already been performed. Although the factor of reduction, .618, is not spectacular, only *one* function evaluation provides that reduction for iterations 2, 3, ...

Compared to this method, any sequential method previously discussed is most prodigal in terms of *information*. The dichotomous search uses two evaluations to decide upon the next interval, but never again uses those pieces of information. The odd-point equal-interval searches do reuse some information, it is true, but all save one value are discarded during any one iteration.

Of course, we have not given a proof that what we claimed for the present search technique always holds—we treated one of two possible results, $f(x_1) > f(x_2)$), and looked only to the next iteration. The reader should examine the case $f(x_2) > f(x_1)$ and should be able to show that the claim will always be true, as one proceeds from any one iteration to the next.

The relationship between L_N and N is simple in this case and is

$$L_N = (.618)^{N-1}, \quad N = 1, 2, \ldots \qquad (6.12)$$

for an initial interval $[0, 1]$, since with N evaluations there are $N - 1$ iterations, each of which produces an interval .618 of the last.

Given an initial interval of length L, we would have, in general,

$$L_N = (.618)^{N-1}L \qquad (6.13)$$

Also, if the initial interval is $[a, b]$, the placement of the observations must reflect the change; for example, one first evaluates at $a + .382(b - a)$ and at $a + .618(b - a)$, and so on. Regardless of the intervals involved, the guiding rule is quite simple: f must, at each iteration, be evaluated at two points symmetric about the midpoint of the interval: the leftmost being .618 of the length of the entire interval from the right end point of the interval, and the rightmost point has the same distance from the left end point of the interval (Figure 6.12).

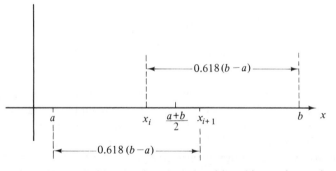

Figure 6.12. Golden-Section Search with arbitrary interval $[a, b]$.

The search technique we have discussed in this subsection is not uniformly designated. It has been called the *golden-section search* [1] and the *golden-mean search* [3]. These names arise from the fact that interest in the bisection of a line segment into two unequal segments such that the ratio of the shorter to the longer equals that of the longer to the whole dates back to the Greek geometers. Further remarks on this subject may be found in References [1] and [2]. Interesting examples of the use of this golden ratio in art and architecture may be found in Reference [2].

The use of the golden section in art is by no means restricted to works from antiquity. The striking "The Sacrament of the Last Supper" by Salvador

Dali, which may be observed in the National Gallery of Art, Washington, D.C., utilizes the golden section in addition to other geometric constructs.

i. *An Optimal Search—The Fibonacci Search*

The last one-dimensional search technique we mention here is, finally, an optimal one. As the golden-section search, the *Fibonacci search*, has ties with the distant past, and the reader would find Wilde's detailed historical notes on these ties of interest [3].

To summarize briefly, a thirteenth-century Italian, one Leonardo de Pisa, was interested in describing the month-by-month growth of a population of rabbits, and the sequence of integers which served as his model for that particular process is known as the *Fibonacci sequence*, $\{F_i\}$, $i = 0, 1, 2, \ldots$. F_i is called the ith Fibonacci number, whose value is given by the recurrence relation

$$F_0 = F_1 = 1$$
$$F_{i+1} = F_i + F_{i-1}, \quad i \geq 1$$

The first few terms of the sequence are

$$1, 1, 2, 3, 5, 8, 13, 21, 34, 55, 89, 144, 233, \ldots$$

Clearly, the numbers begin to grow large very rapidly once beyond the first few terms (given the fecundity of the species, we would anticipate this); that will prove a most favorable property in the search application.

It is also worthwhile to digress a bit and mention that the Fibonacci sequence arises quite naturally in a diversity of areas. For example, many plants whose leaves grow in a spiral fashion exhibit leaf patterns identical to two adjacent terms of the Fibonacci sequence. (See, e.g., Reference [2] for several interesting examples.) In general, the diversity of problems in which the Fibonacci sequence arises naturally is quite remarkable.

The Fibonacci search is such that it can reduce a given extremal interval to $1/F_N$ its original size with just N function evaluations. A particularly clean example is obtained by letting the initial interval be $[0, F_N]$ for some positive integer N, so that N function evaluations will provide an extremal interval of unit length (actually, $1 + \epsilon$, where ϵ is the minimum separation constant). Our strategy in the example will be to reduce the extremal interval to length F_{N-1} after the first iteration, F_{N-2} after the second, F_{N-i} after the ith, and $F_{N-(N-1)} = F_1 = 1$ after the $(N - 1)$st iteration. As in the case of the golden-section search, $N - 1$ iterations will require N evaluations.

Example 6.1

Let us assume our initial interval is $[0, 21] = [0, F_7]$, and to be specific, f will be a unimodal function as in Figure 6.13. To guarantee that the next extremal interval will have length $F_6 = 13$, it is necessary to first evaluate f at $x_1 = 8$ and at $x_2 = 13$ (Figure 6.13), since with $x_1 < x_2$, the next extremal interval will be either $(x_1, 21]$, or $[0, x_2)$, and these must have equal length.

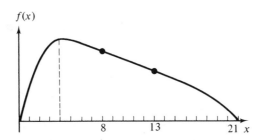

Figure 6.13. First function evaluations of Fibonacci Search.

Clearly, $f(8) > f(13)$, so $[0, 13)$ is next to be considered. Now $F_5 = 8$, so we must next evaluate f at $x_3 = 5$ and at $x_4 = 8$ (Figure 6.14), for only then are we guaranteeing that the next extremal interval will have length 8. But $f(8)$ is already known, so we evaluate at $x_3 = 5$ only.

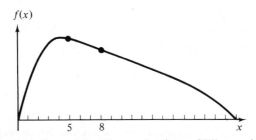

Figure 6.14. Second function evaluations of Fibonacci Search.

Since $f(5) > f(8)$, we next deal with $[0, 8)$. $F_4 = 5$, so we evaluate f at $x_5 = 3$ and at $x_6 = 5$ (Figure 6.15); however, $x_5 = 3$ is the only *new* observation required. $f(5) > f(3)$, and $(3, 8)$ is the extremal interval.

Since $F_3 = 3$, the next iteration demands that $x_7 = 5$ and $x_8 = 6$. Thus $x_8 = 6$ is the one new point, $f(5) > f(6)$, and $(3, 6)$ is the new extremal interval.

To assure that the next extremal interval has length $F_2 = 2$, we must evaluate f at $x_9 = 4$ and at $x_{10} = 5$ (Figure 6.16), the second evaluation hav-

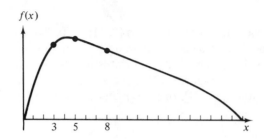

Figure 6.15. Third function evaluations of Fibonacci Search.

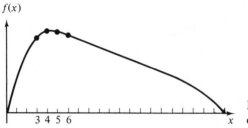

Figure 6.16. Fourth function evaluations of Fibonacci Search.

ing been accomplished. Since $f(4) > f(5)$, we have narrowed the interval to $(3, 5)$ and have $f(4)$.

Finally, we evaluate f at either $4 - \epsilon$ or $4 + \epsilon$, and the last extremal interval has length $1 + \epsilon$. Suppose that we chose $4 + \epsilon$. Altogether, then, f has been evaluated at $x = 8$, $x = 13$, $x = 5$, $x = 3$, $x = 6$, $x = 4$, and $x = 4 + \epsilon$—seven evaluations.

Given an arbitrary initial interval of length L, then, N Fibonacci observations will yield a final extremal interval of length

$$L_N = \left(\frac{1}{F_N}\right)L \tag{6.14}$$

In the general case the first two function evaluations take place at distances $(F_{N-2}/F_N)L$ and $(F_{N-1}/F_N)L$ from the left end point, respectively; that is, starting with an interval $[a, b]$, f is first evaluated at $a + (F_{N-2}/F_N)L$ and $a + (F_{N-1}/F_N)L$, where $b - a = L$ (Figure 6.17).

Thus the next interval of uncertainty will be either $[a, a + (F_{N-1}/F_N)L]$ or $(a + (F_{N-2}/F_N)L, b]$, the first of which has length $(F_{N-1}/F_N)L$, while the second has length

$$b - a - \left(\frac{F_{N-2}}{F_N}\right)L = L\left[1 - \frac{F_{N-2}}{F_N}\right] = L\left[\frac{F_N - F_{N-2}}{F_N}\right]$$

But we have $F_N = F_{N-1} + F_{N-2}$, so the second interval also has length $(F_{N-1}/F_N)L$.

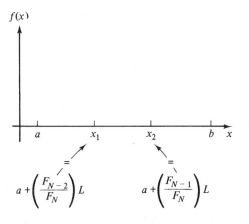

Figure 6.17. Location of first two evaluation points in Fibonacci Search for initial arbitrary interval $[a, b]$.

During iteration 2, we shall need function evaluations at F_{N-3}/F_{N-1} and F_{N-2}/F_{N-1} of the distance from the left end point of the extremal interval resulting from iteration 1. For definiteness, suppose that we had $f(x_2) > f(x_1)$; then

$$\frac{F_{N-3}}{F_{N-1}}\left(\frac{F_{N-1}}{F_N}\right)L = \left(\frac{F_{N-3}}{F_N}\right)L$$

$$\frac{F_{N-2}}{F_{N-1}}\left(\frac{F_{N-1}}{F_N}\right)L = \left(\frac{F_{N-2}}{F_N}\right)L$$

Consequently, we need function values for

$$x_3 = a + \left(\frac{F_{N-2}}{F_N}\right)L + \left(\frac{F_{N-3}}{F_N}\right)L$$

and

$$x_4 = a + \left(\frac{F_{N-2}}{F_N}\right)L + \left(\frac{F_{N-2}}{F_N}\right)L$$

However, $x_3 = a + (F_{N-1}/F_N)L = x_2$, so one of the two points has already been examined, and we find the Fibonacci search, like the golden section, to be frugal with information. The extremal interval after iteration 2 has length $(F_{N-2}/F_N)L$.

At the ith iteration, the points of evaluation needed are located at F_{N-i}/F_{N-i+1}, F_{N-i-1}/F_{N-i+1} of the distance from the left end point of the extremal interval existing after the $(i - 1)$st iteration. These general expressions, for $i = N - 1$, the last iteration, become $F_1/F_2 = \frac{1}{2}$ and $F_0/F_2 = \frac{1}{2}$.

We shall always find, however, that just prior to the last iteration we have argument values w_1, w_2, w_3, where $w_2 - w_1 = w_3 - w_2$ and $f(w_2) > \max \{f(w_1), f(w_3)\}$ (Figure 6.18), so that we have an extremal interval of length $(2/F_N)L$ with the function already evaluated at the midpoint of $[w_1, w_3]$. Our

368 / *Search Techniques*   *Ch. 6*

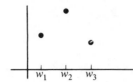

Figure 6.18. Final evaluation points at last iteration of Fibonacci Search.

general expressions indicate that we next evaluate f at w_2, but this has already been done; so the last evaluation occurs at one of $w_2 \pm \epsilon$, and we essentially halve the existing interval, ending the search with an extremal interval of length

$$\frac{1}{2}\left(\frac{2}{F_N}\right)L + \epsilon = \left(\frac{1}{F_N}\right)L + \epsilon$$

We again leave as an exercise the demonstration that one of our two new points—with the obvious exception of the first iteration—is always actually an "old" point. We omit the proof of the minimax optimality of the Fibonacci search procedure. It is suggested, however, that the student, for small values of N, satisfy himself that, in the minimax sense, it is impossible to place the observations in such a way as to improve upon this procedure.

Following is another example of the Fibonacci search procedure in a more general setting.

Example 6.2

Suppose that f is unimodal on $[-1, 5]$ and it is desired to locate the global minimum within an interval of length .1. Since

$$\frac{6}{F_9} = \frac{6}{55} > .1 > \frac{6}{89} = \frac{6}{F_{10}}$$

10 function evaluations will suffice.

Rather than selecting a mathematical expression for our function, we shall merely indicate its shape on the given interval, as in Figure 6.19.

The first evaluation points will be

$$-1 + \frac{F_8}{F_{10}}6 = -1 + \frac{34}{89}6 \simeq 1.3$$

and

$$-1 + \frac{F_9}{F_{10}}6 = -1 + \frac{55}{89}6 \simeq 2.7$$

and our new interval is

$$(1.3, 5]$$

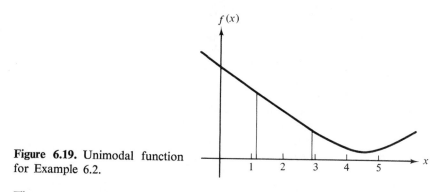

Figure 6.19. Unimodal function for Example 6.2.

The next two points are

$$1.3 + \frac{F_7}{F_9}3.7 = 1.3 + \frac{21}{55}3.7 \simeq 2.7$$

and

$$1.3 + \frac{F_8}{F_9}3.7 = 1.3 + \frac{34}{55}3.7 \simeq 3.5$$

The new interval becomes

$$(2.7, 5]$$

The next points are

$$2.7 + \frac{F_6}{F_8}2.3 = 2.7 + \frac{13}{34}2.3 \simeq 3.5$$

$$2.7 + \frac{F_7}{F_8}2.3 = 2.7 + \frac{21}{34}2.3 \simeq 4.1$$

and so on.

j. COMPARISON OF METHODS

In Table 6.1 we have exhibited relationships between N, the number of function evaluations, and L_N, the length of the extremal interval after N evaluations, for purposes of comparing the various search techniques discussed heretofore. It should be realized that the comparison is not entirely equitable, since the exhaustive search does not benefit from the unimodality assumption, whereas the others are useless without it. For the dichotomous search several different values of ϵ are used. In all other methods, ϵ is assumed to be so small that we never violate its significance in the course of the search. In addition, solving one of the equations relating L_N to N as if N were a continuous variable and then selecting the next highest positive integer may yield an integer that is not feasible for that particular search technique, e.g.,

```
0001          DIMENSION F(500)
0002     1    READ(5,1000)A,B,ALPHA
0003    1000  FORMAT(3F10.5)
0004          WRITE(6,1005)
0005    1005  FORMAT(1H1,50X,'FIBONACCI SEARCH'////)
0006          BB=B
0007          AA=A
0008          UNCIV=B−A
0009          FNPR=UNCIV/ALPHA
0010          F(1)=1.
0011          F(2)=1.
0012          DO 5 I=3,500
0013          F(I)=F(I−1)+F(I−2)
0014          IF(FNPR−F(I))6,6,5
0015     5    CONTINUE
0016     6    IFN=I
0017          WRITE(6,1007)A,B,ALPHA,IFN,IFN,F(IFN)
0018    1007  FORMAT(5X,'THE INTERVAL TO BE SEARCHED EXTENDS FROM  ',F10.5,'  TO
               1 ',F10.5//5X,'THE DESIRED INTERVAL OF UNCERTAINTY IS ',E13.5//5X
               2 ,13,' FIBONACCI NUMBERS WERE GENERATED IN ORDER TO ACHIEVE THIS F
               3 INAL INTERVAL OF UNCERTAINTY'//5X,'THIS ',13,'TH FIBONACCI NUMBER
               4 IS ',E13.5///)
0019    15    P1=AA+(F(IFN−1)/F(IFN))*UNCIV
0020          P2=AA+(F(IFN−2)/F(IFN))*UNCIV
0021          Y1=FUNCTN(P1)
0022          Y2=FUNCTN(P2)
0023          IF(Y1−Y2)10,11,12
0024    12    AA=P2
0025          YMAX=Y1
0026          GO TO 13
```

```
0027    10 BB=P1
0028       YMAX=Y2
0029       GO TO 13
0030    11 BB=P1
0031       AA=P2
0032       YMAX=Y2
0033    13 UNCIV=BB-AA
0034       IFN=IFN-1
0035       IF(IFN-3)20,15,15
0036    20 WRITE(6,1010)AA,BB,YMAX
0037  1010 FORMAT(5X,'THE FINAL INTERVAL OF UNCERTAINTY EXTENDS FROM  ',E13.5
      1' TO  ',E13.5//5X,'THE MAXIMUM VALUE OF THE FUNCTION OBTAINED FO
      2R THIS INTERVAL IS  ',E13.5,/1H1)
0038       GO TO 1
0039       END


0001       REAL FUNCTION FUNCTN(P)
0002       FUNCTN=3.+6.*P-4.*(P**2)
0003       RETURN
0004       END
```

FORTRAN program for Fibonacci search

an odd integer for the dichotomous search or the value 8 for a three-point equal-interval search. In such cases N has been increased to the next highest feasible integer. All initial intervals are assumed to have unit length.

Table 6.1

	Values of L_N				
Technique	*.00001*	*.0001*	*.001*	*.01*	*.1*
Exhaustive	199,999	19,999	1999	199	19
Dichotomous, $\epsilon = 10^{-6}$	34	28	20	14	8
Dichotomous, $\epsilon = 10^{-5}$		28	20	14	8
Dichotomous, $\epsilon = 10^{-4}$			20	14	8
Two Point, equal interval	56	46	36	24	12
Three Point, equal interval	35	29	21	15	9
Four Point, equal interval	52	44	32	24	12
Golden section	26	21	16	11	6
Fibonacci	25	20	16	11	6

k. Golden Section Versus Fibonacci

If the reader will accept that the Fibonacci search is optimal, then he will also accept the statement that the golden-section search is near optimal. In our Table 6.1 the advantage of the former over the latter is in no case greater than a single function evaluation. The reason for this is easily obtained. If we begin computing F_i/F_{i+1}, $i = 0, 1, 2, \ldots$, we develop the sequence $\{1, \frac{1}{2}, \frac{2}{3}, \frac{3}{5}, \frac{5}{8}, \frac{8}{13}, \frac{13}{21}, \frac{21}{34}, \frac{34}{55}, \frac{55}{89}, \frac{89}{144}, \frac{144}{233}, \ldots\}$. These ratios quickly approach, and remain near, .618, which is precisely the factor of reduction of the golden-section search, the ratio F_i/F_{i+1} being the reduction factor in the Fibonacci search.

l. Computer Program for Fibonacci Search

We include a documented listing of a working FORTRAN program for the Fibonacci search. The reader is encouraged to examine the program, understand its operation, and run it on several sample functions of his own choosing. The program is presently set to seek the maximum of the function

$$f(x) = -4x^2 + 6x + 3$$

Treating other functions would necessitate only modifying the subroutine. The inputs are A = the left end point of the initial interval; B = the right end point; and ALPHA = the desired final length of interval.

6.3. Functions of Several Variables

In preceding chapters, more often than not, we found that the degree of difficulty of a problem increases substantially when the number of variables involved increases. This is again the case within the present context.

Let us consider the problem of finding the global maximum of

$$f = f(x_1, x_2, \ldots, x_n)$$

subject to

$$a_j \leq x_j \leq b_j, \quad j = 1, \ldots, n$$

The inequalities may be constraints in our usual sense, or the problem may actually be unconstrained, with the global maximum known, somehow, to lie in the region described. We will see in Chapter 7 that constraints can serve to simplify a solution procedure. In much of the remainder of this chapter, such constraints will be indispensable. In fact, our one-dimensional procedures relied heavily upon the fact that a constraint of that form was given— consider the problem of finding the global max of a unimodal function over the entire real line.

i. *Exhaustive Search in n Dimensions*

Applying the *n*-dimensional exhaustive search to the present problem is perhaps the best way to appreciate the magnitude of our task. Suppose it is desired to locate that hypercube with side δ in which the global maximum is known to lie, where δ is known to be "sufficiently small" to make the procedure valid. Thus it is necessary to partition $[a_j, b_j]$, obtaining a discrete set S_j, and to evaluate f at each point of $S_1 \times S_2 \times \cdots \times S_n$.

In two dimensions, for example, the diagram in Figure 6.20 indicates the procedure. The function f must be evaluated at each point indicated by a dot in the region $a_1 \leq x_1 \leq b_1, a_2 \leq x_2 \leq b_2$. The encircled dot represents that point providing the largest function value; it is the center of the shaded square with side δ, which is the final extremal region.

In three dimensions the final extremal region will be a cube with side of length δ and whose center is that point which provided the largest function value.

We have found previously that to partition $[a_j, b_j]$ requires N_j points, where

$$(N_j + 1)\frac{\delta}{2} = b_j - a_j$$

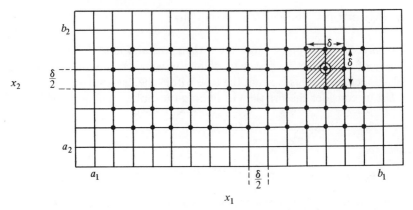

Figure 6.20. Exhaustive search in two dimensions.

or
$$N_j = \frac{2}{\delta}(b_j - a_j) - 1$$

Consequently, the totality of evaluation points required for the n-dimensional exhaustive search numbers

$$\prod_{j=1}^{n} N_j = \prod_{j=1}^{n} \left[\frac{2}{\delta}(b_j - a_j) - 1 \right] \tag{6.15}$$

which for reasonably small δ and even moderate n is a very large number. In fact, for $n = 10$, $\delta = .1$, and $b_j - a_j = 1$ for all j, we obtain 19^{10}.

With this observation the reader will, quite likely, suspect that assumption-making time is upon us again. Let us see what kinds of assumptions on f might be helpful.

ii. *Types of Unimodality in n Dimensions*

One might anticipate some n-dimensional analogues of unimodality, and this is, in fact, conventional.

a. UNIMODALITY

Let us first assume that there exists a unique global maximum $\mathbf{x}^*$, that is $f(\mathbf{x}^*) > f(\mathbf{x})$ for all $\mathbf{x} \in S$, where $S = \{\mathbf{x} \mid a_j \leq x_j \leq b_j, j = 1, \ldots, n\}$. The following definitions are then meaningful.

A function f is said to be *unimodal* (over S) if there exists a path from $\mathbf{x}$ to $\mathbf{x}^*$ over which f is strictly increasing, for all $\mathbf{x} \in S$. We observe that functions that have nonoptimal relative maxima are not unimodal, since if $\mathbf{x}'$ is such a point, then no such path can exist. In fact, if a function has no nonoptimal relative maxima, then it must be possible to construct a path from **x** to **x*** such that f is strictly increasing over that path. Thus f is unimodal if and only if f has no nonoptimal relative maxima.

We shall find contour graphs of functions of two variables quite useful in the sequel. We exhibit the contour graphs of both a unimodal and non-unimodal function of two variables in Figure 6.21.

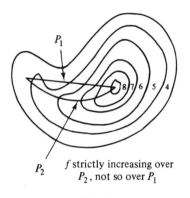

f strictly increasing over
P_2, not so over P_1

(a)

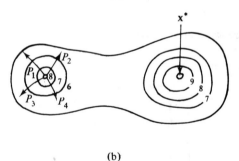

Figure 6.21. Unimodal (a) and
non-unimodal (b) functions.   (b)

The global maximum of a unimodal function may be found by the following conceptual procedure. Choose a point $x_1 \in S$, and locate a path from x_1 over which f begins to increase monotonically (one must exist), and follow that path until f no longer increases, stopping at a point x_2, whereupon another path of strict increase is chosen, and so on. Although the process might require an infinite amount of time, it would end with the finding of **x***, from which no path exists.

b. STRONG UNIMODALITY

A stronger property is possessed by those functions called *strongly unimodal,* by which we term those functions f such that f is strictly increasing over the *straight-line path* from **x** to **x*** for all $x \in S$. Clearly, f is strongly unimodal implies f is unimodal (in one dimension) over every straight line segment that contains x^*. If we now consider conducting our conceptual search, we see that we are allowed to examine only straight-line paths from each successive point. Of course, strong unimodality implies unimodality. Contour maps in Figure 6.22 illustrate both strongly and nonstrongly unimodal functions.

Over the path x' to x^* in Figure 6.22b, f appears as in Figure 6.23.

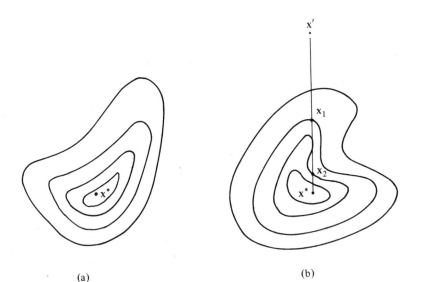

(a)   (b)

Figure 6.22. Strongly (a) and non-strongly (b) unimodal functions.

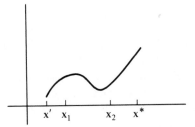

Figure 6.23. f values over the path of Figure 6.22(b).

c. LINEAR UNIMODALITY

A final classification is that of the *linearly unimodal* functions, which are defined as those functions f with the property that they are unimodal (in one dimension) over the straight line segment joining any pair of points x', $x'' \in S$. We observe that linear unimodality implies strong unimodality. We include examples of functions both possessing and lacking this property in Figure 6.24.

The reader should not be disturbed by the references to unimodality in one dimension in the foregoing definitions. The line segments over which we considered f lay in n-dimensional space, it is true, but we know such a segment has a "one-dimensional" representation. Indeed, the segment joining $\mathbf{x}'$, $\mathbf{x}''$ may be generated by $\lambda\mathbf{x}'' + (1 - \lambda)\mathbf{x}'$, $0 \leq \lambda \leq 1$.

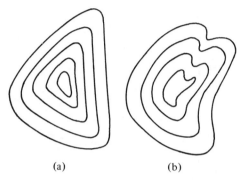

Figure 6.24. Linearly (a) and non-linearly (b) unimodal functions.

(a)   (b)

Also, it should be clear that since we have been speaking in terms of a global maximum, when we have mentioned unimodality in one dimension, we have implied part 1 of our definition of unimodal functions in one variable.

d. CONCEPTUAL SEARCH PROCEDURES

A conceptual search procedure for finding the global maximum of a linearly unimodal function is simpler than for unimodal or strongly unimodal functions. Given an initial point $\mathbf{x}_0 \in S$, select any other point $\mathbf{x}_0' \in S$ and maximize f over that portion of the line segment through $\mathbf{x}_0$, $\mathbf{x}_0'$ which lies in S. This one-dimensional maximization produces a point $\mathbf{x}_1$, whereupon we choose some other point $\mathbf{x}_1'$, maximize over the segment through $\mathbf{x}_1$, $\mathbf{x}_1'$ find $\mathbf{x}_2$, and so on. Of course, a maximization may produce the point with which one began that interation, $\mathbf{x}_i$, whereupon one tries again, replacing $\mathbf{x}_i'$ with some $\mathbf{x}_i'' \neq \mathbf{x}_i'$. This procedure, like the previous conceptual searches, has no stopping criterion; that is, if it arrived at the global maximum, it would continue attempting to find improvement.

Figure 6.25 depicts a realization of our conceptual search with a linearly unimodal function f of two variables. Since each of our one-dimensional searches involves a unimodal function, *we may use any of the sequential strategies for univariate functions previously discussed.* The ability to decompose the n-dimensional search into a sequence of one-dimensional searches will be exploited repeatedly in what follows.

The reader will have noticed that functions which come to mind in defining the various kinds of unimodality bear a certain semblance to concave

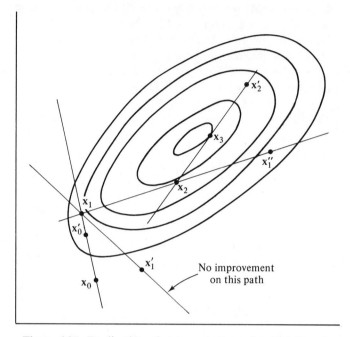

Figure 6.25. Realization of a conceptual search with a linearly unimodal function.

functions and may have wondered what the precise relationships are. First, concave functions need not have a unique global maximum, and if comparisons are to be drawn, we must immediately restrict attention to strictly concave functions.

Exercise 6.11

Prove or give a counterexample to the following: f strictly concave on S implies f is linearly unimodal on S.

Exercise 6.12

Prove or give a counterexample to the converse of the statement in Exercise 6.11.

Exercise 6.13

What additional property might the linearly unimodal function be given in order to make possible the establishment of the converse of Exercise 6.11?

The answers to the questions resulting from the substitution of "unimodal" or "strongly unimodal" for "linearly unimodal" in the three exercises are then apparent.

iii. *A Note on the One-Dimensional Maximizations*

Prior to taking up particular n-dimensional search techniques, let us establish a property the reader may have noticed in our conceptual maximization procedures. Suppose that we have a point x_0, choose a direction, and maximize f over the path defined. Let the maximizing point be denoted x_1.

Any point on the half-line originating at x_0 and passing through x_1 may be expressed as

$$\mathbf{x} = \lambda \mathbf{x}_1 + (1 - \lambda)\mathbf{x}_0, \quad \lambda \geq 0 \tag{6.16}$$

Therefore, on that half-line

$$f(\mathbf{x}) = f(\lambda \mathbf{x}_1 + (1 - \lambda)\mathbf{x}_0) = g(\lambda)$$

Now
$$\frac{dg}{d\lambda} = \sum_{j=1}^{n} \frac{\partial f}{\partial x_j} \cdot \frac{dx_j}{d\lambda} = \sum_{j=1}^{n} \frac{\partial f}{\partial x_j}(x_{1j} - x_{0j}) \tag{6.17}$$

in which

$$\mathbf{x}_0 = (x_{01}, x_{02}, \ldots, x_{0n}), \qquad \mathbf{x}_1 = (x_{11}, x_{12}, \ldots, x_{1n})$$

We must have $dg(1)/d\lambda = 0$, since $\lambda = 1$ is the maximizing value for $g(\lambda)$; thus, from equations (6.16) and (6.17),

$$\sum_{j=1}^{n} \frac{\partial f(\mathbf{x}_1)}{\partial x_j}(x_{1j} - x_{0j}) = 0$$

We now define the gradient of f, ∇f, as

$$\nabla f = \left(\frac{\partial f}{\partial x_1}, \frac{\partial f}{\partial x_2}, \ldots, \frac{\partial f}{\partial x_n} \right)$$

and write this last condition as

$$\nabla f(\mathbf{x}_1)(\mathbf{x}_1 - \mathbf{x}_0) = 0$$
or
$$\nabla f(\mathbf{x}_1)(\mathbf{x}_0 - \mathbf{x}_1) = 0$$

For any real number C, then,

$$\nabla f(\mathbf{x}_1)(C(\mathbf{x}_0 - \mathbf{x}_1)) = 0 \tag{6.18}$$

Considering $\mathbf{x}$ on the half-line in question, we obtain from (6.18)

$$\begin{aligned}
\nabla f(\mathbf{x}_1)(\mathbf{x} - \mathbf{x}_1) &= \nabla f(\mathbf{x}_1)[\lambda \mathbf{x}_1 + (1 - \lambda)\mathbf{x}_0 - \mathbf{x}_1] \\
&= \nabla f(\mathbf{x}_1)[(\lambda - 1)(\mathbf{x}_1 - \mathbf{x}_0)] \\
&= 0
\end{aligned}$$

But we recognize $\nabla f(\mathbf{x}_1)(\mathbf{x} - \mathbf{x}_1) = 0$ as the equation of the tangent hyperplane to the level surface $f(\mathbf{x}) = f(\mathbf{x}_1)$ at the point $\mathbf{x} = \mathbf{x}_1$. Therefore, the path over which we maximize will be tangent to a level surface at the point of maximization. If the reader will reflect upon the process of maximization over a particular path, this property will be seen to be intuitively obvious. In Figure 6.26 we exhibit this property. P is the path over which we maximize, $\mathbf{x}_1$ is the maximizing point for f over the path P, and $\mathbf{x}_2$ is the maximizing point over P'.

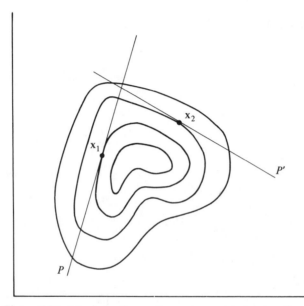

Figure 6.26. Maximization paths are tangent to level surfaces.

iv. *Sequential Searches*

a. ONE-AT-A-TIME SEARCH

The first n-dimensional search we shall formally describe is the *one-at-a-time or univariate search*.

A point $\mathbf{x}_0 \in S$ is first selected and a coordinate direction specified, say x_{i_1}, $i_1 \in \{1, \ldots, n\}$. The first task is to maximize f over that portion of the line through $\mathbf{x}_0$ parallel to the x_{i_1} axis which lies within S. This maximization produces a point $\mathbf{x}_1$, whereupon a direction x_{i_2}, $i_2 \neq i_1$, is chosen, and a maximization in the x_{i_2} direction performed, and so on. The last portion of the first iteration consists of maximizing f over that portion of the line through $\mathbf{x}_{n-1}$ and parallel to the x_{i_n} axis which lies in S, and $\mathbf{x}_n$ is found as a result.

The second iteration begins with $\mathbf{x}_n$ as initial point and successive one-dimensional maximizations in the directions $x_{i_1}, x_{i_2}, \ldots, x_{i_n}$.

A portion of the process is illustrated in Figure 6.27 for a bivariate function. We have arbitrarily specified $i_1 = 2$, $i_2 = 1$. In general, the sequence of directions used could be changed from iteration to iteration.

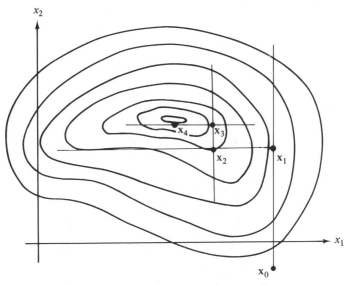

Figure 6.27. A one-at-a-time search.

When is the search stopped? One might allow the search to continue so long as improvement in objective function values occurs. However, one might find that a large number of additional iterations—perhaps a large quantity of time or money—were devoted to obtaining a very slight improvement beyond a value already obtained. Thus one might rephrase the rule as continuation so long as "worthwhile" improvement is achieved, our subjective terminology implying that the problem solver can weigh the advantages of slight improvement against an increased searching cost. Such a dilemma will only arise when it is noticed that a succession of one-dimensional searches is producing new values only slightly greater than their immediate predecessors. Here again, as in the case of the exhaustive search in one dimension, some knowledge of the behavior of the objective function can be invaluable. For example, knowledge that no sharp "spikes" exist on the surface being searched would lead one to terminate the search when, for example, each search of one complete iteration produced slight improvement—perhaps improvement less than some predetermined constant obtained from the knowledge of the function's behavior.

b. Some Complications

In the previous example the search worked quite well; two complete iterations—four maximizations—served to provide x_4 quite close to x^*. However, the process breaks down at x_2 in the example of Figure 6.28; movement in either coordinate direction produces only function values smaller than $f(x_2)$. Let us examine the corresponding surface in three dimensions illustrated in Figure 6.29.

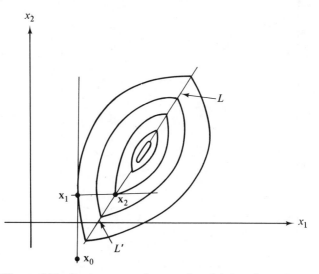

Figure 6.28. Contour map for a unimodal function with a ridge.

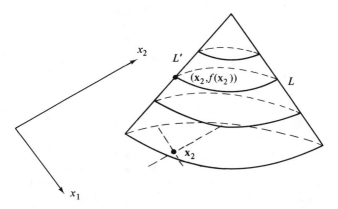

Figure 6.29. Surface representation of a unimodal function with a ridge.

The line segments L, L' of Figure 6.28, which we have transcribed into Figure 6.29, are called ridges, and Figure 6.29 illustrates the topographic appeal possessed by this terminology. We have become stuck on a ridge— movement from x_2 in the coordinate directions, indicated by the dashed lines through x_2, produces only smaller function values.

Ridges in general plague n-dimensional search techniques, but let us return to Figure 6.28 and *rotate the coordinate axes somewhat* (Figure 6.30).

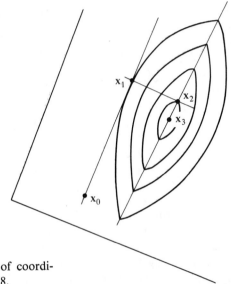

Figure 6.30. Rotation of coordinate axes in Figure 6.28.

Now things work much better. By aligning the ridges in the direction of one of the coordinate axes, we not only avoid stalling on the ridge, we actually use the ridge to advantage, moving along it to quite near the global maximum.

A common variation on the univariate search, the method of rotating coordinates, accomplishes this rotation by selecting a new set of n mutually orthogonal† search directions at the outset of each iteration. We shall not pursue the details of this refinement, but the reader will understand that selecting new search directions might provide one with a means for escaping entrapment on a ridge. It should be noticed that our one-at-a-time method is not guaranteed to lead to the global maximum, even if the objective function is assumed linearly unimodal. The function in Figure 6.28, for example, is linearly unimodal, but its ridge confounds the procedure.

It should also be observed that unless f is linearly unimodal there is no

---

†A set $\{\alpha_1, \alpha_2, \ldots, \alpha_k\}$ *is orthogonal if* $\alpha_i \alpha_j = 0$ *when and only when* $i \neq j$. In two and three dimensions, orthogonal implies perpendicular.

guarantee that the one-dimensional maximizations can be carried out by sequential search methods; f need not be unimodal over those paths if f is unimodal or strongly unimodal.

c. PRACTICAL CONSIDERATIONS FOR ONE-DIMENSIONAL SEARCHES

To utilize any of the sequential searches in one variable previously discussed, it is necessary to know over what interval the search is to take place. In the univariate search, for example, this knowledge is not explicitly available—one has a point $\mathbf{x}_i \in E^n$ and a coordinate direction, say x_k.

The line through $\mathbf{x}_i$ in the direction x_k is exactly the set

$$L = \{\mathbf{y} \,|\, \mathbf{y} = \mathbf{x}_i + \lambda\mathbf{e}_k,\ \lambda \text{ real}\}$$

where $\mathbf{e}_k$ is the kth unit vector. A typical $\mathbf{y} \in L$ has the form

$$\mathbf{y} = \begin{pmatrix} y_1 \\ y_2 \\ \cdot \\ \cdot \\ \cdot \\ y_n \end{pmatrix} = \begin{pmatrix} x_{i1} \\ x_{i2} \\ \cdot \\ \cdot \\ \cdot \\ x_{in} \end{pmatrix} + \lambda \begin{pmatrix} 0 \\ \cdot \\ \cdot \\ 1 \\ \cdot \\ \cdot \end{pmatrix}$$

As λ is increased from zero, the points generated eventually reach and depart from the set S. The boundary is reached for $\lambda = \lambda_1$ such that $x_{ik} + \lambda_1 = b_k$, that is, for $\lambda = \lambda_1 = b_k - x_{ik}$. Similarly, decreasing λ from zero results in eventually leaving S in the negative x_k direction; so the value of λ corresponding to the other end point of the segment of interest is

$$\lambda = \lambda_2 = a_k - x_{ik}$$

Thus, and it is obvious without the foregoing, the end points of our segment in E^n are

$$\mathbf{x}_i' = \begin{pmatrix} x_{i1} \\ x_{i2} \\ \cdot \\ \cdot \\ \cdot \\ x_{i,k-1} \\ a_k \\ x_{i,k+1} \\ \cdot \\ \cdot \\ \cdot \\ x_{in} \end{pmatrix} \quad \text{and} \quad \mathbf{x}_i'' = \begin{pmatrix} x_{i1} \\ x_{i2} \\ \cdot \\ \cdot \\ \cdot \\ x_{i,k-1} \\ b_k \\ x_{i,k+1} \\ \cdot \\ \cdot \\ \cdot \\ x_{in} \end{pmatrix}$$

and the distance between these points is

$$d = \| \mathbf{x}_i' - \mathbf{x}_i'' \| = \left[\sum_{j=1}^{n} (x_{ij}' - x_{ij}'')^2 \right]^{1/2}$$
$$= b_k - a_k$$

Points $\mathbf{x}$ along the segment may be generated by varying λ over $[0, 1]$ and finding

$$\mathbf{x} = \lambda \mathbf{x}_i'' + (1 - \lambda)\mathbf{x}_i' \tag{6.19}$$

Thus points of function evaluation needed in the one-dimensional search technique may be found by varying λ; for example, the midpoint of the segment corresponds to $\lambda = \frac{1}{2}$, the point at distance .618 the length of the interval from $\mathbf{x}_i'$ corresponds to $\lambda = .618$, and so on. However, in the case of movement in a coordinate direction, one need not bother with expression (6.19). Points of interest such as in the examples immediately preceding are more easily constructed than with the use of (6.19). (How?)

Next let us imagine that the direction of interest is not necessarily a coordinate direction; rather, suppose that it is described by some arbitrary n vector $\mathbf{v}$. The corresponding line may be expressed as

$$L = \{\mathbf{y} \,|\, \mathbf{y} = \mathbf{x}_i + \lambda \mathbf{v}, \, \lambda \text{ real}\}$$

Consider now

$$\mathbf{y} = \begin{pmatrix} x_{i1} \\ x_{i2} \\ \cdot \\ \cdot \\ \cdot \\ x_{in} \end{pmatrix} + \lambda \begin{pmatrix} v_1 \\ v_2 \\ \cdot \\ \cdot \\ \cdot \\ v_n \end{pmatrix}$$

Let λ_1 be that value of λ for which $x_{ik} + \lambda v_k \le b_k$, $k = 1, \ldots, n$, and $x_{ir} + \lambda v_r = b_r$ for at least one $r \in \{1, \ldots, n\}$. Similarly, let λ_2 be that value of λ for which $x_{ik} + \lambda v_k \ge a_k$, $k = 1, \ldots, n$, and $x_{ip} + \lambda v_p = a_p$ for at least one $p \in \{1, \ldots, n\}$.

Then
$$\mathbf{x}_i' = \mathbf{x}_i + \lambda_1 \mathbf{v}, \qquad \mathbf{x}_i'' = \mathbf{x}_i + \lambda_2 \mathbf{v}$$

are the end points of the line segment of interest. The segment has length $|\lambda_2 - \lambda_1| \, \|\mathbf{v}\|$, so that if instead of $\mathbf{v}$ we used direction $\mathbf{v}/\|\mathbf{v}\|$, the segment would have length $|\lambda_2 - \lambda_1|$. Again, the search may be conducted on the segment in terms of a parameter λ that varies over $[0, 1]$.

Also of interest is the concept of the half-line from $\mathbf{x}_i$ in the direction $\mathbf{v}$.

This is defined to be the set

$$H = \{\mathbf{y} \,|\, \mathbf{y} = \mathbf{x}_i + \lambda \mathbf{v}, \, \lambda \geq 0\}$$

The reader should sketch an H in two dimensions; the motivation for the terminology half-line will then be evident.

Example 6.3

We wish to illustrate the univariate search by means of a simple example. Let

$$f(x_1, x_2, x_3) = -x_1^2 + 2x_1x_2 - 3x_2^2 + 4x_2x_3 - 5x_3^2$$

Let the region of interest S be defined by the rectangular parallelepiped

$$-2 \leq x_1 \leq 3, \quad -3 \leq x_2 \leq 3, \quad -1 \leq x_3 \leq 2$$

We have

$$\nabla f(x_1, x_2, x_3) = (-2x_1 + 2x_2, 2x_1 - 6x_2 + 4x_3, 4x_2 - 10x_3)$$

Let an initial point $\mathbf{x}_0$ be

$$\mathbf{x}_0 = (-2, 2, 1)$$

For implementing the one-at-a-time or univariate search, let us treat the directions in the order by which they are subscripted, that is, first the x_1 direction, and so on.

First iteration

The line segment over which we search is that defined by

$$L = \{(-2 + \lambda, 2, 1) \,|\, 0 \leq \lambda \leq 5\}$$

For a typical point $(-2 + \lambda, 2, 1)$ on L, we have

$$
\begin{aligned}
f(-2 + \lambda, 2, 1) &= -(-2 + \lambda)^2 + 2(-2 + \lambda)2 - 3(2^2) \\
&\quad + 4(2)(1) - 5(1^2) \\
&= -\lambda^2 + 8\lambda - 21 = g_1(\lambda)
\end{aligned}
$$

Thus the first iteration problem is to maximize $g_1(\lambda)$ over $[0, 5]$.

In deciding upon the proper way to solve the subproblem, we might ask if $g_1(\lambda)$ is unimodal. We have

$$\frac{d^2 g_1}{d\lambda^2} = -2$$

which implies that $g_1(\lambda)$ is strictly concave (see Chapter 2), and thus is uni-modal. An optimal search is thus possible for the subproblem, and could be applied to obtain a result to whatever accuracy is desired.

In the present example, however, let us proceed analytically; the solution to the subproblem is $\lambda = 4$, and we obtain a point $\mathbf{x}_1 = (2, 2, 1)$.
Second iteration

The line segment over which we optimize f is

$$L = \{(2, 2 + \lambda, 1) \mid -5 \le \lambda \le 1\}$$

that is, we maximize $g_2(\lambda)$ over $-5 \le \lambda \le 1$, where

$$g_2(\lambda) = f(2, 2 + \lambda, 1)$$

which, again, is unimodal, and has its maximum for $\lambda = -\frac{2}{3}$, and we arrive at $\mathbf{x}_2 = (2, \frac{4}{3}, 1)$.
Third iteration

The next line segment over which we optimize (with respect to x_3) is

$$L = \{(2, \tfrac{4}{3}, 1 + \lambda) \mid -2 \le \lambda \le 1\}$$

and the function of concern is

$$g_3(\lambda) = f(2, \tfrac{4}{3}, 1 + \lambda)$$

As may be verified by other means (Chapter 2), the example function f is a negative definite quadratic form with maximum (over all E^3) $\mathbf{x}^* = (0, 0, 0)$. The reader should perform several additional iterations.

d. GRADIENT METHOD

For functions having continuous first partials, the gradient direction at a point $\mathbf{x}$ is the direction of maximal rate of increase of f from $\mathbf{x}$ at least in a small neighborhood of $\mathbf{x}$. It is reasonable to assume then that there are search techniques which seek to exploit this property.

It must be remembered, simultaneously, that in the most general situations we have imagined there may be no concise mathematical expression for the objective function and, consequently, none for the gradient. As a result, one would of necessity utilize numerical procedures for approximating gradients and, consequently, be concerned about the errors in such approximations and their effect upon the search procedures. Thus search procedures requiring the evaluation of the gradient might, in practice, have broader implications than we choose to consider here; we assume that the gradient may be evaluated or approximated in some fashion and dismiss questions of its accuracy.

The search technique that we shall call the *method of steepest ascents* presumes the existence of a predetermined, probably small real number δ and an initial point $\mathbf{x}_0$. The gradient of f at $\mathbf{x}_0$, $\nabla f(\mathbf{x}_0)$, is evaluated, and the search moves to a point $\mathbf{x}_1$ a distance d from $\mathbf{x}_0$ in the direction $\nabla f(\mathbf{x}_0)$; that is,

$$\mathbf{x}_1 = \mathbf{x}_0 + \frac{d\nabla f(\mathbf{x}_0)}{\|\nabla f(\mathbf{x}_0)\|}$$

In general, one finds

$$\mathbf{x}_{i+1} = \mathbf{x}_i + \frac{d\nabla f(\mathbf{x}_i)}{\|\nabla f(\mathbf{x}_i)\|}$$

Now, *if d is sufficiently small*, selection of these gradient directions will assure $f(\mathbf{x}_{i+1}) > f(\mathbf{x}_i)$. Thus, if one has some knowledge of the behavior of f, it might be possible to specify d such that $f(\mathbf{x}_{i+1}) < f(\mathbf{x}_i)$ never occurs. On the other hand, for some choice of step size d, one might at some time locate $\mathbf{x}_{i+1}$ such that $f(\mathbf{x}_{i+1}) < f(\mathbf{x}_i)$, in which case the step size would be decreased—perhaps halved—and a shorter step in the same direction attempted. If worse function values were occurring frequently with step size d, one might be led to reduce the step size for all the remainder of the process.

Thus there are two conflicting forces at work: on the one hand, the smaller the d the more likely we are to always realize function improvement, but the longer the process will take; on the other hand, choosing a larger d would seem to imply a more rapid approach to the global maximum, except that the process may have to devote much time to backtracking—reducing step size and attempting again—thus nullifying the advantages of the larger d.

With f having level contours as in Figure 6.31 and d "small," the gradient search proceeds as indicated in Figure 6.31.

Here the process could be substantially accelerated by means of a larger d. It is equally simple to construct instances in which d is too large to be of use.

The diagram in Figure 6.32 illustrates a few steps taken with a too great value of d, in which case finding acceptable points, those with primes, entails decrementing d.

If the function f is unimodal, the gradient search will proceed to a neighborhood of the global maximum. Only a nonglobal relative maximum could defeat the procedure, although, as we have seen, convergence may be very slow. In general, the procedure will be more effective for functions which are strongly or linearly unimodal. Notice that no one-dimensional maximizations are required.

The remarks made regarding stopping criteria for the univariate search also apply to the method of steepest ascents.

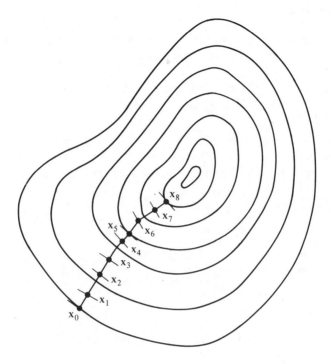

Figure 6.31. Illustrated use of the gradient search technique.

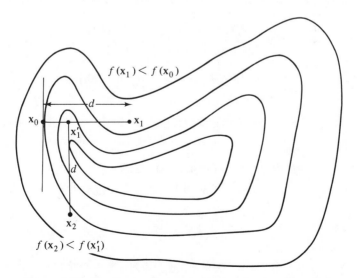

Figure 6.32. Gradient search using a too large value of d.

Example 6.4

To illustrate the gradient search, let us apply that method over the region $0 \le x \le 15$, $0 \le y \le 12$, to maximize the function

$$f(x, y) = -2x^4 + 44x^3 - 282x^2 + 440x - y^4$$
$$+ 28y^3 - 277y^2 + 1140y - 1900$$

The reader is correct in observing that our $f(x, y)$ here has the form $f(x, y) = f_1(x) + f_2(y)$ and hence could be maximized by maximizing f_1, f_2 separately. However, we treat it here as a function of two variables.

Let us choose as a starting point $\mathbf{x}_0 = (x_0, y_0) = (2, 8)$, and let the step size be $d = 1$. We have

$$\nabla f(2, 8) = (-224, 36)$$
$$f(2, 8) = -196$$

and $\qquad\qquad \|\nabla f(2, 8)\| \cong 226.87$

Our next point $\mathbf{x}_1$ is

$$\mathbf{x}_1 = (2, 8) + \frac{1}{226.87}(-224, 36) = (1.02, 8.16)$$

Hence $\qquad\qquad f(\mathbf{x}_1) \cong -68 > -196$

Thus $\mathbf{x}_1$ is better than $\mathbf{x}_0$, and we proceed to find $\mathbf{x}_2$ and $f(\mathbf{x}_2)$. We have

$$\nabla f(1.02, 8.16) \cong (-6.5, 39.1)$$

and $\qquad\qquad \|\nabla f(1.02, 8.16)\| \cong 39.63$

so $\qquad\qquad \mathbf{x}_2 = (1.02, 8.16) + \frac{1}{39.63}(-6.5, 39.1)$
$$\cong (.86, 9.15)$$

and $\qquad\qquad f(\mathbf{x}_2) \cong -23.14 > -68$

We have

$$\nabla f(.86, 9.15) = (45.4, 39.4)$$
$$\mathbf{x}_3 = (.86, 9.15) + \frac{1}{60.11}(45.4, 39.4)$$
$$\cong (1.6, 9.8)$$

and $\qquad\qquad f(\mathbf{x}_3) < f(\mathbf{x}_2)$

Thus it is necessary to decrease the step size and make another attempt at moving from x_2. With a step size of $\frac{1}{2}$ from x_2, we obtain

$$x_3' \cong (1.3, 9.5)$$
$$f(x_3') \cong -14.6 > f(x_2)$$

so x_3' yields some improvement over x_2.

$$\nabla f(1.3, 9.5) \cong (-94, 34)$$
$$x_4 \cong (1.3, 9.5) + \tfrac{1}{100}(-94, 34)$$

Motivated by the fact that a unit length step was too long for the previous iteration, let the step size now be $\frac{1}{3}$, so that

$$x_4' \cong (1.3, 9.5) + \tfrac{1}{300}(-94, 34)$$
$$\cong (.99, 9.6)$$

The reader should verify that $f(x_4') > f(x_3')$. The point near which the process should halt is $x^* = (1, 10)$.

The reader should then, for the same function, choose the starting point $x_0 = (11, 11)$ and execute several iterations, so that the gradient search will now converge to a different point. (Which one?) For the given function the second point will provide the *same* objective function value as does $(1, 10) = x^*$; however, the example should serve to illustrate that a nonglobal relative extremum can quite easily defeat the gradient search and lead to a point not nearly optimal.

e. METHOD OF OPTIMAL STEEPEST ASCENTS

The choice of d being such a problem in the steepest ascents method, the optimal steepest ascents method circumvents that problem by eliminating the need for a choice of d or a scheme for varying d within a search.

Again assuming a starting point x_0, one calculates $\nabla f(x_0)$ and conducts a one-dimensional maximization of f over the line segment from x_0 to the boundary of S in the direction $\nabla f(x_0)$. The result of this maximization is a point x_1, whereupon $\nabla f(x_1)$ is calculated and a one-dimensional maximization of f performed over the line segment from x_1 to the boundary of S in the direction $\nabla f(x_1)$. A maximizing point x_2 is found, and the procedure continues.

The boundary point associated with x_i and direction $\nabla f(x_i)$, or to employ a normalized direction

$$\frac{\nabla f(x_i)}{\| \nabla f(x_i) \|}$$

is for some positive λ'

$$\mathbf{x}_i' = \mathbf{x}_i + \lambda' \frac{\nabla f(\mathbf{x}_i)}{\|\nabla f(\mathbf{x}_i)\|}$$

Obtaining the points $\mathbf{x}_i'$ has been previously discussed. To illustrate a few successively obtained points with this search, we include Figure 6.33. Stopping criteria would be those for the method of steepest ascents.

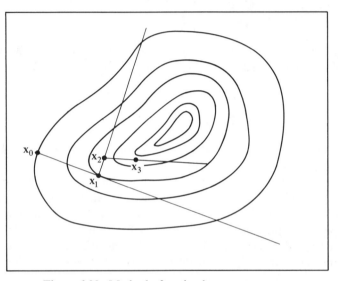

Figure 6.33. Method of optimal steepest ascents.

The optimal steepest ascents method will work whenever f is unimodal, but here we have another instance where the one-dimensional maximizations cannot be guaranteed always realizable by sequential methods unless f is linearly unimodal.

Example 6.5

For an example of optimal steepest ascents, let us consider the maximization of the function

$$f(x, y, z) = y(x - 1)^2 - zx + z^2 - y^2$$

over the region $-3 \leq x \leq 2, -2 \leq y \leq 4, 0 \leq z \leq 15$.

Let
$$\mathbf{x}_0 = (-1, 0, 4)$$
$$\nabla f(x, y, z) = (2y(x - 1) - z, (x - 1)^2 - 2y, 2z - x)$$
$$\nabla f(-1, 0, 4) = (-4, 4, 9)$$

The half-line generated is

$$H = (-1, 0, 4) + \lambda(-4, 4, 9), \quad \lambda \geq 0$$

which *leaves* the region of interest for $\lambda > \frac{1}{2}$. Thus we wish to maximize

$$g_1(\lambda) = f(-1 - 4\lambda, 4\lambda, 4 + 9\lambda)$$
$$= 4\lambda(-2 - 4\lambda)^2 - (4 + 9\lambda)(-1 - 4\lambda) + (4 + 9\lambda)^2 - (4\lambda)^2$$

for $0 \leq \lambda \leq \frac{1}{2}$. Expanding $g_1(\lambda)$, we find

$$g_1(\lambda) = 64\lambda^3 + 165\lambda^2 + 113\lambda + 20$$

which for $0 \leq \lambda \leq \frac{1}{2}$ is strictly increasing. Thus we find x_1 by letting $\lambda = \frac{1}{2}$

and
$$x_1 = (-3, 2, 8\tfrac{1}{2})$$
$$\nabla f(x_1) = (-24\tfrac{1}{2}, 12, 20)$$

f. ANOTHER COMPLICATION

Clearly, any movement in the gradient direction from x_1 results in departure from the region of interest. What do we conclude? Is it that x_1 is optimal over the region defined? The answer to that question is in general, "no." There may be "less than steepest" paths from x_1 that would give some improvement.

When, in any of these search techniques, an iteration would take the process outside the region of interest S, a problem arises. Of course, one could temporarily disregard the question of feasibility, move outside the region S, and, hopefully, future iterations would again bring one back into S. However, this would not seem a very sound procedure, since there is absolutely no guarantee that the unconstrained maximum will, indeed, lie within S.

Another possibility would be to abandon the sequence of points developed, choose a new initial x_0, and begin anew. However, if the unconstrained maximum lies outside S and if the technique being used is effective, we would expect to again be led outside S, and, consequently, to be faced with the same dilemma later.

Let the last feasible point of the sequence be x_k, and let x'_{k+1} be the last feasible point of the step that would provide $x_{k+1} \notin S$; in other words, x'_{k+1} is a boundary point of S. There would seem to be two reasonable approaches: (1) Back up to x_k and employ a *different* technique for locating x_{k+1}; (2) use x'_{k+1} as the new point.

We would like to assume that $f(x'_{k+1}) \geq f(x_k)$ (otherwise, what could be said about f?) and employ (2). It may be that computing an x_{k+2} using the given technique from the point x'_{k+1} will provide $x_{k+2} \in S$. If so, we can carry on. If not, we must decide upon some procedure for driving the next point back into S.

Suppose in the present example that we locate an x_1' by moving back into S in a direction perpendicular to the boundary violated. Let $x_1' = (-\frac{5}{2}, 2, 8\frac{1}{2})$, where the retreat distance of $\frac{1}{2}$ was chosen rather arbitrarily.

$$\nabla f(x_1') = (-\tfrac{45}{2}, \tfrac{33}{4}, \tfrac{39}{2})$$

From x_1', the next point discovered is

$$x_2 = (-3, 2.2, 8.9)$$

and
$$f(x_2) > f(x_1)$$

Thus, for the increased effort, we have a feasible point x_2 better than x_1, but, again, a boundary point. $\nabla f(x_2)$ is such that any movement from x_2 in that gradient direction eliminates feasibility. Once again, we can move away from the boundary $x_1 = -3$, locate x_2', take another gradient step, and so on.

The reader should compute the next several points and determine, by other means, the true x^* over S by identifying the type of unimodality if any of $f(x, y, z)$.

Leaving the feasible region and methods for overcoming that problem are most important when dealing with non-linear programming problems by search techniques. Our treatment has been, of necessity, cursory.

g. PARallel TANgents (PARTAN) Methods

The stronger the properties possessed by the objective function, the easier it is to contrive efficient search procedures. A class of functions for which this is true is the definite quadratic functions, or, more generally, those whose level surfaces are definite quadratics (see Chapter 2). Since maximization has been our goal, let us consider those functions whose level surfaces are negative definite quadratics.

For functions of two variables $f(x_1, x_2)$, we consider those whose level contours or level surfaces are negative definite quadratics; that is,

$$Q(x_1, x_2) = a_{11}x_1^2 + a_{12}x_1x_2 + a_{22}x_2^2 + b_1x_1 + b_2x_2 + c = z_0$$

Now
$$a_{11}x_1^2 + a_{12}x_1x_2 + a_{22}x_2^2 + b_1x_1 + b_2x_2 = z_0 - c$$

may be rewritten,

$$Q(x_1, x_2) = a_{11}(x_1 - u_1)^2 + a_{22}(x_2 - u_2)^2 + a_{12}(x_1 - u_1)(x_2 - u_2) - u_3$$
$$= z_0 - c \tag{6.20}$$

where u_1, u_2 are uniquely determined from

$$-2a_{11}u_1 - a_{12}u_2 = b_1$$
$$-2a_{22}u_2 - a_{12}u_1 = b_2$$

and $\qquad\qquad u_3 = a_{11}u_1^2 + a_{22}u_2^2 + a_{12}u_1u_2$

It is seen from equation (6.20) that the present level contours are concentric ellipses, as we shall observe, and, owing to this property, very efficient optimization procedures are possible.

Consider a line through $\mathbf{x}^*$, our global max, and let $\mathbf{p}$, $\mathbf{x}_1$, $\mathbf{x}_2$ be three points which lie on that line. The point $\mathbf{p}$ may be represented as

$$\mathbf{p} = \lambda'\mathbf{x}_2 + (1 - \lambda')\mathbf{x}^* \qquad (6.21)$$

and as $\qquad\qquad \mathbf{p} = \lambda''\mathbf{x}^* + (1 - \lambda'')\mathbf{x}_1 \qquad (6.22)$

for appropriate values of λ'', λ'.

Equating (6.21) and (6.22), we have

$$\lambda'\mathbf{x}_2 + (1 - \lambda')\mathbf{x}^* = \lambda''\mathbf{x}^* + (1 - \lambda'')\mathbf{x}_1$$
$$\lambda'(\mathbf{x}_2 - \mathbf{x}^*) = (\lambda'' - 1)\mathbf{x}^* + (1 - \lambda'')\mathbf{x}_1$$

Thus $\qquad\qquad \mathbf{x}_2 - \mathbf{x}^* = \dfrac{1 - \lambda''}{\lambda'}[\mathbf{x}_1 - \mathbf{x}^*] \qquad (6.23)$

Streamlining the representation with $u = (1 - \lambda'')/\lambda'$,

$$\mathbf{x}_2 - \mathbf{x}^* = u[\mathbf{x}_1 - \mathbf{x}^*] \qquad (6.24)$$

Substituting (6.24) into (6.20), we find

$$Q(\mathbf{x}_2) = u^2 Q(\mathbf{x}_1) \qquad (6.25)$$

or $\qquad\qquad u = \left(\dfrac{Q(\mathbf{x}_2)}{Q(\mathbf{x}_1)}\right)^{1/2}$

As we have seen, the tangents to the level surfaces at $\mathbf{x}_1$, $\mathbf{x}_2$ have the equations

$$\nabla Q(\mathbf{x}_2)[\mathbf{x} - \mathbf{x}_2] = 0$$
$$\nabla Q(\mathbf{x}_1)[\mathbf{x} - \mathbf{x}_1] = 0$$

But from (6.25)

$$\nabla Q(\mathbf{x}_2) = u^2 \nabla Q(\mathbf{x}_1)$$

that is, the tangents have the same normal and, consequently, are parallel.

We may then reverse our manipulations and show that (6.25) implies that x_1, x_2 are colinear with x^*, the global maximum. Furthermore, the two tangents being parallel implies (6.25). In addition, as was previously observed, the points of tangency, x_1 and x_2, are the maxima of Q over the lines corresponding to the contour tangents.

A quite efficient method for locating the global maximum in this bivariate case is then evident. One selects two parallel lines and performs one-dimensional maximizations of Q over those portions of the lines that lie within S. Two maxima, x_1 and x_2, result from these operations, and we know the desired point x^* lies on the line L through x_1 and x_2. Thus one maximizes Q over that portion of L within S, and the result of this one-dimensional maximization is chosen as x^*.

Figure 6.34 illustrates the method. The illustration should not be construed as to imply that x^* must lie "between" x_1, x_2. Selection of the two parallel lines is entirely arbitrary.

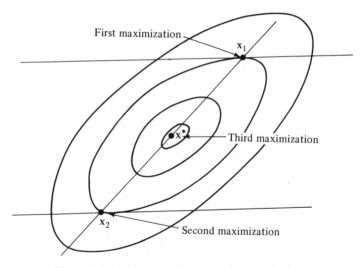

Figure 6.34. The use of the PARTAN method.

Since but three linear maximizations are required in all cases, the superiority of this method to those previously described is evident. We emphasize that this performance is guaranteed only for the restricted class of functions described.

It is also true, however, that in some vicinity of the global maximum a quadratic approximation to the actual surface is frequently accurate. Thus the possibility of utilizing PARTAN to locate rapidly the global max within a *reduced* extremal region provided by the application of some other technique presents itself in the event of a nonquadratic function.

A different, but equivalent, method for conducting PARTAN in two dimensions is available, and will facilitate an intuitive understanding of the *n*-dimensional counterpart of PARTAN. Given an initial point x_0, one conducts *two iterations* of an optimal steepest ascents search, obtaining the successive maximizing points x_2, x_3. The desired global maximum x^* is then found on the line that passes through x_0 and x_3. The equivalence of this alternative procedure with PARTAN as previously described is easily seen.

By the very nature of the optimal steepest ascents method, that is, it searches always in a direction orthogonal to the present contour tangent, the lines L_1, L_2, contour tangents at x_0 and x_3, respectively, are parallel, and x_0, x_3 are the respective maxima of the function f over L_1, L_2. Figure 6.35 illustrates these facts.

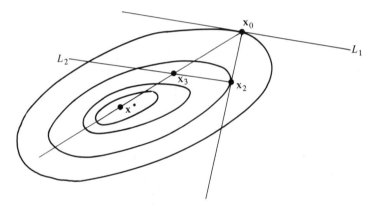

Figure 6.35. An alternative PARTAN procedure in two dimensions.

One would normally prefer the first method to this alternative, since the former does not require any computation of the gradient.

For f a negative definite quadratic of n variables, the PARTAN method assumes the nature of a combination of two different search procedures, as did the second definition of PARTAN in two dimensions. (Here it is worthwhile to point out that many hybrid methods, combinations of particular search techniques, have been used with success. In addition, persons frequently utilize variations on existing procedures to obtain more efficient methods for specific kinds of problems.)

As usual, denote a given starting point by x_0. A more compact general decription of the method is available if, as in Figure 6.35, the second point found is labeled $\mathbf{x}_2$. To itemize the first few iterations, $\mathbf{x}_2$ is found by conducting an iteration with the optimal steepest ascents method from $\mathbf{x}_0$. The point $\mathbf{x}_3$ is found in the same manner, beginning with $\mathbf{x}_2$. $\mathbf{x}_4$ is found with a PARTAN step, that is, maximizing over the line through $\mathbf{x}_0$, $\mathbf{x}_3$. From $\mathbf{x}_4$, $\mathbf{x}_5$ is obtained with another gradient move; $\mathbf{x}_6$ is obtained from $\mathbf{x}_5$ by a PARTAN step using $\mathbf{x}_5$, $\mathbf{x}_2$, and so on.

In general, then, for $i \geq 3$, with i odd, $\mathbf{x}_i$ is obtained by an optimal gradient search from $\mathbf{x}_{i-1}$; with i even, $\mathbf{x}_i$ is obtained by a PARTAN iteration with $\mathbf{x}_{i-1}$, $\mathbf{x}_{i-4}$, that is, by maximizing over the line through $\mathbf{x}_{i-1}$, $\mathbf{x}_{i-4}$.

As in the case of two-dimensional PARTAN, the number of iterations that this method will require is known in advance; for if $Q(x_1, x_2, \ldots, x_n)$ is our quadratic objective function, the global maximum will be provided by $\mathbf{x}_{2n}$. That is, $2n - 1$ one-dimensional maximizations will yield the global max. We shall not undertake verification of this claim, but an example may make it intuitively appealing.

Example 6.6

Let us now examine an application of the PARTAN search in three dimensions. Let

$$f(x_1, x_2, x_3) = (x_1, x_2, x_3) \begin{pmatrix} -1 & 0 & 1 \\ 0 & -1 & 1 \\ 1 & 1 & -5 \end{pmatrix} \begin{pmatrix} x_1 \\ x_2 \\ x_3 \end{pmatrix} = \mathbf{x}^T A \mathbf{x}$$

The function f is a negative definite quadratic form, so the unique global maximum over the entirety of E^3 is $\mathbf{x}^* = (0, 0, 0)$.

We shall perform each step of the PARTAN search "analytically" and shall not, therefore, specify a particular region. Let

$$\mathbf{x}_0 = (4, 3, -3)$$

$$\nabla f(\mathbf{x}) = 2A\mathbf{x} = 2 \begin{pmatrix} -1 & 0 & 1 \\ 0 & -1 & 1 \\ 1 & 1 & -5 \end{pmatrix} \begin{pmatrix} x_1 \\ x_2 \\ x_3 \end{pmatrix}$$

$$= (-2x_1 + 2x_3, -2x_2 + 2x_3, 2x_1 + 2x_2 - 10x_3)$$

$$\nabla f(4, 3, -3) = (-14, -12, 44)$$

Here we need not work with a normalized gradient direction, whence $\mathbf{x}_2$ is found according to

$$\max_{0 \leq \lambda} \{h(\lambda) = f(4 - 14\lambda, 3 - 12\lambda, -3 + 44\lambda)\}$$

Now

$$h(\lambda) = (4 - 14\lambda, 3 - 12\lambda, -3 + 44\lambda) \begin{pmatrix} -1 & 0 & 1 \\ 0 & -1 & 1 \\ 1 & 1 & -5 \end{pmatrix} \begin{pmatrix} 4 - 14\lambda \\ 3 - 12\lambda \\ -3 + 44\lambda \end{pmatrix}$$

$$= -(4 - 14\lambda)^2 + 2(4 - 14\lambda)(-3 + 44\lambda) - (3 - 12\lambda)^2$$
$$+ 2(3 - 12\lambda)(-3 + 44\lambda) - 5(-3 + 44\lambda)^2$$

Thus

$$h'(\lambda) = -2(4 - 14\lambda)(-14) + 2[(4 - 14\lambda)44 + (-3 + 44\lambda)(-14)]$$
$$- 2(3 - 12\lambda)(-12) + 2[(3 - 12\lambda)44 + (-3 + 44\lambda)(-12)]$$
$$- 10(-3 + 44\lambda)44$$
$$= -24{,}616\lambda + 2276$$

Hence $h'(\lambda) = 0$, for $\lambda = 2276/24{,}616 \cong .09$.

Now
$$\mathbf{x}_2 \cong (2.7, 1.9, 1)$$

From $\mathbf{x}_2$ another gradient step is taken, and the result of that gives

$$\mathbf{x}_3 = (-1, -.1, .1)$$

where
$$\nabla f(\mathbf{x}_2) = (-3.4, -1.8, -.8)$$

Now $\mathbf{x}_4$ is found using $\mathbf{x}_0, \mathbf{x}_3$, as described.

$$(\mathbf{x}_3 - \mathbf{x}_0) = (-1, -.1, .1) - (4, 3, -3)$$
$$= (-5, -3.1, 3.1)$$

and the subproblem to be solved is

$$\max_{\lambda} \{f(-1 - 5\lambda, - .1 - 3.1\lambda, .1 + 3.1\lambda)\}$$

Expanding the function of interest and setting the derivative to zero, we find

$$\lambda = -.08$$

As a result we obtain the point

$$\mathbf{x}_4 \cong (-.6, -.75, -.15)$$

The reader should establish that our sequence of points has provided an in-

creasing set of function values. Furthermore, he should proceed to obtain
x_5, x_6 and observe the deviation from the true $x^* = (0, 0, 0)$.

Exercise 6.14

Write a FORTRAN program for executing PARTAN in two variables.

REFERENCES

[1] BELLMAN, R., and S. DREYFUS, *Applied Dynamic Programming*, Princeton University Press, Princeton, N.J., 1962.

[2] BERGAMINI, D., et al., *Mathematics*, Time, Inc., New York, 1963.

[3] WILDE, D., *Optimum Seeking Methods*, Prentice-Hall, Inc., Englewood Cliffs, N.J., 1964.

[4] WILDE, D., and C. BEIGHTLER, *Foundations of Optimization*, Prentice-Hall, Inc., Englewood Cliffs, N.J., 1967.

PROBLEMS

1. Suppose that you were interested in a particular problem in constructing a sequential search procedure which minimized the *expected number* of function evaluations needed to achieve proximity L_N to the true global extremum. In general terms, how would you proceed, and what, if anything, would you require in the way of additional information?

2. Choose a relatively small k, say $k = 4$ or $k = 5$, and begin the Fibonacci search over an interval of length F_k, with a unit length interval the desired objective. Suppose that prior to the last iteration we compare $f(x_i)$, $f(x_{i+1})$ and find them equal. What is the effect on the duration of the search, assuming that equality does not occur thereafter? Can you generalize the observation? What if equality occurs $p > 1$ times? How do these results relate to the concept of a minimax optimal search?

3. In Table 6.1, compute the numbers N for
Dichotomous

$$\epsilon = 10^{-6}, L_N = 10^{-5}, L_N = 10^{-4}$$

Two-point equal interval

$$L_N = 10^{-5}, L_N = 10^{-4}, L_N = 10^{-3}$$

Four-point equal interval

$$L_N = 10^{-5}, L_N = 10^{-4}$$

Three-point equal interval

$$L_N = 10^{-3}, L_N = 10^{-2}, L_N = 10^{-1}$$

Golden section

$$L_N = 10^{-5}, L_N = 10^{-3}$$

Fibonacci

$$L_N = 10^{-5}, L_N = 10^{-4}$$

4. For each of the following functions, initial and desired intervals, and techniques, carry out the search for the duration given. State the final interval if it is not specified.

   (a) On $[-.2, \pi/2]$, $\sin x$, two-point equal interval, three iterations.

   (b) Same as (a), but with three-point equal interval.

   (c) On $[-1, \frac{3}{4}]$, $x^2 + 1$, golden section, final interval length $\leq \frac{1}{4}$.

   (d) Same as (c), but with four-point equal-interval search.

   (e) Same as (c), but with Fibonacci.

   (f) $(x^3/3) + \frac{3}{2}x^2 - 2x$ on $[\frac{3}{4}, 4]$, dichotomous search with $\epsilon = 10^{-3}$, three iterations.

   (g) Same as (f), but Fibonacci search with final interval length ≤ 1.5.

5. Let c_1, c_2 be the positive and negative solutions, respectively, to the golden-section equation. Verify that

$$F_k = \frac{c_1^{k+1} - c_2^{k+1}}{c_1 - c_2}$$

   for a few small values of k. Then show the equation holds for $k = 0, 1, 2, \ldots$.

6. As in Problem 5, treat the equation $F_{i+j} = F_i F_j - F_{i-1}F_{j-1}$ for $j > 0$ and $i \geq 0$. What is the coefficient of x^n in $F(x) = 1/(1 - x - x^2)$?

7. Explain why the Fibonacci search is slightly better than the golden section. Show that in both the golden-section and Fibonacci searches one of the two "new" points is, except for the first step, always an "old" point.

8. For each of the following, treat the function on the region given with the technique and for the duration specified.

   (a) Minimize $x_1^2 + x_2^2 + x_3^2 + x_4^2 - 4x_1 - 2x_2 + 4x_3 - 2x_4 + 10$ on $0 \leq x_1 \leq 4$; $-1 \leq x_2 \leq 3$; $-4 \leq x_3 \leq 4$; $-3 \leq x_4 \leq 2$. Use the univariate search and continue until a global minimum is found, with $(0, 0, 0, 0)$ an initial point.

   (b) Same as (a), but use the gradient search with step size 2 units, beginning at point $(4, 3, 4, 2)$.

   (c) Same as (b), but using $(0, -1, -4, -3)$ as an initial point.

   (d) Same as (a), but using optimal steepest descents beginning with $(0, 0, 0, 0)$.

   (e) Same as (d) but beginning with $(4, 3, 4, 2)$.

   (f) Same as (d), but beginning with $(0, -1, -4, -3)$.

   (g) For the function $-(x_1^2 + 3x_2^2 + 9x_3^2) + 2x_1x_2 + 4x_1x_3 + 2x_2x_3 + 20$ find the maximum on $-3 \leq x_1 \leq 3$; $-2 \leq x_2 \leq 2$; $-4 \leq x_3 \leq 4$, with $x_0 = (-3, 2, 3)$, using the PARTAN search.

   (h) Same as (g), but beginning with $(1, 1, 1)$.

9. Develop a recursive expression for the number of n-tuples $Z(n)$ over the set $\{0, 1\}$ in which there are not two consecutive 0s. Define $Z(0) = 1$.

DYNAMIC PROGRAMMING

7.1. Introduction

As we set out to discuss dynamic programming, it would be good if the reader would reflect upon the nature of the methods of linear programming. One is interested in methods for solving optimization problems of a particular variety, and various algorithms for solving such problems may be obtained. The name "dyanamic programming" might lead one to believe that he will encounter here algorithms for the solution of other classes of problems, but the terminology is misleading, and the methods of dynamic programming are not algorithmic in nature.

Dynamic programming might best be described as a technique of problem formulation, and this should be kept in mind as we proceed.

7.2. Introductory Concepts

Let us consider an *n-stage decision process*, as shown in Figure 7.1, where S_j, $j = 0, \ldots, n$, are sets of elements called *states*

7

D_j, $j = 1, \ldots, n$, are sets of elements called *decisions*

T_j, $j = 1, \ldots, n$, are mappings that with each possible pair (s_j, d_j) where s_j denotes an arbitrary element of S_j and d_j an arbitrary element of D_j, associate a single element s_{j-1} of S_{j-1}

R_j, $j = 1, \ldots, n$, are mappings that associate with each pair (s_j, d_j) a single real number.

$R_0(s_0)$ is a mapping that associates a single real number with each element s_0 in S_0

We also assume the existence of a function

$$\varphi(R_n(s_n, d_n), R_{n-1}(s_{n-1}, d_{n-1}), \ldots, R_1(s_1, d_1), R_0(s_0))$$

which associates with any set of $(n + 1)$ R_j values a single real number.

Figure 7.1 is to be interpreted as follows, then. The process begins at some state within the set S_n at stage n. The decision maker selects a decision from the set D_n. The function R_n then determines the payoff or return or cost resulting from the process's having been in that particular state and the making of the particular decision. The function T_n, then, determines the state in which the process is to be found at stage $n - 1$, whereupon the decision mak-

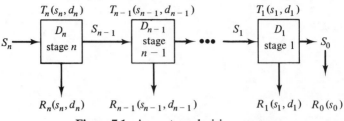

Figure 7.1. An *n*-stage decision process.

er elects a decision from the set D_{n-1}, realizes a payoff computed by R_{n-1}, and T_{n-1} carries the process into some state at stage $n-2$, and so on. Finally, at stage 1, the process is in some state, the decision maker chooses a decision from D_1, and realizes a payoff R_1. The transformation T_1 then determines a terminal state in the terminal set S_0, a return $R_0(s_0)$, and the process ends.

The problem is to select a sequence of decisions

$$d_n^*, d_{n-1}^*, \ldots, d_1^*$$

so that the total return function φ is maximized or minimized over all possible decision sequences.

We observe that

$$\varphi(R_n(s_n, d_n), R_{n-1}(s_{n-1}, d_{n-1}), \ldots, R_1(s_1, d_1), R_0(s_0))$$

may be written

$$\varphi(R_n(s_n, d_n), R_{n-1}(T_n(s_n, d_n), d_{n-1}),$$
$$R_{n-2}(T_{n-1}(T_n(s_n, d_n), d_{n-1}), d_{n-2}), \ldots, R_0(\cdots)) \qquad (7.1)$$
$$= h(s_n, d_n, d_{n-1}, \ldots, d_1)$$

where we have eliminated all the state variables $s_{n-1}, \ldots, s_1, s_0$ by utilizing the transformation equations

$$T_j(s_j, d_j) = s_{j-1}$$

successively.

The complicating factor in such a problem is, of course, the "hereditary" nature; that is, decisions made affect the future of the process, since future states and, consequently, future returns depend upon present decisions.

The reader should not be disturbed by the numbering of the stages in reverse order of their occurrence. This numbering, of course, in no manner affects the results to follow, but will allow a more homogeneous notation.

The reader should, in addition, be certain he understands the formulation and the workings of our n-stage decision process before proceeding.

Let us suppose now that $\varphi(R_n, R_{n-1}, \ldots, R_0) = \sum_{i=0}^{n} R_i$. This is a common and important total return function and will facilitate the present analysis. It will be assumed, for the sake of specificity, that maximization of the sum of the stage returns is the objective.

Symbolically, the problem is

$$\max_{D_n D_{n-1} \cdots D_1} [R_n + R_{n-1} + \cdots + R_1 + R_0]$$

or

$$\max_{d_n d_{n-1} \cdots d_1} [R_n + R_{n-1} + \cdots + R_1 + R_0]$$

In words, we wish to maximize over all possible sequences of decisions the total return function, in this case, the sum of the stage returns.

In light of equation (7.1), we might write

$$\max_{d_n d_{n-1} \cdots d_1} [h(s_n, d_{n-1}, \ldots, d_1)]$$

and this maximum, clearly, *is a function of s_n alone;* we recall that the process may begin in any one of a set S_n of states, and it is to be expected that an optimal decision sequence, or optimal policy, will change as the starting state changes.

Therefore, we have

$$\max_{d_n \cdots d_1} [h(s_n, d_{n-1}, \ldots, d_1)] = f_n(s_n)$$

where f_n is some function.

Prior to investigating solution procedures for the kind of optimization problems posed, let us observe examples of serial decision processes, which arise quite naturally.

Example 7.1

A smooth curve is defined on the interval $[I_n, I_0]$. It is desired to approximate this curve by a piecewise-linear curve, where the linear segments are to be defined on the subintervals $[I_n, I_{n-1}], [I_{n-1}, I_{n-2}], \ldots, [I_1, I_0]$, respectively, with $I_n < I_{n-1} < \cdots < I_1 < I_0$ (see Figure 7.2). Here $P(t)$ is an approximating curve of the stated variety for the curve $f(t)$, $P(t)$ has linear segments

$$P_1(t), P_2(t), \ldots, P_5(t)$$

and we have a five-stage decision process, decisions being the choices of $P_5, P_4, \ldots, P_1$.

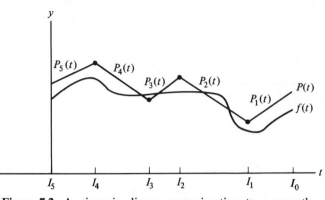

Figure 7.2. A piecewise linear approximation to a smooth curve.

Furthermore, it is desired to construct the approximating polygonal path $P(t)$ so that

$$\int_{I_n}^{I_0} |P(t) - f(t)| \, dt$$

is minimized.

Let us then define an *n*-stage process, as in Figure 7.1, along with all sets and transformations involved.

Let $S_i = \{$all finite real numbers$\}$, since at $I_n, \ldots, I_1$ the approximating curve is allowed to assume any finite value. Let $D_i = \{$all finite real numbers$\}$, since by specifying a real number we specify a slope. Then, given the state at that stage, which is just a point in the plane, selection of a slope actually specifies a line in the plane, and, consequently, a line segment $P_i(t)$. The transformation T_i is available from the line equation itself, which uniquely determines the state of the process at the next stage.

The stage returns are most conveniently defined as

$$R_j = \int_{I_j}^{I_{j-1}} |P_j(t) - f(t)| \, dt$$

for then the sum of the stage returns is the total return.

The reader should complete the details and show that the problem illustrated in Figure 7.2 may be represented as an *n*-stage process as sketched in Figure 7.1.

Example 7.2

For another example of a sequential decision process, let us choose a common, but important, problem in economic decision making.

Suppose that at the beginning of some span of time T a particular organization has available a sum of capital earmarked for expenditure on research and development (R&D) projects and a set of such candidate projects. The decision maker(s) must decide which projects will be funded and how much capital to allocate to each of those selected, given the budgetary constraint.

At the end of a time interval of length $t \ll T$, progress on the projects previously funded is assessed, any new candidate projects are appended to the list of possibilities, some old ones perhaps deleted, and the capital available for R&D projects during the next time t is allocated to a chosen subset of projects.

Thus the decision-making process continues—at the end of each t-length period the R&D capital for the next period is allocated to the current set of possible projects, the degree of progress of those currently funded being assessed. It is unnecessary that all R&D resources be expended on R&D projects; some may be invested at the going rate of return.

Now, as money and other resources are allocated to a given project, it (hopefully) approaches successful completion. During this time, the project represents a loss to the company. Once brought to fruition it (hopefully) begins to produce a return, perhaps slowly at first. The organization's problem is to make the periodic allocations in such a way as to maximize some function of the projects' net returns over the long term T.

The inherent sequential decision-making nature of the problem is clear, although we would encounter difficulties if we attempted to explicitly define the various sets and transformations necessary to fit this problem into the general framework of Figure 7.1.

It may be beneficial, however, to touch upon several of these difficulties. At any stage j, the state of the system would seem to be accurately represented by a collection of pairs {(project i, status of project i)} and the amount of capital available for R&D expenditure during period j. A decision at stage j would be a vector

$$(A_{1j}, A_{2j}, \ldots),$$

where A_{kj} is the amount to be spent upon project k during period j. As described, however, the problem does not admit explicit definition of these entities, since the set of alternative projects at some future point of time and the amount of capital available then are not known at the outset, nor are they necessarily functions of preceding states and decisions, a fact that makes deterministic specification of the stage transformations impossible. Furthermore, knowledge of the state of progress of project i after period j and the amount of capital to be allocated to project i during period $j + 1$ does not determine the state of progress of project i after period $j + 1$, by the nature of R & D work. The foregoing observation, coupled with the fact that even upon completion the return a project will yield during subsequent periods is not certain, makes specification of stage return functions a hazy proposition also.

A troublesome fact, at this stage of our inspection of n-stage decision processes, is the nondeterministic aspect of the problem as stated. We obviously need to know more. Decision making that is nondeterministic is conventionally dichotomized into (1) decision making under risk, meaning that the occurrences of the nondeterministic events are described by probability distributions, and (2) decision making under uncertainty, meaning that no such distributions are known.

The setting for our sample problem is clearly, then, one of uncertainty. A first step toward obtaining a more tractable form might be the use of approximating probability distributions for the "uncertain" aspects of the prolem (these might be acquired from past experience, technological knowledge, etc.). Dealing then in expected returns, probabilities of project progress, and the like, more workable forms could be obtained. But discussing decision processes possessed of stochastic elements is inappropriate at this point. We wish at this time only to point out the existence of important serial decision-making problems in business and industrial situations, as well as certain of the complications attending their formulation in our new framework. We neglect pursuing the details of this kind of problem.

Example 7.3

By making several simplifying assumptions, we can construct a similar kind of problem and one also representative of an important class of problems, although here we remove the interesting complications by assumption.

Suppose that an individual is designing an n-year investment program for himself. There are p investment alternatives $A_1, A_2, \ldots, A_p$ available to him at the start, and these will continue to be available throughout the n years. A given alternative will possess in general different annual rates of return through the years, and for the sake of simplicity we endow our investor with complete knowledge of these. The rate of return of A_i during year j is, say, r_{ij}. Furthermore, income from A_i during period j is taxed according to a known table t_{ij}. Let us assume money not invested is untaxed, and that our investor's initial funds total C dollars.

Thus at the outset of year 1, the investor allocates up to C dollars among the p alternatives. At the end of the year, he collects his investment plus interest, pays his taxes, and ends the year with C' dollars, whereupon he allocates some or all of this amount among the p alternatives for year 2, and so on. Let us suppose the investor's objective is to maximize the present worth of his total profit over the 10 years.

We ask the reader to describe all sets, transformations, and return functions that arise when considering this problem as a serial decision process.

7.3. Solving Dynamic-Programming Problems

i. *Recursion Relations and Optimization*

In the n-stage decision process let us assume the process has run all the way to the point where a decision from D_1 must be selected; that is, the process has just arrived at stage 1. Regardless of what has taken place up to that point, a d_1 must be chosen so that

$$R_1(s_1, d_1) + R_0(s_0) = R_1(s_1, d_1) + R_0(T_1(s_1, d_1))$$

is maximized; for if this is *not* done, then certainly the *entire* decision sequence cannot be optimal. Also, even if the sequence of decisions up to this point is nonoptimal, but we are interested in making the best of the situation *given these circumstances*, we are led to the same strategy.

Therefore, we consider

$$\max_{d_1} [R_1(s_1, d_1) + R_0(T_1(s_1, d_1))]$$

The bracketed quantity is a function of the state variable s_1 and the decision variable d_1. The maximum, over all possible d_1 of the expression is a function of s_1 alone, and we may write

$$\max_{d_1} [R_1(s_1, d_1) + R_0(T_1(s_1, d_1))] = f_1(s_1)$$

where f_1 is some function.

The function f_1 may be a continuous function of the variable s_1; it may be a tabular relationship that associates, with each element $s_1 \in S_1$, a number. All we are demanding is that we have obtained a relationship between elements of S_1 and the corresponding optimal return from stages 1 and 0. Implicit is the fact that from each s_1 we would also identify the decision or decisions which provide that optimal return.

When the maximization has been performed, and here the fact that no statement is made about *how* the maximization is to be done must be emphasized, we have, in effect, solved the problem arising from the one-stage process[†] of Figure 7.3. This single-stage optimization problem involved the single decision variable d_1.

[†]Although we consider the terminal stage to be stage zero, Figure 7.3 represents but *one* decision-making stage.

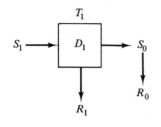

Figure 7.3. A one-stage decision process.

The next step is to consider the two-stage problem

$$\max_{d_2 d_1} [R_2(s_2, d_2) + R_1(s_1, d_1) + R_0(s_0)]$$

or $$\max_{d_2 d_1} [R_2(s_2, d_2) + R_1(T_2(s_2, d_2), d_1) + R_0(T_1(T_2(s_2, d_2), d_1))] \qquad (7.2)$$

It is convenient to write this last expression as

$$\max_{d_2} \{\max_{d_1} [R_2(s_2, d_2) + R_1(T_2(s_2, d_2), d_1) + R_0(T_1(T_2(s_2, d_2), d_1))]\} \quad (7.3)$$

which indicates that we first find

$$\max_{d_1} [R_2(s_2, d_2) + R_1(T_2(s_2, d_2), d_1) + R_0(T_1(T_2(s_2, d_2), d_1))]$$

and then maximize that result over all possible d_2.

a. VALIDITY OF THE SEQUENTIAL OPTIMIZATIONS

That (7.3) and (7.2) are equivalent maximization problems is not difficult to illustrate. For purposes of compact representation and to generalize the foregoing statement to an arbitrary function of two variables, let us suppose that $g(x, y)$ is defined for all pairs (x, y) with $x \in X, y \in Y$, where X, Y are arbitrary sets.

We shall show that

$$\max_{X, Y} \{g(x, y)\} = \max_{X} \{\max_{Y} \{g(x, y)\}\}$$

where we assume that the sets X, Y together with the function g are such that it is legitimate to speak of *maxima;* that is, the maxima do exist.

The equality can be seen intuitively; consider that for each $x \in X$ we compute

$$\max_{Y} \{g(x, y)\}$$

obtaining a set S of numbers, one for each element of X. Choosing the largest element of S, then, accomplishes

$$\max_x \{\max_Y \{g(x, y)\}\}$$

and, clearly, one also obtains

$$\max_{x,Y} \{g(x, y)\}$$

Now let

$$\max_{x,Y} \{g(x, y)\} = g(x^*, y^*)$$

For any x, we have

$$\max_Y \{g(x, y)\} \geq g(x, y^*)$$

So

$$\max_x \{\max_Y \{g(x, y)\}\} \geq \max_x \{g(x, y^*)\} = g(x^*, y^*)$$

But it must be, then, that

$$\max_x \{\max_Y \{g(x, y)\}\} = g(x^*, y^*) = \max_{x,Y} \{g(x, y)\}$$

Thus (7.3) and (7.2) are equivalent representations.

b. Recursive Nature of the Optimizations

Having established the validity of expression (7.3) as a representation of the two-stage problem expressed by (7.2), the next claim is that (7.3) may be written

$$\max_{d_2} \{R_2(s_2, d_2) + \max_{d_1} [R_1(T_2(s_2, d_2), d_1) + R_0(T_1(T_2(s_2, d_2), d_1))]\} \quad (7.4)$$

This is easily seen, for $R_2(s_2, d_2)$ is *not* a function of d_1, since the events at stage 2 are not contingent upon what transpires at stage 1. Therefore, we maximize the entire expression over d_1 by maximizing only that portion of it which is a function of d_1.

The next observation that can be made is of central importance, and is, actually, the heart of dynamic programming. We have *already* calculated

$$\max_{d_1} [R_1(T_2(s_2, d_2), d_1) + R_0(T_1(T_2(s_2, d_2), d_1))]$$

and it is simply $f_1(T_2(s_2, d_2))$. Recall that when the one-stage problem was solved, we found an optimal decision and optimal return for the one-stage process. Hence we have

$$\max_{d_2 d_1} [R_2(s_2, d_2) + R_1(s_1, d_1) + R_0(s_0)]$$
$$= \max_{d_2} [R_2(s_2, d_2') + f_1(T_2(s_2, d_2))] = f_2(s_2) \tag{7.5}$$

where f_2 is some function of s_2, and the maximization is with respect to a *single* variable.

c. Recursion Relations for Higher-Numbered Stages

Next let us calculate

$$\max_{d_3 d_2 d_1} \{R_3(s_3, d_3) + R_2(s_2, d_2) + R_1(s_1, d_1) + R_0(s_0)\}$$

which becomes, upon utilizing T_3, T_2, and T_1,

$$\max_{d_3 d_2 d_1} \{R_3(s_3, d_3) + R_2(T_3(s_3, d_3), d_2)$$
$$+ R_1(T_2(T_3(s_3, d_3), d_2), d_1) \tag{7.6}$$
$$+ R_0(T_1(T_2(T_3(s_3, d_3), d_2), d_1))\}$$

(Fortunately, solving problems is frequently simpler than writing such expressions.)

As before, we treat the three-dimensional maximization sequentially, and, for the same reasons as before, we are justified in bringing the maximizations over d_2 and d_1 inside the braces in (7.6), which becomes

$$\max_{d_3} \{R_3(s_3, d_3) + \max_{d_2 d_1} [R_2(T_3(s_3, d_3), d_2)$$
$$+ R_1(T_2(T_3(s_3, d_3), d_2) d_1) \tag{7.7}$$
$$+ R_0(T_1(T_2(T_3(s_3, d_3), d_2), d_1))]\}$$

From (7.5), however, we recognize the maximization over all d_2, d_1 of the bracketed term of (7.7) as being simply $f_2(T_3(s_3, d_3))$. Therefore, the maximum return from the last three stages of our process, clearly a function of s_3 *only* after the maximization over all d_3, d_2, and d_1, may be written

$$\max_{d_3} \{R_3(s_3, d_3) + f_2(T_3(s_3, d_3))\} = f_3(s_3) \tag{7.8}$$

where f_3 is some function, and again we have a one-dimensional subproblem.

Clearly, for $1 < i \leq n$, we shall find, using the same sort of arguments used to obtain (7.8) and (7.5), that

$$
\max_{d_i d_{i-1} \cdots d_1} \{R_i(s_i, d_i) + R_{i-1} + \cdots + R_1 + R_0\}
$$
$$
= \max_{d_i} \{R_i(s_i, d_i) + f_{i-1}(T_i(s_i, d_i))\} = f_i(s_i)
\tag{7.9}
$$

In (7.9), f_i is *some* function and f_{i-1} is the counterpart of f_i which we obtained when carrying out the analysis at stage $i - 1$. For notational simplicity, the arguments of $R_{i-1}, R_{i-2}, \ldots, R_0$ have been omitted in (7.9). When $i = n$, (7.9) becomes

$$
\max_{d_n d_{n-1} \cdots d_1} \{R_n(s_n, d_n) + R_{n-1} + \cdots + R_1 + R_0\}
$$
$$
= \max_{d_n} \{R_n(s_n, d_n) + f_{n-1}(T_n(s_n, d_n))\} = f_n(s_n)
\tag{7.10}
$$

The first term in the double equality (7.10) is immediately recognized as the problem of interest, that is, *the maximization of the total return over all possible decision sequences* as a function of the state at the initial stage.

d. PRINCIPLE OF OPTIMALITY AND FUNCTIONAL EQUATIONS

Consider again the thinking that has lead to a typical equation of the form (7.9). It is certainly the case that any optimizing sequence of decision variables has the following property. Given the state attained at any stage, the subsequence of decisions beginning at that stage must be optimal for the subprocess beginning at that stage, given the state mentioned.

That this is true is obvious from a simple contradiction argument. If there were a better decision subsequence, then there would be, contrary to hypothesis, a better decision sequence for the entire process.

The property described was named *the principle of optimality* by R. Bellman. Since the typical recursive optimization expression (7.9) is not valid in every problem, as we shall see, the choice of the noun "principle" might be questioned, but the expression "principle of optimality" is in common use, and it is not unusual to see an equation of the type (7.9) presented as the result of an application of the "principle of optimality."

Regarding terminology, equations of the type (7.9) are frequently called "functional equations." In light of even our brief mention of functionals in Section 2.6, this nomenclature is understandable. In the present context we are defining functions of other functions.

ii. *Remarks on the Nature, Advantages, and Applicability of Dynamic Programming*

The reader can now better understand what dynamic programming is. It says nothing at all about how the subproblems, the optimizations at individual stages, are to be solved. Each of these could conceivably be an extremely difficult problem, but the dynamic-programming technique is oblivious to how these are solved and offers no assistance toward their solution. What it does offer is a technique for casting a problem involving n-decision variables into n subproblems, each involving a single decision variable. Each of these, we imagine, should be more tractable than the original. It is the recursive nature of the formulation that has made this decomposition possible, and, in addition, the recursive expression obtained at each stage allows the use of results from the previous stage optimization to facilitate optimization at the present stage.

On the basis of these heuristic comments, it is at least plausible that the dynamic-programming approach has merit. Let us then, prior to working some sample problems, offer a few general comments on the range of applications and several remarks on variations of the model of Figure 7.1.

Is dynamic programming limited to problems that exist as n-stage decision processes? Many important problems, of course, do have this form; many chemical engineering problems, for example, have exactly that form, and the reader should be able to think of other examples. Still, the applicability of the technique would be sharply curtailed if this were true. Fortunately, however, dynamic programming can be applied to any problem, regardless of the form in which it is found to arise, so long as it may be equivalently formulated as an n-stage decision process as described and diagrammed at the outset of this chapter, and found to satisfy the equations such as (7.9). The technique is truly no better than its user. This burden of proper formulation—and not all problems admit such a formulation in terms of various sets, transformations, and returns—rests entirely on the problem solver, as does the burden of solving the subproblems thereby created. Thus the possibility of utilizing dynamic programming to good advantage is enhanced by one's familiarity with other mathematical optimization techniques, a knowledge of the structure of the given problem, and, perhaps, a certain amount of cleverness. There will be more on the feasibility of dynamic programming later.

Processes will be encountered in which there is no terminal set S_0, and consequently no mapping T_1. Here the process ends within the state set S_1, with the selection of a d_1 and the realization of a return R_1. The nature of the process itself, of course, will determine this. Also, we may find processes with a terminal set S_0 but no return associated with that set; that is, $R_0(s_0) = 0$ for all $s_0 \in S_0$. If, for example, the process is required (constrained) to terminate in one of some particular set of states with no associated return, we could find an S_0 with $R_0 \equiv 0$.

In some problems, the *only* objective is to make the system terminate in one of some set of terminal states. Variations such as these, however, fit within the general process sketched in Figure 7.1.

Let us now formulate and solve a problem that does not have the "look" of an n-stage decision process.

Example 7.4

$$\max \sum_{i=1}^{n} \sqrt{x_i}$$

subject to

$$\sum_{i=1}^{n} x_i = c, \quad c > 0,$$

$$x_i \geq 0, \quad i = 1, \ldots, n$$

The first step toward solving this problem, which could be approached by the method of Lagrange multipliers (Chapter 2), is to formulate it as an n-stage decision process. For this, there are no rules or algorithms, but let us attempt a logical development of such a formulation.

The form

$$\sum_{i=1}^{n} \sqrt{x_i}$$

is our total return function, so if a return at the ith stage is $\sqrt{x_i}$, then the sum of the stage returns is the total return function; that is the only return function φ we have treated thus far, in fact.

A reasonable definition for a decision d_i is that it is the specification of an x_i which is consistent with the constraints. D_i, therefore, would be the set of all such real numbers x_i. We may also consider the act of specifying x_i, which we have denoted d_i, to be equivalent to x_i, and write $x_i = d_i$. Also, we have $R_i = \sqrt{x_i}$ a function of $d_i = x_i$, as required. The state sets and transformations remain to be specified. The problem itself is to divide c units into n parts so as to achieve a certain objective; that is, it is an allocation problem, and we may consider it to be a sequential one—we first select x_n, then $x_{n-1}, \ldots$, and finally x_1. Suppose then that we let $s_i = $ *the quantity as yet unallocated* as we approach stage i, that is, just prior to specifying x_i. Then the amount as yet unallocated after specifying x_i, $s_{i-1} = s_i - x_i$, and the transformations T_i are immediately defined:

$$T_i(s_i, d_i) = T_i(s_i, x_i) = s_i - x_i$$

S_i will be the set of all possible values of the quantity as yet unallocated at stage i.

We must be certain to allocate all c units, so we could insert a terminal state set S_0, consisting of the element 0 itself, and demand that the process end there. Or we could have the process end at stage 1, with no terminal constraints, and take care in the solution process to see that the constraint

$$\sum_{i=1}^{n} x_i = c$$

is satisfied.

The second alternative will be employed here, and the set S_n will be, as expected, the set $\{c\}$.

Our process appears as in Figure 7.4. The stage 1 block illustrates how the constraint will be satisfied; whatever is left, if anything, at that stage must be used. *There is no choice of x_1 at stage 1.*

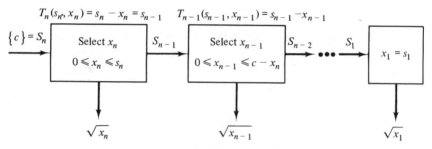

Figure 7.4. The n-stage decision process for the problem

$$\max \sum_{i=1}^{n} \sqrt{x_i}, \text{ s.t. } \sum_{i=1}^{n} x_i = c, \, x_i \geq 0, \, c > 0.$$

We may then begin the dynamic-programming solution.

$$f_1(s_1) = \max_{x_1 = s_1} \{\sqrt{x_1}\} = \sqrt{s_1}$$

As the maximization indicates, there was no choice of x_1, and the optimal x_1, $x_1^* = s_1$.

$$f_2(s_2) = \max_{0 \leq x_2 \leq s_2} \{\sqrt{x_2} + f_1(T_2(s_2, x_2))\}$$

$$= \max_{0 \leq x_2 \leq s_2} \{\sqrt{x_2} + f_1(s_2 - x_2)\}$$

$$= \max_{0 \leq x_2 \leq s_2} \{\sqrt{x_2} + \sqrt{s_2 - x_2}\}$$

The last expression consists of a maximization of a function of the single variable x_2 subject to the constraint $0 \leq x_2 \leq s_2$. If we consider at first the unconstrained problem, this subproblem appears amenable to ordinary calculus methods. Denoting

$$Q_2(x_2) = \{\sqrt{x_2} + \sqrt{s_2 - x_2}\}$$

we let

$$\frac{\partial Q_2}{\partial x_2} = \frac{1}{2}(x_2)^{-1/2} + \frac{1}{2}(s_2 - x_2)^{-1/2}(-1) = 0\dagger$$

Thus an optimal solution may exist at the point $x_2 = s_2/2$, which simultaneously satisfies the constraint $0 \leq x_2 \leq s_2$. We also observe that

$$\frac{\partial^2 Q_2}{\partial x_2^2} = -\frac{1}{2}\left(\frac{s_2}{2}\right)^{-3/2} < 0$$

if $s_2 > 0$

so

$$x_2 = \frac{s_2}{2}$$

is a relative maximum. Furthermore, as the reader may verify, $\sqrt{x_2} + \sqrt{s_2 - x_2}$ is concave on $0 \leq x_2 \leq s_2$, so

$$x_2^* = \frac{s_2}{2}$$

is a global maximum. The optimal return then for the two-stage process is

$$\sqrt{\frac{s_2}{2}} + \sqrt{\frac{s_2}{2}} = 2\sqrt{\frac{s_2}{2}} = \sqrt{2s_2} = f_2(s_2)$$

Next we find

$$\begin{aligned}
f_3(s_3) &= \max_{0 \leq x_3 \leq s_3} \{\sqrt{x_3} + f_2(T_3(s_3, x_3))\} \\
&= \max_{0 \leq x_3 \leq s_3} \{\sqrt{x_3} + f_2(s_3 - x_3)\} \\
&= \max_{0 \leq x_3 \leq s_3} \{\sqrt{x_3} + \sqrt{2(s_3 - x_3)}\}
\end{aligned}$$

Omitting the differentiations, we find

$$x_3^* = \frac{s_3}{3}$$

and

$$f_3(x_3) = \sqrt{3s_3}$$

---

†We have indicated partial differentiation, owing to the presence of s_2, but if we consider it fixed, we could write ordinary derivatives as well.

The three-stage process has been solved, and here we emphasize that in solving the *n*-stage problem the $1, 2, \ldots, (n-1)$-stage problems are solved as a by-product.

Were *n* large, there would be a great deal of work involved in proceeding to the *n*th stage. A pattern has begun to develop, it appears, and if an inductive proof is possible, the intermediate stages may be omitted. It appears that for any i, $x_i^* = s_i/i$ and $f_i(s_i) = \sqrt{is_i}$. The statements are true for $i = 1, 2, 3$.

Let us assume that they are true for j. Then

$$x_j^* = \frac{s_j}{j}, \qquad f_j(s_j) = \sqrt{js_j}$$

and
$$f_{j+1}(s_{j+1}) = \max_{0 \leq x_{j+1} \leq s_{j+1}} \{\sqrt{x_{j+1}} + f_j(s_{j+1} - x_{j+1})\}$$

$$= \max_{0 \leq x_{j+1} \leq s_{j+1}} \{\sqrt{x_{j+1}} + \sqrt{j(s_{j+1} - x_{j+1})}\}$$

by the assumption of the form of $f_j(s_j)$. Let

$$Q_{j+1} = \sqrt{x_{j+1}} + \sqrt{j(s_{j+1} - x_{j+1})}$$

Let
$$\frac{\partial Q_{j+1}(x_{j+1})}{\partial x_{j+1}} = 0$$

Then
$$s_{j+1} = (j+1)x_{j+1}$$

Thus we have $x_{j+1}^* = s_{j+1}/(j+1)$. It is then easily verified that this is a relative and global maximum, and that $f_{j+1}(s_{j+1}) = \sqrt{(j+1)s_{j+1}}$. The inductive proof is thus complete.

Therefore, for our *n*-stage problem we have that $f_n(c) = \sqrt{nc} =$ the maximum of the form

$$\sum_{i=1}^{n} \sqrt{x_i}$$

subject to the constraint

$$\sum_{i=1}^{n} x_i = c$$

and that the optimal choice of x_n is $x_n^* = c/n$. We must, however, determine the optimal values for $x_{n-1}, \ldots, x_1$. To do this, one backtracks through the work, using the T_j. Since

$$s_{n-1} = s_n - x_n$$
$$s_{n-1}^* = c - x_n^*$$

where s_{n-1}^* is to be interpreted as the state of the process at stage $n - 1$, resulting from an optimal choice of x_n. Thus

$$s_{n-1}^* = c - \frac{c}{n} = \frac{(n-1)c}{n}$$

and since

$$x_{n-1}^* = \frac{s_{n-1}^*}{n-1}$$

$$x_{n-1}^* = \frac{c}{n}$$

As indicated, we have obtained the optimal total return by working "right to left" in the process. The optimal decisions corresponding to this return are then obtained by working "left to right," utilizing the expressions for optimal decisions, $x_j^* = s_j/j$, which were determined as we worked toward $f_n(s_n)$.

Since

$$s_{n-2}^* = s_{n-1}^* - x_{n-1}^* = \frac{n-2}{n}c$$

we have that $x_{n-2}^* = s_{n-2}^*/(n-2) = c/n$, and so on, and we observe that

$$x_j^* = \frac{c}{n}, \quad j = 1, \ldots, n$$

The problem just solved is unquestionably simpler to attack by the method of Lagrange multipliers, but affords a good example of the mechanics of dynamic programming.

We wish to point out that in solving the problem we have solved a *family* of problems, one for each nonnegative value that c could assume. In this problem, c is a sort of "variable constant"; for a particular problem from the family, it is a constant, but, on the other hand, if it is allowed to vary as a parameter, one can solve any new problem from the family generated without any additional work.

Let us now solve a different and actually more important type of problem.

Example 7.5

An individual must travel from a city S_4 to a city S_0. Upon departing from S_4 he must pass through one of a set of four cities $S_3 = \{s_{31}, \ldots, s_{34}\}$, each of whose distance from S_4 is known. From the member of S_3 he selects, he must pass through one of a set of four cities $S_2 = \{s_{21}, \ldots, s_{24}\}$ whose distances from each city in S_3 are known. From the city in S_2 he must pass through one of a set of five cities $S_1 = \{s_{11}, \ldots, s_{15}\}$ with all distances from S_2 cities again known, and, finally, he travels from the city chosen in S_1 to his destination S_0 over a route of known length.

We assume that there is *one and only one route* from s_{mi} to $s_{m-1,j}$ for $m = 1, 2, 3, 4$ and all appropriate i, j. Again, the distance from s_{mi} to $s_{m-1,j}$ is known. The problem is to select a route that minimizes the total distance traveled. Omitting some of the paths, a map of the possibilities might appear as in Figure 7.5. There are $4 \cdot 4 \cdot 5 = 80$ possible routes, so enumeration of all possibilities and selection of the best would require a fair amount of computation.

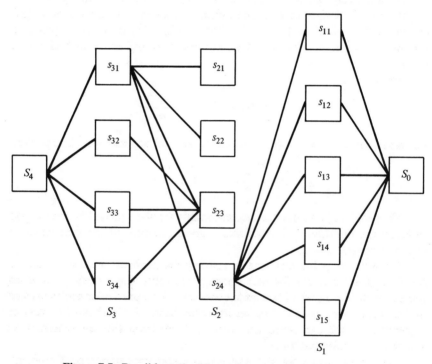

Figure 7.5. Possible paths taken when travelling from city S_4 to city S_0.

Let the distances of the various paths be defined in Table 7.1. If the driver is in s_{mk} and chooses to drive to $s_{m-1,j}$, we shall say that decision d_{mj} has been made.

Figure 7.6 illustrates the notation for decisions. The formulation of our problem as an *n*-stage decision process should be apparent.

In this particular problem, the states at a given stage are the cities at that stage. The decision set at each stage is the set of possible routes, and the

Table 7.1. DISTANCES BETWEEN CITIES OF S_4, S_3, S_2, S_1, AND S_0

		To S_3			
		s_{31}	s_{32}	s_{33}	s_{34}
From	S_4	3	4	5	6

		To S_2			
		s_{21}	s_{22}	s_{23}	s_{24}
	s_{31}	6	2	9	3
From S_3	s_{32}	3	1	4	8
	s_{33}	5	7	1	3
	s_{34}	2	3	1	4

		To S_1				
		s_{11}	s_{12}	s_{13}	s_{14}	s_{15}
	s_{21}	2	1	3	5	4
From S_2	s_{22}	4	2	6	5	6
	s_{23}	3	1	8	2	3
	s_{24}	4	5	1	3	2

		To	S_0
	s_{11}	2
	s_{12}	4
From S_1	s_{13}	3
	s_{14}	5
	s_{15}	1

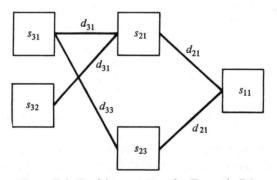

Figure 7.6. Decision notation for Example 7.5.

transformations are obviously defined, for given a city at a particular stage and the specification of a route, a unique city at the next stage is determined. The return at a given stage is the distance associated with a starting city and the route chosen. P_{ij} will denote the distance that results in traveling from S_i to S_j.

The process might then be depicted as in Figure 7.7, where the minimization of

$$\sum_{i=1}^{4} P_{i,i-1} = P_{40}$$

over all possible routes is the objective.

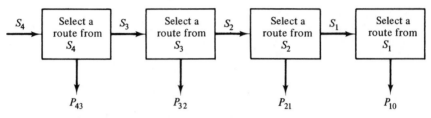

Figure 7.7. The 4-stage decision process for Example 7.5.

To solve this problem by dynamic programming, we first find

$$f_1(s_1) = \min_{d_1} \{P_{10}\}$$

where s_1 is a discrete variable ranging over $s_{11}, \ldots, s_{15}$, and $\min_{d_1}$ implies the minimization over all possible routes from S_1 cities.

Once again, the first stage does not afford a choice, since there is but one route from s_{1j} to $S_0, j = 1, 2, \ldots, 5$. There is actually no reason to tabulate any information from this stage.

At stage 2, we have

$$f_2(s_2) = \min_{d_2 d_1} \{P_{20}\} = \min_{d_2 d_1} \{P_{21} + P_{10}\}$$

$$= \min_{d_2} \{P_{21} + \min_{d_1} P_{10}\}$$

In this case, $f_2(s_2)$ will be a table, relating optimal strategies to each possible starting city in S_2.

If in s_{21} and the route to s_{11} is chosen, total trip length from S_2 is 2 (from s_{21} to s_{11}) + 2(from s_{11} to S_0), or 4. If the route to s_{12} is chosen, total trip length is $1 + 4 = 5$. If s_{13}, the total is 6; if s_{14}, 10; and if s_{15}, 5. Therefore, regardless of the distance covered previously, if the driver finds himself in s_{21} at stage 2, he should drive to s_{11}, that is, make decision d_{21}.

If at stage 2, the driver finds himself in s_{22}, he may minimize his total distance *from that point* by driving to either s_{11} or s_{12}. Starting in s_{23}, he should

Table 7.2. VALUES OF THE OPTIMAL RETURN FUNCTION $f_2(s_2)$

Stage 2 State	Optimal Decision(s)	Total Return $f_2(s_2)$
s_{21}	d_{21}	4
s_{22}	d_{21} or d_{22}	6
s_{23}	d_{25}	4
s_{24}	d_{25}	3

first drive to s_{15}. From s_{24}, the best choice is s_{15}. Thus $f_2(s_2)$ is as shown in Table 7.2.

Next we find

$$f_3(s_3) = \min_{d_3} \{P_{32} + \min_{d_2 d_1} P_{20}\}$$
$$= \min_{d_3} \{P_{32} + f_2(T_3(s_3, d_3))\}$$

Here we have used the usual notation for the transformation, although it was not explicitly introduced into the problem formulation. Notice that in solving the two-stage problem, we have had to enumerate all the possibilities.

Let us now compute f_3. Suppose that at stage 3 the driver finds himself in s_{31}. If he drives to s_{21}, he must go distance 6 and, being in s_{21}, the best choice, from Table 7.2, is d_{21} (i.e., drive to s_{11}), for which the return is 4. Therefore, if in s_{31} at stage 3 and if decision d_{31} is made, the minimum total distance is 10. If from s_{31} our driver goes to s_{22}, the minimum distance is 2 plus 6 (from Table 7.2), which equals 8. From s_{31} to s_{23} is distance 9, and the table for f_2 shows that upon reaching s_{23}, d_{25} is the best decision, and the overall distance is $9 + 4 = 13$. From s_{31} to s_{24} is a distance 3 and the best overall distance then, again from the table, is $3 + 3 = 6$.

Therefore, we conclude that if the driver is in s_{31}, regardless of where he has been previously, the minimum distance remaining is 6, which he can realize by driving first to s_{24}, then to s_{15}.

Performing these calculations for *every* city in S_3 and choosing the best decision *pair* for each city gives the function $f_3(s_3)$, using $f_2(s_2)$, and these are presented in Table 7.3. The advantage of the recursive optimization becomes apparent from the ease of solving the three-stage problem.

To find the *overall* optimal route it is only necessary to compute

$$\min_{d_4} \{P_{43} + f_3(T_4(s_4, d_4))\} = \min \{3 + 6, 4 + 7, 5 + 5, 6 + 5\} = 9$$

Thus the minimum distance is 9, and the optimal decision sequence is d_{41}, d_{34}, d_{25}; in other words, our hypothetical driver should travel from S_4 to s_{31} to s_{24} to s_{15}, and thence to S_0.

It should be noticed that we could have approached the problem of re-

Table 7.3. VALUES OF THE OPTIMAL RETURN FUNCTION $f_2(s_3)$

Stage 3 State	Optimal Decision(s)	Total Return $f_3(s_3)$
s_{31}	d_{34} (to s_{24})	6
s_{32}	d_{31} (to s_{21}) or d_{32} (to s_{22})	7
s_{33}	d_{33} (to s_{23})	5
s_{34}	d_{33} (to s_{23})	5

trieving an optimizing decision sequence differently in this example. Rather than first finding the optimal total return, then backtracking to find the optimizing decisions, we could maintain at each stage an optimizing decision *sequence* for each state. These would be kept within the tables for f_1, f_2, and so on, and no backtracking would be necessary, so we economize on the amount of storage used—the only table required in storage, while a new one is being constructed, is the one from the previous stage. If one is to backtrack, however, all previous tables must be retained, so more storage is required.

Having solved a discrete example by dynamic programming, it is advisable that we compare this approach to total enumeration, that is, to the strategy of forming all possible decision sequences and selecting the one which provides the best value of the objective function.

In the foregoing example, it was required to evaluate five possibilities for each stage 2 state. The work at stage 2, then, consisted of 20 evaluations, each the addition of two numbers.

At stage 3, for each of four states, it was necessary to evaluate but four alternatives; again, each evaluation consists of the addition of two numbers. Hence at stage 3 we performed 16 additions. Here the computational saving afforded by the recursive optimization, the use of the previously computed f_2, becomes evident.

At stage 4, four additions were performed. Thus $20 + 16 + 4 = 40$ additions were required to solve the problem by dynamic programming.

Total enumeration requires three additions for each of the 80 alternatives, or 240 additions.

Our comparison has not been complete, since in both procedures values must be compared in order that certain minima be selected. If the reader will determine the number of comparisons that must be executed for each approach, he will find the dynamic-programming solution even more desirable than evidenced by our analysis. In what follows, such comparisons will be neglected.

In the general model of the foregoing type, suppose that we have p_i states and q_i decisions at stage i, $i = 1, \ldots, n$. At stage i, $p_i q_i$ operations are required, so the total number of additions is, including a stage one contribution

$$\sum_{i=1}^{n} p_i q_i$$

In the case of total enumeration, there are

$$\prod_{j=1}^{n} q_j$$

possible decision sequences, evaluating each of which requires $n - 1$ operations of addition.

Thus we can compare the amount of work required by the two approaches:

$$\text{Dynamic programming} \quad \sum_{i=1}^{n} p_i q_i \tag{7.11}$$

$$\text{Enumeration} \quad (n - 1) \prod_{j=1}^{n} q_j \tag{7.12}$$

Clearly, values for p_i, q_i can be contrived for which the value (7.11) exceeds that of (7.12), but will those values of p_i correspond to a reasonable model? The state variables are, after all, an artifact of the model selected. For the present model, the p's and q's are certainly not independent; we have, in fact, $p_{i-1} = q_i$ for $i = 2, \ldots, n$. With that relationship in mind, the reader should examine the magnitudes of the expressions (7.11), (7.12) and convince himself of the general advantage of dynamic programming.

Another interesting observation is the manner in which the number of stages affects the amount of work. As n increases, (7.11) increases linearly, whereas (7.12) increases exponentially. For one thing, this observation points up that the feasibility of dynamic programming is not so much a matter of the number of stages. The more critical parameters will be taken up a little later.

In comparing dynamic programming to total enumeration, it is helpful to realize that dynamic programming, itself, is an enumeration procedure—we account for all possible decision sequences; the difference is that the recursive nature of the optimization (7.9) allows many of the alternatives to be eliminated without being explicitly treated. Thus we might characterize dynamic programming as a method of *implicit enumeration*. That is, *one accounts for all possibilities without the necessity of explicitly evaluating them.*

The reader may notice that the dynamic-programming approach to problem solving is, roughly speaking, a strategy he himself has used in solving, or at least in approximating solutions to, decision problems he may have faced. It was, however, Richard Bellman who first formalized and made rigorous the method as a problem-solving device and explored its application to a vast variety of problems.

7.4. More General Measures of Total Return

Let us now consider total return functions of a more general nature. In particular, let the return from the terminal stage remain $R_0(s_0)$.
Let the return from stages 1 and 0 be

$$\varphi_1(R_1(s_1, d_1), R_0(s_0))$$

Beginning with stage 2, the return is

$$\varphi_2(R_2(s_2, d_2), R_1(s_1, d_1), R_0(s_0))$$

And beginning with stage i, the return is

$$\varphi_i(R_i(s_i, d_i), \ldots, R_1(s_1, d_1), R_0(s_0))$$

and the total return is

$$\varphi_n(R_n(s_n, d_n), \ldots, R_1(s_1, d_1), R_0(s_0))$$

Let us examine a rather simplified problem, in which a more general function arises.

A man is going to make k wagers, an equal amount each time, on n horse races. He has discovered that the probability of not winning any bets on race i is a function only of the number of bets made on race i; that is, if m bets are made on race i the probability of losing them all is $g_i(m)$. The g_i are known, and the man's objective is to allocate bets to races so as to maximize his probability of winning at least one bet (a modest objective). Here we should minimize

$$\varphi_n(R_n, R_{n-1}, \ldots, R_1) = \prod_{j=1}^{n} R_j = \prod_{j=1}^{n} g_j(x_j)$$

i. Assumption on the Nature of the φ_i

In general, we shall assume that our functions φ_i have the form

$$\varphi_i(R_i, \ldots, R_1, R_0) = \alpha_i(R_i, \varphi_{i-1}(R_{i-1}, \ldots, R_1, R_0)), \quad i = 1, \ldots, n \quad (7.13)$$

In other words, the return from the i-stage process is just a function α_i of the ith stage return and the return from the $(i - 1)$-stage process. The only composite return functions mentioned thus far and having this particular form were

$$\varphi_i = \sum_{j=0}^{i} R_j \quad \text{and} \quad \varphi_i = \prod_{j=0}^{i} R_j$$

ii. Developing the General Recursion Relations

If we begin writing the recursion relations in this more general case, we have, assuming maximization,

$$
\begin{aligned}
f_1(s_1) &= \max_{d_1} \{\varphi_1(R_1(s_1, d_1), R_0(s_0))\} \\
&= \max_{d_1} \{\varphi_1(R_1(s_1, d_1), R_0(T_1(s_1, d_1)))\} \\
f_2(s_2) &= \max_{d_2 d_1} \{\varphi_2(R_2(s_2, d_2), R_1(s_1, d_1), R_0(s_0))\} \qquad (7.14) \\
&= \max_{d_2} \{\max_{d_1} \{\varphi_2(R_2(s_2, d_2), R_1(s_1, d_1), R_0(s_0))\}\} \\
&= \max_{d_2} \{\max_{d_1} \{\alpha_2(R_2(s_2, d_2), \varphi_1(R_1(s_1, d_1), R_0(s_0)))\}\}
\end{aligned}
$$

Just as when comparable expressions were obtained for the case of the sum of the stage returns, we would *like* to observe that this last expression (7.14) is equal to

$$\max_{d_2} \{\alpha_2(R_2(s_2, d_2), \max_{d_1} \varphi_1(R_1(s_1, d_1), R_0(s_0)))\}$$

If that manipulation is legitimate in this general case, then the recursive power of the method will be preserved, for

$$\max_{d_1} \varphi_1(R_1(s_1, d_1), R_0(s_0)) = f_1(s_1) = f_1(T_2(s_2, d_2))$$

has already been obtained at that point, and, again, a maximization in a single variable, d_2, results.

At stage i, then, we would obtain

$$f_i(s_i) = \max_{d_i} \{\alpha_i(R_i(s_i, d_i), f_{i-1}(T_i(s_i, d_i)))\} \qquad (7.15)$$

quite analogous to the expression (7.9), where φ_i is the sum of the stage returns $R_i, \ldots, R_0$.

428 | *Dynamic Programming*

iii. *On the Validity of the General Relations—*
A Sufficiency Condition

Unfortunately, equation (7.15) is not always legitimate. Consider the following counterexample:

$$\max_{x,y} g(x, y) = x \cdot y$$

subject to

$$a \leq x \leq b < 0$$
$$0 < c \leq y \leq d$$

With respect to the function g, the question we ask is whether

$$\max_{x,y} x \cdot y = \max_{x}(x \cdot \max y) = \max_{x} (x \cdot d)$$

The answer is no, clearly, since

$$\max_{x} x \cdot d = b \cdot d \quad \text{and} \quad \max_{x,y} x \cdot y = b \cdot c$$

Since it is now obvious that the recursion relations will not always be valid, it would be desirable to be able to know in advance when a particular return function will admit a valid recursive formulation and when it will not.

Conditions necessary for the recursion to hold have not been discovered, but a sufficient condition is easily obtained, and this we now do.

Consider a function of two variables, $\phi(x, y)$, where x ranges over some set X, and y ranges over a set Y. Let us assume ϕ is a monotonically nondecreasing function of y; that is, if $y_1 < y_2$, then

$$\phi(x_0, y_1) \leq \phi(x_0, y_2)$$

for any $x_0 \in X$ and for any $y_1, y_2 \in Y$. Then

$$\max_{y} \phi(x, y) = \phi(x, \max_{y} y)$$

and consequently

$$\max_{x} \max_{y} \phi(x, y) = \max_{x} \phi(x, \max_{y} y)$$

More generally, suppose that ϕ is a function of a variable x and a function β, itself a function of the variables $y_1, y_2, \ldots, y_m$. That is, we have $\phi(x, \beta(y_1, y_2, \ldots, y_m))$. Suppose that ϕ is monotonically nondecreasing in $\beta(y_1, y_2, \ldots, y_m)$. Then

$$\max_{y_1, \cdots, y_m} \phi(x, \beta(y_1, \ldots, y_m)) = \phi(x, \max_{y_1, \cdots, y_m} \beta(y_1, \ldots, y_m))$$

In terms of our usual notation, equation (7.15) will hold if the function α_i is a monotonic, nondecreasing function of φ_{i-1}.

iv. *Application to a Reliability Problem*

We consider a particular serial system, which gives rise to a multiplicative objective function, represented as follows:

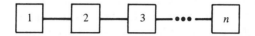

Each box is a component, and failure of a single component causes the entire system to fail. We would like to investigate the possibility of adding backup components in parallel at some stages in order to maximize the probability of successful operation of the entire system. A particular such configuration might appear as

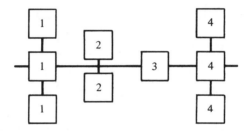

Thus, if a component fails and there is one in parallel with it, a switch of some sort allows the parallel component to begin operating and prevent system failure.

Denote by x_j the number of *additional* components to be used at stage j and by $p_j(x_j)$ the probability of failure of stage j if x_j backup components are used there. Each component of kind j has cost c_j, and the allocation of parallel components is constrained by the amount of money available, c dollars.

The problem, then, is to find $x_1, x_2, \ldots, x_n$ so as to maximize

$$\prod_{j=1}^{n} [1 - p_j(x_j)]$$

subject to

$$\sum_{j=1}^{n} c_j x_j \leq c$$

and x_j a nonnegative integer, $j = 1, \ldots, n$.

The exact formulation of this problem as an n-stage decision process, including the specification of all sets, transformations, and functions involved, is straightforward and is left as an exercise.

7.5. Reversing the Direction of the Problem

In general, there is no real "direction" associated with the problems we formulate as serial decision processes, although there is a direction, as indicated by the arrows, associated with the process as diagrammed in Figure 7.1.

Assuming that a problem has been formulated so as to appear as Figure 7.1, we inquire whether it is possible to "reverse" the flow of the system, obtaining the diagram in Figure 7.8, and then proceed to solve the same problem.

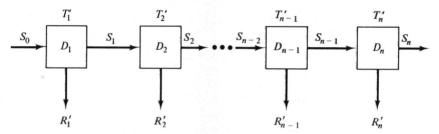

Figure 7.8. Reversing the direction of the process in an *n*-stage problem.

If this is possible, we notice that the optimal return will be expressed as a function of an element of S_0. Also, it will be necessary that there exist one-to-one transformations $T'_1, T'_2, \ldots, T'_n$ such that

$$s_i = T'_i(s_{i-1}, d_i), \quad i = 1, \ldots, n$$

Since we originally had $T_1, \ldots, T_n$ with the property that

$$s_{i-1} = T_i(s_i, d_i), \quad i = 1, \ldots, n,$$

then if these equations can be solved for s_i, the existence of the T'_i will be guaranteed. They may not always exist.

But if they do, we are essentially saying, with respect to Figure 7.1, that the input state is a function of the output state and the decision.

Assuming the existence of the T'_i and, for notational simplicity, that the total return function is the sum of the stage returns, write

$$R_n(s_n, d_n) = R_n(T'_n(s_{n-1}, d_n), d_n) = R'_n(s_{n-1}, d_n)$$

Let
$$f_1(s_{n-1}) = \max_{d_n} \{R'_n(s_{n-1}, d_n)\}$$

For the two-stage process, we obtain

$$f_2(s_{n-2}) = \max_{d_{n-1}} \{R'_{n-1}(s_{n-2}, d_{n-1}) + f_1(T'_{n-1}(s_{n-2}, d_{n-1}))\}$$

In general, we find

$$f_i(s_{n-i}) = \max_{d_{n-i+1}} \{R'_{n-i+1}(s_{n-i}, d_{n-i+1}) + f_{i-1}(T'_{n-i+1}(s_{n-i}, d_{n-i+1}))\}$$

and, finally,

$$f_n(s_0) = \max_{d_1} \{R'_1(s_0, d_1) + f_{n-1}(T'_1(s_0, d_1))\}$$

Exercise 7.1

The reader should derive several of the lower-order recursion relations and then solve Example 7.4 using this new approach.

i. *Practical Advantages to Direction Reversal*

There are several potential advantages to the ability to solve a problem in either of two directions. First, the transformations may be more easily handled in one direction than the other, assuming they exist in both directions. For example, suppose that

$$s_{i-1} = T_i(s_i, d_i) = \sqrt{s_i - d_i}$$

where we may be interested only in the positive square root and where $s_i - d_i \geq 0$ always. Solving for s_i, we obtain

$$s_i = T'_i(s_{i-1}, d_i) = s_{i-1}^2 + d_i$$

which is more rapidly calculated than would be the square root.

Another advantage is that the assumption regarding the nature of the return functions (7.13) may be valid in one direction and not so in the other.

For example, if the total return from a four-stage process is $R_1 \cdot R_2(R_3 + R_4)$ and the serial process has R_4, R_3, R_2, R_1 occurring from left to right, we cannot satisfy (7.13). If we consider the process in the other direction, however, with returns appearing R_1, R_2, R_3, R_4 left to right, we have

$$\varphi_1(R_4) = R_4$$
$$\varphi_2(R_3, R_4) = R_3 + R_4 = R_3 + \varphi_1(R_4) = \alpha_2(R_3, \varphi_1(R_4))$$
$$\varphi_3(R_2, R_3, R_4) = R_2(R_3 + R_4) = R_2 \cdot \varphi_2(R_3, R_4) = \alpha_3(R_2, \varphi_2(R_3, R_4))$$
$$\varphi_4(R_1, R_2, R_3, R_4) = R_1 R_2(R_3 + R_4) = R_1 \cdot \varphi_3(R_2, R_3, R_4)$$
$$= \alpha_4(R_1, \varphi_3(R_2, R_3, R_4))$$

If our returns are all positive, the monotonicity assumption will also hold. A little later we shall see another use for an expression of the optimal return as a function of final state.

7.6. Extensions of the Model and the Methods

Having generalized upon the kinds of total return functions our models may possess, we may turn to extensions of our basic model in other directions. We have found, in general, when dealing with optimization problems that unconstrained problems are more easily solved than constrained problems. In the example problems we have cast as n-stage decision processes thus far, however, constraints have lent themselves quite nicely to providing the stage transformations T_i, which must exist.

i. *Multiple Constraints*

An obvious extension, then, is the treatment of problems having more than one constraint. To be sure we have already considered problems with more than a single constraint; for example, nonnegativity restrictions or integer restrictions on variables have been considered; but now we consider more formal kinds of constraints. Consider, then, the following problem.

Example 7.6

The cargo hold in a particular ship has a volume V and may carry a maximum weight W. There are available n kinds of items that might be shipped, item j having a volume v_j and a weight w_j. There is a sufficient number of each item that the hold could be filled entirely with items of kind j, $j = 1$, $\ldots$, n. Furthermore, carrying x_j items of kind j results in a profit $g_j(x_j)$, and we would like to specify the composition of a cargo that maximizes profit. The problem, often referred to as the *knapsack problem*, may be expressed

$$\max \sum_{j=1}^{n} g_j(x_j)$$

subject to

$$\sum_{j=1}^{n} v_j x_j \leq V \qquad (7.16)$$

$$\sum_{j=1}^{n} w_j x_j \leq W$$

$$x_j = \text{a nonnegative integer}, \ 1 \leq j \leq n$$

For the most part, the formulation as an n-stage serial decision process is apparent. The question is, how to handle two constraints.

One possibility is to consider state variables as being two dimensional; for example, let $s_j = \mathbf{s}_j = (V_j, W_j)$, where V_j, W_j are the volume and weight, respectively, still unused at the beginning of the jth stage, at which point we specify x_j. Then define

$$T_j(\mathbf{s}_j, x_j) = (V_j - v_j x_j, W_j - w_j x_j) = (V_{j-1}, W_{j-1}) = \mathbf{s}_{j-1}$$

For illustrative purposes in the example, we would like V, W, v_j, and w_j all positive integers, so that our state vectors will always be ordered pairs of nonnegative integers.

What is obtained in this problem, which we shall verify by completing the formulation and observing the first two or three recursive maximization problems, is a rather large number of maximizations to perform at each stage—a direct result of the two-dimensional state variable.

The first computation involves that of calculating

$$f_1(\mathbf{s}_1) = \max (g_1(x_1))$$

where $$x_1 \le \left[\frac{V_1}{v_1}\right],^\dagger \qquad x_1 \le \left[\frac{W_1}{w_1}\right]$$

Since V_1 can range between 0 and V and W_1 between 0 and W, there are $(V + 1)(W + 1)$ possibilities for $\mathbf{s}_1$,‡, and consequently that number of maximizations at stage 1.

At stage 2 we calculate

$$f_2(\mathbf{s}_2) = \max (g_2(x_2) + g_1(x_1))$$

where $$x_2 \le \left[\frac{V_2}{v_2}\right], \qquad x_2 \le \left[\frac{W_2}{w_2}\right]$$

or $$f_2(\mathbf{s}_2) = \max (g_2(x_2) + f_1(T_2(\mathbf{s}_2, x_2)))$$

where $$x_2 \le \left[\frac{V_2}{v_2}\right], \qquad x_2 \le \left[\frac{W_2}{w_2}\right]$$

and $$(V + 1)(W + 1)$$

maximizations are required.

---

†$[Z]$ = greatest integer not exceeding Z, i.e., the "integer part" of Z.

‡This is not a totally accurate statement. From the nature of the problem, for example, we cannot have used all the volume and none of the weight. Neglecting physical considerations, however, this is the number of state vectors that needs to be examined in the general problem. Also, not all the listed values of V_j or W_j alone may be possible, but the computations to eliminate these may be as much work as that sustained by including them.

Continuing in this fashion, by the time $f_n(s_n)$ has been calculated, we have performed

$$n(V + 1)(W + 1) \qquad (7.17)$$

maximizations.

For purposes of comparison, suppose that we had solved the example problem with but a single state variable—say the second constraint of (7.16) were eliminated. The reader will verify that the number of maximizations for the one-dimensional problem is

$$n(V + 1) \qquad (7.18)$$

and comparison of the forms of (7.17) and (7.18) indicates the effect on the size of the problem brought about by the increase in the dimension of the state variable.

The effect of further increases in dimensionality can be inferred by the considerations that led to (7.17). Simply stated, the higher the dimension of the state-variable space, the greater the number of combinations of state *variables* at each stage. Furthermore, this number of combinations, which is also the number of optimizations that must be performed, increases *exponentially* with dimension.

The topic of multiple constraints will be taken up again later.

Exercise 7.2

Solve the problem (7.16) for $v_1 = 2$, $v_2 = 5$, $v_3 = 1$, $V = 24$, $w_1 = 3$, $w_2 = 7$, $w_3 = 2$, $W = 33$, and

(a) $g_1(x_1) = 2x_1$, $g_2(x_2) = 3x_2$, $g_3(x_3) = \frac{3}{2}x_3$

Then

(b) $g_1(x_1) = \frac{1}{2}x_1$, $g_2(x_2) = x_2$, $g_3(x_3) = 2x_3$

ii. *Multidimensional Decision Variables— A Production Inventory Problem*

It is natural to consider next problems in which the *decision* variables are vectors. In such problems, generally, each optimization at each stage will be in terms of several variables. Previous experience with optimization involving a higher-dimensional space of decision variables implies that one may expect a great deal of computation, in general, when solving such a problem by dynamic programming. It is not difficult, then, to anticipate that problems in which one or both decision and state variables are multidimensional can grow out of hand quite rapidly. Without certain approximation techniques and other trickery, dynamic programming is an infeasible approach for many higher-dimensional problems.

For the moment, though, let us consider an idealized sort of problem in which both state and decision variables are multidimensional. We include a portion of the problem formulation.

We envision a simple production-inventory system over a time span of n periods. The organization uses raw materials m_1, m_2, m_3, m_4 beginning the time span with $q_{01}, q_{02}, q_{03}, q_{04}$ units of each, respectively. The costs of these materials are $c_{1j}, c_{2j}, c_{3j}, c_{4j}$ dollars per unit, respectively, during period $j, j = 1, \ldots, n$; that is, we assume the materials' costs possibly vary in time, but in a manner known in advance. Two products, P_1 and P_2, are manufactured, and the composition and selling price of each is summarized as follows:

Product	Units of Raw Material Required				Selling Price During Period j
	m_1	m_2	m_3	m_4	
P_1	p_{11}	p_{12}	p_{13}	p_{14}	c'_{1j}
P_2	p_{21}	p_{22}	p_{23}	p_{24}	c'_{2j}

We assume the per-period demand for the products is known for all n periods, and b_{ij} = demand for product i in period $j, i = 1, 2, j = 1, \ldots, n$.

Suppose that the problem is to determine an ordering–manufacturing program through the n periods so as to maximize the total profit.

Let us see now one way to cast this problem in the desired form. (The desired formulation is not unique in this problem.)

A decision $\mathbf{x}_j \in D_j$ will be a vector $(x_{1j}, x_{2j}, \ldots, x_{8j})$ where

x_{kj} = number of units of m_k purchased for period j,
$\quad k = 1, \ldots, 4, j = 1, \ldots, n$

$\quad$ = number of units of product P_{k-4} produced in period j,
$\quad k = 5, 6, j = 1, \ldots, n$

$\quad$ = number of units of product P_{k-6} sold in period j,
$\quad k = 7, 8, j = 1, \ldots, n$

Also, $x_{7j} \leq b_{1j}, x_{8j} \leq b_{2j}$.

We allow the firm to sell less of its products than the known demand. Owing to the time-varying costs and prices, it is a possibility that the firm would willingly incur a penalty for not meeting demand in order to obtain a better future profit. (Of course, we have removed the risk in such a maneuver by specifying all future costs and prices.) The penalty for each unit of demand not met will be f_i for product $P_i, i = 1, 2$. In addition, there may be a holding cost associated with storing finished product. Let the per-period holding cost charged at beginning of period be h_i for product $P_i, i = 1, 2$.

A state variable $\mathbf{s}_j \in S_j$ will be a vector $(s_{1j}, s_{2j}, \ldots, s_{6j})$, where $s_{kj} =$ number of units of m_k on hand at the beginning of period j, $k = 1, \ldots, 4$, $j = 1, \ldots, n$, and $s_{kj} =$ number of units of P_{k-4} on hand at the beginning of period j, $k = 5, 6, j = 1, \ldots, n$.

We want $T_j(\mathbf{s}_j, \mathbf{x}_j) = \mathbf{s}_{j-1}$, and such a T_j is easily defined. Clearly,

$$T_j(s_{1j}, s_{2j}, s_{3j}, s_{4j}, s_{5j}, s_{6j}, x_{1j}, x_{2j}, x_{3j}, x_{4j}, x_{5j}, x_{6j}, x_{7j}, x_{8j})$$
$$= (s_{1,j-1}, \ldots, s_{6,j-1})$$
$$= (s_{1j} + x_{1j} - p_{11}x_{5j} - p_{21}x_{6j}, s_{2j} + x_{2j} - p_{12}x_{5j} - p_{22}x_{6j},$$
$$\ldots, s_{4j} + x_{4j} - p_{14}x_{5j} - p_{24}x_{6j}, z_5, z_6)$$

where

$$z_5 = x_{5j} + s_{5j} - x_{7j}$$

and

$$z_6 = x_{6j} + s_{6j} - x_{8j}$$

The stage return R_j will be simply the net profit during period j. In fact,

$$R_j = x_{7j}c'_{1j} + x_{8j}c'_{2j} - \sum_{i=1}^{4} x_{ij}c_{ij} - \sum_{i=1}^{2} h_i s_{4+i,j} - \sum_{i=1}^{2} f_i[b_{ij} - x_{i+6,j}]$$

In words, the profit during period j is simply the income from units sold during the period less the total of the cost of raw materials purchased during the period, penalties for not meeting demands, and holding costs for products kept during the period.

We leave for the reader the construction of several recursion relations and the investigation of the effects of dimensionality.

iii. Performing the Individual Optimizations

At a given point in solving a problem by dynamic programming one is faced with finding

$$\max_{\mathbf{d}_j} \{\alpha_j[R_j(\mathbf{s}_j^0, \mathbf{d}_j), f_{j-1}(T_j(\mathbf{s}_j^0, \mathbf{d}_j))]\} = f_j(\mathbf{s}_j^0) \qquad (7.19)$$

where $\mathbf{s}_j^0$ is some particular state at stage j.

As mentioned previously and exhibited in our examples, dynamic programming offers no solution procedure for this optimization.

Now at this point the reader has been exposed to a number of techniques for performing single optimizations of the kind (7.19). In particular, the methods of Chapter 6 can be very important in relation to dynamic programming. These invariably require something in the way of special properties from the objective function—for example, convexity or unimodality—

but, if applicable, could mean the difference between an efficiently obtained, accurate approximation and the opposite.

It must be realized that it is not the convexity or unimodality of all the stage returns that is important; rather it is the function inside the braces of equation (7.19), that is, the objective function of a typical optimization.

For an arbitrary function α_i in equation (7.19) it may not be immediate that it possesses or lacks one of these desirable properties.

In some relatively simple, but common, instances desirable properties of α_i can be deduced from the nature of the stage returns. For example, if the sum of stage returns is of interest and each of these is concave, then α_i is concave, if T is linear.

However, if the objective function of a typical optimization within the dynamic-programming solution is not adequately behaved for the application of one of the more efficient search techniques, the dynamic-programming approach need not be considered defeated, even if decision variables are of higher dimension. One can adopt the technique of successively refining the grid in exhaustive search, as was mentioned in the Chapter 6. Let us now apply an exhaustive search approach.

iv. *Discrete Approximations to Continuous Problems*

Let us consider discretizing a simple continuous problem. Some further computational impediments will be observed. Let us suppose that, in the given problem,

$$D_j = [a_j, b_j], \qquad S_j = [u_j, v_j]$$

where $a_j, b_j, u_j, v_j, j = 1, \ldots, n$, are real numbers with $a_j < b_j, u_j < v_j$ for all j; that is, the D_j, S_j are merely closed intervals of the real line.

D_j, S_j are partitioned by δ_j, Δ_j, respectively, as follows.†

$$[a_j, a_j + \delta_j, a_j + 2\delta_j, \ldots, b_j - \delta_j, b_j] = D'_j$$
$$[u_j, u_j + \Delta_j, u_j + 2\Delta_j, \ldots, v_j - \Delta_j, v_j] = S'_j$$

D'_j, S'_j are the decision and state sets to be used in solving the problem, which is but an *approximate* version of the given problem.

As always, the first task is to find $f_1(s'_1)$, which in this case, means somehow computing

$$\max_{d'_1} \{R_1(s'_1, d'_1)\}$$

for each element $s'_1 \in S'_1$. Thus, at the end of the first phase of the work, Table 7.4 may be constructed.

---

†It is not essential that the points be equally spaced.

Table 7.4. OPTIMAL DECISIONS ASSOCIATED WITH THE FIRST STAGE RETURN
($k_1, \ldots, k_p$ ARE SOME NONNEGATIVE INTEGERS)

State	Optimal Decision	Optimal R_1 Value
u_1	$a_1 + k_1\delta_1$	$R_1(u_1, a_1 + k_1\delta_1)$
$u_1 + \Delta_1$	$a_1 + k_2\delta_1$	$R_1(u_1 + \Delta_1, a_1 + k_2\delta_1)$
$u_1 + 2\Delta_1$	$a_1 + k_3\delta_1$	.
.	.	.
.	.	.
.	.	.
$v_1 - \Delta_1$	$a_1 + k_{p-1}\delta_1$	
v_1	$a_1 + k_p\delta_1$	

At stage 2, we need to calculate

$$f_2(s_2') = \max_{d_2'} \{R_2(s_2', d_2') + f_1(T_2(s_2', d_2'))\}$$

A typical operation then is to compute

$$R_2(u_2 + q\Delta_2, a_2 + r\delta_2) + f_1(T_2(u_2 + q\Delta_2, a_2 + r\delta_2))$$

where q, r are some given nonnegative integers. There is no difficulty with the first term in the expression since R_2 was originally defined on $S_2 \times D_2$, one member of which is $(u_2 + q\Delta_2, a_2 + r\delta_2)$. Consider, however,

$$T_2(u_2 + q\Delta_2, a_2 + r\delta_2) = \bar{u}_1$$

We know this is an element of S_1, but there is no guarantee that $T_2(u_2 + q\Delta_2, a_2 + r\delta_2) \in S_2'$, and, consequently, f_1 may be *undefined* for $T_2(u_2 + q\Delta_2, a_2 + r\delta_2)$. So, what does one do? Clearly, some kind of interpolation is implied. One could assume f_1 linear over the subintervals of length Δ_1 of S_1' and interpolate linearly. The smaller Δ_1, the more closely f_1 would be linear on the subintervals; but, on the other hand, the magnitude of the computational job will increase (linearly, as a matter of fact) as the number of partition points increases.

If one is interested in greater accuracy than linear interpolation can be expected to provide, then he can fit a higher-degree polynomial $p_f(u)$ to the data in some neighborhood of $\bar{u}_1$ and thus approximate $f_1(\bar{u}_1)$ with $p_f(\bar{u}_1)$. We shall not pursue any of the numerous methods for polynomial interpolation, but the interested reader will find a clear exposition of the principal techniques as well as some account of the error of the various methods in any text on elementary numerical methods. The more sophisticated schemes will be more time consuming, in general—for example, in calculating the coefficients of the interpolating polynomials.

The proper approximation method in a given instance can only be decided by knowledge of the accuracy required and the amount of computation time that can be expended; of course, some knowledge of the form of f_1 would be invaluable.

However, there remains the problem of determining the value of the decision variable d_1' corresponding to the interpolated value of $f_1(\bar{u}_1)$. An approximation to this decision variable may be obtained by interpolation also. One passes an approximating polynomial $p_d(u)$ through several of the pairs $(u_1 + z\Delta_1, a_1 + k_{z+1}\delta_1)$ and approximates the optimal decision corresponding to $\bar{u}_1$ by $p_d(\bar{u}_1)$.

The information from Table 7.4 has been expressed as the graph of Figure 7.9. Here both optimal decisions and f_1 values are measured on the vertical

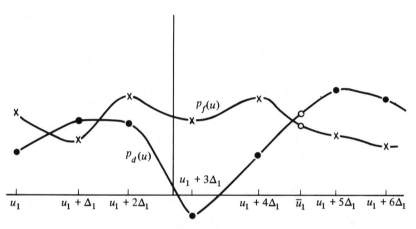

Figure 7.9. Polynomial approximations.

axis. We now discuss the interpolation methods. For notational simplicity, the optimal decisions are shown as dots and the f_1 values as crosses. Both approximating polynomials have been constructed, we assume, in order to match the data *precisely* at several points on either side of $\bar{u}_1$.

The small circles above $\bar{u}_1$ represent the values of $f_1(\bar{u}_1)$ and the corresponding optimizing decision provided by the approximating polynomials $p_f(u)$, $p_d(u)$, respectively, when evaluated at $u = \bar{u}_1$.

The reader is correct in observing that the accuracy of the approximations may leave much to be desired, especially if the Δ_j, δ_j are somewhat large. (How well behaved f_1 and the optimizing decision function are as functions of the state variable also bears upon the accuracy, for given Δ_1, of course.)

If, for example, the optimal decision as a function of the state variable had the appearance of Figure 7.10, interpolation results would be of question-

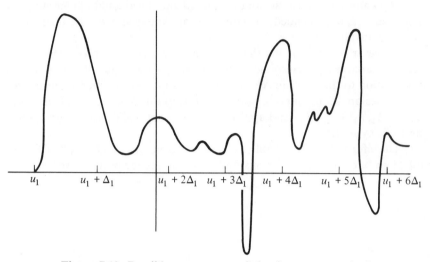

Figure 7.10. Possible appearance of the first stage optimal decision as a function of the state variable.

able value, and the implication is that Δ_1 is much too large—for this function. Too many of the pertinent features of the function have been overlooked.

Observations such as the foregoing are reminiscent of those on search techniques in Chapter 6, and, indeed, we shall be examining the utility of some of these methods in dynamic programming.

Let us interject two more observations at this point. First, in some cases it is simple to discretize the sets S_j, D_j such that there is no problem about $T_j(s'_j, d'_j)$ belonging to S'_{j-1}. Second, we have not pursued beyond the second stage the solution of the problem in which interpolation was required. It should be clear that the work would proceed, stage by stage, as usual, except that we would expect to find interpolation necessary at each stage in order to find values of f_j and the corresponding optimal decisions for state variable values not included in the table.

In discrete examples, we can work so as to maintain, at each stage, an optimizing decision sequence for each value of the state variable at that stage. The advantages are mainly two: first, as soon as the table for f_j is completed, the table for f_{j-1} can be discarded, thus conserving storage space; second, having computed f_n (an n-stage process), an optimizing decision sequence is available for any value of the n-stage state variable. Thus it is unnecessary to make a final pass down through the tables for $f_{n-1}, f_{n-2}, \ldots, f_1$ (which would be destroyed, anyway).

Suppose, however, that we are working in our discrete approximation environment. At stage $k + 1$, utilizing the table for f_k, suppose that we must interpolate, which we do as before, to calculate an intermediate k-stage return. But, if we are maintaining optimizing decision sequences as we go, then we are left with the problem of interpolating between two decision *sequences* to obtain a decision sequence corresponding to our newly found return, as illustrated:

$$
\begin{array}{ccc}
s_{k,i} & f_k(s_{k,i}) & d_{ka}, d_{k-1,b}, \ldots, d_{1w} \\
\bar{s}_k \xrightarrow{\text{interpolate}} & & \xrightarrow{\text{interpolate}} \\
s_{k,i+1} & f_k(s_{k,i+1}) & d_{ka'}, d_{k-1,b'}, \ldots, d_{1w'}
\end{array}
$$

One possibility of course is to use interpolation *component by component in the decision sequences*. This tactic, however, tends to assume that a best decision at one stage is unrelated to a best decision at another, which is entirely untrue.

While the problem may at first appear to be one involving multivariate interpolation, for which methods do exist, it is not. Rather than asking what value of f_k corresponds to a vector of k components **d** for which we have no f_k value, we are finding the function value and asking what vector **d** will provide that value.

When interpolation is necessary, it becomes evident that to enhance the accuracy of the approximating decisions one would be better off to retain all the tables corresponding to $f_1, f_2, \ldots, f_{n-1}$, and to retrieve an optimal decision sequence by means of a final pass back through the tables, obtaining *one* element of an optimal sequence from each table.

A natural question, then, is "What happens if the state and/or decision variable happen to be of dimension greater than 1 at the outset?" Any interpolation scheme then must surely take place in more than one dimension.

The dimensionality of problems having been established as a valid concern, both with regard to interpolation above and previously with regard to computational effort and even feasibility, it is necessary that methods for alleviating these difficulties be sought.

Exercise 7.3

Construct a discrete approximating problem to Example 7.4 so that interpolation is not required.

v. *Recursion Relations Not Involving Stage Returns*

The typical recursion relation need not have, as the function to be optimized at each stage, a contribution from a single stage and a contribution from all the remainder of the process considered up to that time. Consider the following problem.

A man places even money wagers on a series of n events. To decide how to bet, he takes the advice of an acquaintance who supposedly possesses inside information. The acquaintance, however, has only a probability p of being correct each time, and although the bettor is aware of this, he bets accordingly each time, and would like to determine the optimal bet size at each stage so as to maximize the expected value of his bankroll at the end of the n wagers, where his capital is c dollars at the outset.

If s_j is the bankroll at stage j, it is clear that the maximum expected return from a j-stage process beginning with s_j dollars is given by

$$f_j(s_j) = \max_{0 \le d_j \le s_j} \{pf_{j-1}(s_j + d_j) + (1 - p)f_{j-1}(s_j - d_j)\}$$

The reader is encouraged to solve this problem in detail and to discover that, at every stage, the optimal policy is to bet the entire bankroll if $p > 1/2$, and to bet nothing if $p \le 1/2$.

To the more conservative strategist, of course, being advised to always "shoot the works" does not smack of security, even though $p > 1/2$, for an optimal strategy admits the possibility of losing everything.

There is nothing inherently wrong with the expected value, of course, but suppose a person's total wealth were two hundred dollars. Would he equate the absolute value of the negative good of having zero dollars with the positive good of having four hundred dollars? Probably not, and thus one is confronted by questions of the utility of resources, and the reader should be aware that *utility theory* is an existing topic of study. At any rate, it may be that the expected value is not always the quantity which a player may desire to maximize.

A player may, for example, seek to maximize the minimum possible return, a conservative strategy which should be reasonable after the considerations of Chapter 6. (How would such a player play the game in question?)

Other players are willing to risk some of their capital to obtain a gain, but find the chance of being completely wiped out unacceptable or disproportionately adverse. For this more conservative philosoply, it is possible to guarantee that the gambler never loses all his capital. If one maximizes the expected value of the logarithm of the total capital, then the process of maximization will guarantee an objective function value greater than $-\infty$, which in turn guarantees that the probability of the log of total capital being $-\infty$ is zero and equivalently the probability of total capital being zero is then zero.

This model gives rise to an optimizing strategy where, at each state j, the fraction $(p-q)s_j$ is wagered if $p > q$ and the bet is zero if $p \le q$ with q the probability of losing. The reader should verify this result. Both these examples were originally treated by Bellman [2].

7.7. Treating Problems in Several Dimensions

Let us consider the following situation: A city has n precincts and has M policemen and W policewomen available for duty during one particular shift. If m_j men and w_j women are assigned to precinct j during this shift a measure of effectiveness within the precinct, $g_j(m_j, w_j)$, is obtained, and g_j may encompass crime prevention, rapidity of response possible to emergency calls, and the like. (We must emphasize that the successful application of quantitative problem-solving methods to many pertinent problems of today is hindered most by a lack of meaningful and accurate measures of performance. Neither is this surprising, for the primary ingredient in many of these pertinent problems is man himself. In our example, however, we have circumvented this greatest obstacle and assumed that we know g_j, $j = 1, \ldots, n$.)

The problem thus becomes

$$\max \sum_{j=1}^{n} g_j(m_j, w_j)$$

subject to

$$\sum_{j=1}^{n} m_j \leq M$$

$$\sum_{j=1}^{n} w_j \leq W$$

$$m_j, w_j \text{ a nonnegative integer, } 1 \leq j \leq n$$

The problem has, immediately, a serial decision process formulation in which both state and decision variables are two dimensional; but we seek an alternative, since we expect the magnitude of the required effort to increase exponentially over the one-dimensional allocation problem.

i. *One-at-a-Time Method*

The following alternative was suggested by Bellman [2]. Denote by $P(\mathbf{m}_0)$ the problem

$$\max \sum_{j=1}^{n} g_j(m_{0j}, w_j)$$

subject to

$$\sum_{j=1}^{n} w_j \leq W$$

where the vector $\mathbf{m}_0 = (m_{01}, m_{02}, \ldots, m_{0n})$ is given and satisfies the constraint

$$\sum_{j=1}^{n} m_{0j} \leq M$$

Denote by $P(\mathbf{w}_0)$ the problem

$$\max \sum_{j=1}^{n} g_j(m_j, w_{0j})$$

subject to

$$\sum_{j=1}^{n} m_j \leq M$$

where the vector $\mathbf{w}_0 = (w_{01}, w_{02}, \ldots, w_{0n})$ is given and satisfies the constraint

$$\sum_{j=1}^{n} w_{0j} \leq W$$

Thus in problem $P(\mathbf{m}_0)$ the constraint on the total number of policemen was removed while retaining the constraint on the total number of police-women. A similar procedure was used to define problem $P(\mathbf{w}_0)$. We now select an arbitrary feasible $\mathbf{m}_0$ and solve $P(\mathbf{m}_0)$, an n-stage problem in one dimension, obtaining an optimal $\mathbf{w}_1^*$. Then solve $P(\mathbf{w}_1^*)$, obtaining $\mathbf{m}_1^*$, and so on, solving

$$P(\mathbf{m}_1^*), P(\mathbf{w}_2^*), P(\mathbf{m}_2^*), P(\mathbf{w}_3^*), P(\mathbf{m}_3^*), \ldots$$

If it occurs that

$$\sum_{j=1}^{n} g_j(m_j, w_j)$$

is sufficiently well behaved, we will have $\{\mathbf{m}_j^*\} \to \mathbf{m}^*$ and $\{\mathbf{w}_j^*\} \to \mathbf{w}^*$, where

$$\mathbf{m}^* = (m_1^*, m_2^*, \ldots, m_n^*)$$

and

$$\mathbf{w}^* = (w_1^*, w_2^*, \ldots, w_n^*)$$

are optimal solutions to the given problem.

It was exactly this strategy that we employed in the *univariate search* (Chapter 6). This becomes clearer if we consider the present problem to have but two variables, $\mathbf{m}$ and $\mathbf{w}$, as in Figure 7.11, although each is an n vector, and we choose this representation for pictorial advantage only. Having previously examined the procedure, we shall not here mention conditions that will cause it to fail, guarantee its convergence, and so forth.

It is interesting to note that two levels of problem decomposition have taken place here. A given maximization in $2n$ variables was first reduced to n maximizations, each in two variables. The latter problem itself was then reduced to a sequence of problems, each member of which required n maximizations of one variable each. The number of elements in the sequence neces-

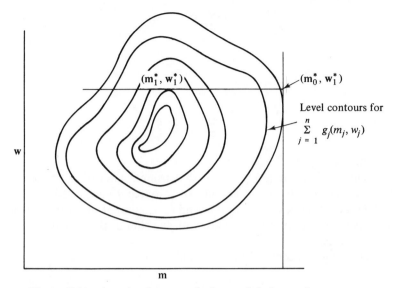

Figure 7.11. A univariate search for a global max in two dimensions.

sary to get one "close" to the true global max is not known in advance, but one does know that the total computational effort required will vary only *linearly* by this number.

a. GENERALIZING THE METHOD

Consider now problems of the form

$$\max f(g_1(x_{11}, \ldots, x_{1m}), g_2(x_{21}, \ldots, x_{2m}), \ldots, g_n(x_{n1}, \ldots, x_{nm}))$$

subject to

$$h_1(x_{11}, x_{21}, \ldots, x_{n1}) \leq b_1$$
$$h_2(x_{12}, x_{22}, \ldots, x_{n2}) \leq b_2$$
$$.$$
$$.$$
$$.$$
$$h_m(x_{1m}, x_{2m}, \ldots, x_{nm}) \leq b_m$$

a particular example of which might be

$$\max \sum_{j=1}^{n} g_j(x_{j1}, x_{j2}, \ldots, x_{jm})$$

subject to

$$\sum_{j=1}^{n} x_{j1} \le b_1$$

$$\sum_{j=1}^{n} x_{j2} \le b_2$$

.

.

.

$$\sum_{j=1}^{n} x_{jm} \le b_m$$

Denote $(x_{1r}, x_{2r}, \ldots, x_{nr})$ by $\mathbf{x}_r$.

This problem might be an m-dimensional allocation problem with n serial stages, so that x_{pq} = amount of resource q allocated in period p, g_p is the return function for period p, and b_q = amount of resource q available for expenditure over the n periods, with the sum of the period returns being the total return.

One might specify arbitrary feasible vectors

$$\mathbf{x}_2^0, \mathbf{x}_3^0, \ldots, \mathbf{x}_m^0$$

(which is simplified considerably by the fact that each variable appears in exactly one constraint), maximize, and obtain $\mathbf{x}_{11}^*$. Then with

$$\mathbf{x}_1 = \mathbf{x}_{11}^*, \qquad \mathbf{x}_3 = \mathbf{x}_3^0, \ldots, \mathbf{x}_m = \mathbf{x}_m^0,$$

obtain $\mathbf{x}_{21}^*, \ldots$; then with

$$\mathbf{x}_1 = \mathbf{x}_{11}^*, \qquad \mathbf{x}_2 = \mathbf{x}_{21}^*, \ldots, \mathbf{x}_{m-1} = \mathbf{x}_{m-1,1}^*$$

obtain $\mathbf{x}_{m1}^*$. (Here $\mathbf{x}_{ij}^*$ denotes the jth approximation to $\mathbf{x}_i^*$, the actual optimal vector for $\mathbf{x}_i$.)

Then one begins anew with

$$\mathbf{x}_2 = \mathbf{x}_{21}^*, \qquad \mathbf{x}_3 = \mathbf{x}_{31}^*, \ldots, \mathbf{x}_m = \mathbf{x}_{m1}^*,$$

$\mathbf{x}_{12}^*$ is obtained, and so on. Again the univariate search character of the method is evident, and, under the proper conditions

$$\{\mathbf{x}_{1j}^*\} \longrightarrow \mathbf{x}_1^*, \{\mathbf{x}_{2j}^*\} \longrightarrow \mathbf{x}_2^*, \ldots, \{\mathbf{x}_{mj}^*\} \longrightarrow \mathbf{x}_m^*$$

To trace the evolution of these decompositions, we have first reduced the maximization in nm variables to n maximizations each with m variables. This problem, in turn, is reduced to a sequence of *sets* of problems, each *set* consisting of nm maximizations in a single variable, or, to state it another way, each set consisting of m subproblems, and each subproblem composed of n maximizations of one variable each.

Thus, if m is large, we are still proposing a great deal of computational effort, and we consider the possibility of reducing m.

ii. *Reducing the Number of Constraints*

Fortunately, we have such a method, and that is the generalized multiplier method of Everett, as discussed in Chapter 2, Section 2.5-v.

There we saw that, under certain conditions, the problem

$$\max f(x_1, x_2, \ldots, x_n)$$

subject to

$$h_i(x_1, x_2, \ldots, x_n) \leq b_i, \quad 1 \leq i \leq m \tag{7.20}$$

$$(x_1, x_2, \ldots, x_n) = \mathbf{x} \in S$$

could be solved by solving the problem

$$\max f(x_1, x_2, \ldots, x_n) + \sum_{i=1}^{m} \lambda_i h_i(x_1, \ldots, x_n)$$

subject to $$\tag{7.21}$$

$$\mathbf{x} \in S$$

successively, until for a set of nonpositive λ_i, $i = 1, \ldots, m$, determined by interpolation and extrapolation, one obtains an $\mathbf{x}^*$ such that $h_i(\mathbf{x}^*) = b_i$, $i = 1, \ldots, m$. Such an $\mathbf{x}^*$ was shown to be optimal for the original problem.

Thus we have a possible way to reduce the number of constraints, and consequently the dimensionality, in a dynamic-programming problem like (7.20).

For notational simplicity, assume that (7.20) without $\mathbf{x} \in S$ is the given problem. Suppose that we choose to solve

$$\max f(x_1, \ldots, x_n) + \sum_{i=1}^{k} \lambda_i h_i(x_1, \ldots, x_n)$$

subject to $$\tag{7.22}$$

$$h_j(x_1, \ldots, x_n) \leq b_j, \quad j = k+1, \ldots, m$$

by dynamic programming, using Everett's method. The set of $\mathbf{x}$ satisfying the constraints of (7.22) corresponds to the set S of (7.21). Therefore, we select $(\lambda_1, \lambda_2, \ldots, \lambda_k) \leq \mathbf{0}$ and solve (7.22) by dynamic programming. If $\mathbf{x}^*(\lambda_1, \lambda_2, \ldots, \lambda_k)$ satisfies $h_i(\mathbf{x}^*) = b_i$, $i = 1, \ldots, k$, we are finished; otherwise, the λ's are perturbed in appropriate fashion and (7.22) is solved by dynamic programming with the new λ's, and so on. Thus we reduce the number of constraints and the dimensionality of the problem—but at the expense of having to solve the reduced problem a potentially large number of times. In addition, the generalized multiplier technique is not guaranteed to work, although even when it fails some approximations are possible [3]. The reason one might hesitate to incorporate $m - 1$ constraints into the objective function, obtaining a "one-dimensional" problem, is that the larger the number of λ's one has to vary, the more times the dynamic-programming problem will likely need to be solved. This number of iterations could become sufficiently large to make the solution of a higher-dimensional problem a *more* desirable alternative.

7.8. Introduction to Nonserial Decision Processes

We have observed that many practical problems may be cast into the form of n-stage serial decision processes. There are decision processes that are nonserial, and recursive optimization techniques can be used to optimize some of these.

i. *Divergent Allocation Process*

Let us consider the following allocation problem: We have c units of some resource that is to be allocated to some n activities, with the usual return functions defined. Furthermore, one of the activities, say the mth, is actually composed of k subactivities, among which the amount of resource allocated to activity m must be distributed, and returns are also defined for these subactivities. Denoting returns from the activities by R_j, $j = 1, \ldots,$ n; returns from subactivities R'_i, $i = 1, \ldots, k$; the allocation to activity j, x_j; the allocation to subactivity i, x'_i, $i = 1, \ldots, k$; and the total return function φ, we may express the problem as

$$\max \varphi(R_1, R_2, \ldots, R_n, R'_1, R'_2, \ldots, R'_k)$$

subject to

$$\sum_{j=1}^{n} x_j \leq c$$

$$\sum_{i=1}^{k} x'_i \leq x_m$$

$x_j \geq 0$, $x'_i \geq 0$, $1 \leq j \leq n$, $1 \leq i \leq k$.

Neglecting the subactivities temporarily, we represent the process as in Figure 7.12. If we consider only the subactivities, on the other hand, we can represent that allocation process as in Figure 7.13.

We want to assume that the total return from the subprocess may be measured somehow, say with a function $\gamma(R'_1, R'_2, \ldots, R'_k)$. With respect to the problem as given, we are asking that we have a measure of the total return from the subactivities.

$$T_n(s_n) = s_n - x_n = s_{n-1}$$

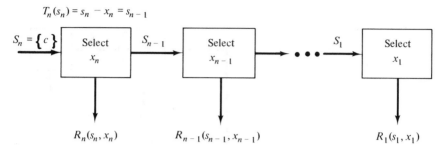

Figure 7.12. A serial stage system for an allocation problem.

$$T'_k(s'_k) = s'_k - x'_k = s'_{k-1}$$

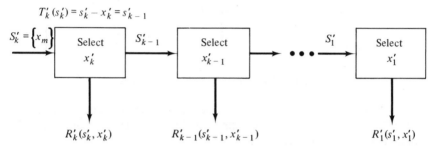

Figure 7.13. A serial stage process for a subactivity of an allocation problem.

a. Solving the Problem

If we wished to assume we had a similar measure for the return from the "main" process, we might consider solving that dynamic-programming problem first, and then optimizing the subprocess with $\{x_m^*\} = S'_k$; but this cannot be expected, in general, to lead to an optimal solution for the entire problem.

Furthermore, finding s'_k so as to optimize the subprocess, and then optimizing the main process with x_m constrained to equal s'_k will not, in general, provide an overall optimal solution. But we can find the optimal return from the subprocess as a function of s'_k, of course, and thus one *could* incorporate the subprocess return into the main process return at the mth stage, since both s'_k and s_{m-1} are functions of s_m, x_m; that is, *at stage m one maximizes both the total contribution from the subprocess and the contribution from main process stages 1 through m, as a function of s_m.*

Before illustrating this strategy, let us observe that these two separate processes may be coupled together in simple fashion and in a way which gives an accurate pictorial representation of the given problem (Figure 7.14).

As previously, our notations reflect the identity between the set of possible x values at a stage and the D at that stage.

At the mth stage we require, for the general diverging branch process, as it has been called by Nemhauser [4], two transformations; $T_m: S_m \times D_m$

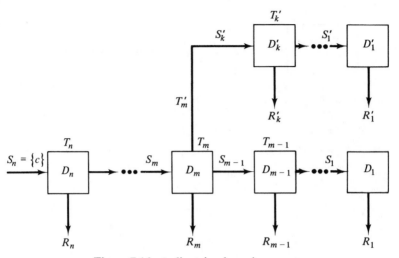

Figure 7.14. A diverging branch process.

$\rightarrow S_{m-1}$ and $T'_m: S_m \times D_m \rightarrow S'_k$. T_m defines the next state of the main process, whereas T'_m defines the first state of the subprocess. In our particular example allocation problem, we would have

$$T_m(s_m, x_m) = s_m - x_m \quad \text{and} \quad T'_m(s_m, x_m) = x_m$$

Also

$$T'_i(s'_i, x'_i) = s'_i - x'_i, \quad \text{for } i = 2, \ldots, k$$

The recursion relations, then, as well as the solution procedure are fairly simple.

One works in the diverging branch, obtaining

$$f'_1(s'_1), f'_2(s'_2), \ldots, f'_k(s'_k)\dagger$$

in the usual manner. But we see that

$$f'_k(s'_k) = f'_k(T'_m(s_m, d_m))$$

Thus the total return from the branch is measured in terms of the state and decision at stage m of the the main process.

Independently, $f_1(s_1), f_2(s_2), \ldots, f_{\tilde{m}-1}(s_{m-1})$ are obtained in the usual way from the main process, and we have

$$f_{m-1}(s_{m-1}) = f_{m-1}(T_m(s_m, d_m))$$

---

†Primes on functions here do not indicate derivatives, of course, and are meant only to distinguish the diverging branch portion of the process.

At stage m of the main process we calculate, assuming for notational simplicity that the total return is the sum of all returns,

$$f_m(s_m) = \max_{d_m} \{R_m(s_m, d_m) + f_{m-1}(T_m(s_m, d_m)) + f'_k(T'_m(s_m, d_m))\}$$

This is the only point in the work where a deviation from the serial process procedure appears—here we have three expressions rather than the usual two. It is here that the branch return is incorporated. f_m, then, measures the total return from the branch and stages 1 to m of the main process. Thereafter, one calculates $f_{m+1}, f_{m+2}, \ldots, f_n$ in the usual way, and the problem is solved.

ii. *Convergent Allocation Process*

A slight variation on the diagram of Figure 7.14 yields a similar, but different, type of nonserial process. Let us synthesize a problem that gives rise to the variation.

Suppose that at time period n two firms agree upon a plan whereby firm 2 will absorb firm 1 at time period m.

Firm 1 begins in some state s'_n, say, and at each stage $n, n-1, \ldots,$ $m+1$ has available decision sets $D'_n, \ldots, D'_{m+1}$, which we envision as broad decisions that govern the overall state of the firm. The sets $S'_n, \ldots, S'_{m+1}$ are possible "state of the firm" sets. (We are assuming deterministic situations.) Also, the necessary transformations $T'_n, \ldots, T'_{m+1}$ exist, as well as the return functions $R'_n, \ldots, R'_{m+1}$.

Firm 2 meanwhile begins in a state s_n and has state and decision sets, transformations, and return functions, labeled in the usual way, through time period 1, the time through which the *entire* process is to be considered. At time period m, when firm 1 is absorbed by firm 2, the return R_m will be a function of an element of S'_m as well as of elements from S_m, D_m. Furthermore, the transformation T_m will be a function of an element from S'_m. The object is to maximize some function of the returns $R'_n, \ldots, R'_{m+1}, R_n, \ldots, R_m, \ldots, R_1$.

The process might be diagrammed as in Figure 7.15.

Example 7.7

Consider a physical setting as diagrammed in Figure 7.16, which is a map of two rivers, the smaller of which feeds into the larger at the indicated point. The small squares represent sites at each of which there exist dams and reservoirs. The reservoirs, according to some schedule, release water to neighboring areas for use therein—say for irrigation purposes, for drinking water, or the like.

For purposes of simplification we might imagine that decisions at each

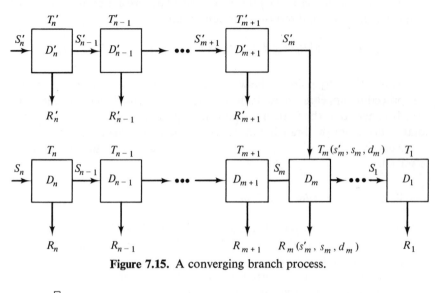

Figure 7.15. A converging branch process.

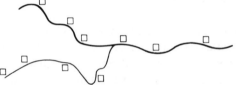

Figure 7.16. Water control system of Example 7.7.

site are merely at what rate to remove water from the particular stream. State variables might be vectors whose components include, for example, rate of stream flow below the site, average water depth there, and so forth.

The site returns might be some measure of goodness for the surrounding area served by the site, as well as for the stream itself—for example, considerations of wildlife, stream pollution, and so on, might arise, and there would unquestionably be constraints on what could be done at the sites.

And in this particular example there would likely be a terminal state set S_0, since the state of the main river below the final site would certainly be a consideration.

A diagram of this process would again appear with the form of Figure 7.15. Thus our general "converging branch process" [4] appears as in Figure 7.17.

What would be desirable, of course, would be to treat stages 1 through $m - 1$ of the main process in the usual way, and then, analogous to the diverging branch tactic, draw the optimal total branch return into the main process at stage m, having optimized the branch return as a function of a state variable.

However, as Figure 7.17 shows, if we find the optimal branch return,

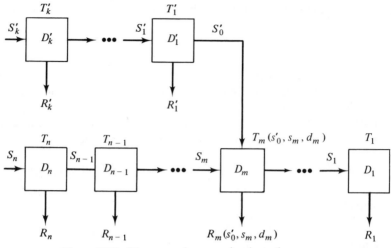

Figure 7.17. The general converging branch process.

$f'_k(s'_k)$, there is a problem in absorbing that into the main process—nothing in the latter process is explicitly a function of s'_k.

But, as we saw in Section 7.5, *it may be possible to reverse the flow of the branch and optimize it in terms of its final state*, that is, to find the optimal branch return as $f'_k(s'_0)$.

As implied by Figure 7.17, it is assumed that we have

$$s_{m-1} = T_m(s'_0, s_m, d_m)$$
$$R_m(s'_0, s_m, d_m)$$

To absorb the branch return at stage m, then, we have, assuming an additive total return function,

$$R_m(s'_0, s_m, d_m) + f_{m-1}(T_m(s'_0, s_m, d_m)) + f'_k(s'_0) \tag{7.23}$$

To maximize expression (7.23) and obtain the usual $f_m(s_m)$ implies that (7.23) must be maximized not only over D_m but also over S'_0.

Now, consider just the branch of Figure 7.17 and imagine that we isolated and treated it in the usual forward fashion; we obtain

$$f_k(s'_k)$$

which gives optimal return as a function of $s'_k \in S'_k$. If the choice of the element of S'_k is available, we can perform

$$\max_{s_{k'}} f'_k(s'_k)$$

and, indeed, select the best initial state.

If this is possible, then it is certainly possible to specify which element of S'_0 it is most desirable to end in, in which case

$$\max_{s'_0} f'_k(s'_0)$$

is meaningful in the converging branch computation.

If, on the other hand, there is no control over initial state, the final state is not entirely controllable.

With respect to the stage-m optimization, then, and the expression (7.23), if the element of S'_0 can be fixed, we should compute

$$\max_{d_m, s'_0} \{R_m(s'_0, s_m\, d_m) + f_{m-1}(T_m(s'_0, s_m, d_m)) + f'_k(s'_0)\} = f_m(s_m)$$

and thereafter

$$f_{m+1}, \ldots, f_n$$

in the usual way.

If the element of S'_0 may not be fixed, then we obtain

$$\max_{d_m} \{\text{expression (7.23)}\} = f_m(s_m, s'_0)$$

and, successively,

$$f_{m+1}(s_{m+1}, s'_0), \ldots, f_n(s_n, s'_0)$$

Having found $f_n(s_n, s'_0)$, suppose we seek to optimize the process given particular initial state elements s_{n0}, s'_{k0}. But our optimal function value has been discovered in terms of s'_0, s_n, and optimal decisions in the branch have been discovered as functions of s'_0, not s'_k.

This difficulty may be overcome, but we choose not to discuss tactics for so doing. If we may optimize over S'_0 at stage m, though, there is no conceptual difficulty; obtaining the optimizing decisions proceeds as usual, and in the branch are obtained in ascending order $d'_1, d'_2, \ldots, d'_k$.

iii. *Some Model Extensions*

We shall mention two other varieties of nonserial systems and simple problems that lead to them, but we shall not pursue their solution or analysis.

Consider, as before, that there exists a set of n activities to which we allocate funds, having c dollars to spend altogether. The return from a given activity is a certain amount of "success," say, as we might have in the early stages of a research and development program for example—before any tangible profit can be made. Here, as an additional but realistic twist, we have that the degree of success in activity j is a function of the amount of money spent on activity i (for some given i, j) as well as the amount of money spent directly on project *j*. For example, in the design of a new jet aircraft it may be that the wing design cannot be completed until tests of the strength and high-temperature properties of a new alloy have been computed. Thus success toward completing the wing design is a function of success toward completing the alloy test, and consequently of the amount of money spent on testing the alloy.

Depending upon whether $i < j$ or $i > j$ in the sequence of allocations, we might diagram this process as in Figure 7.18 or 7.19.

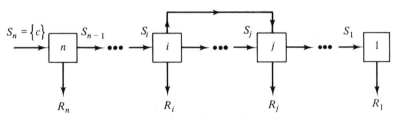

Figure 7.18. A feedforward loop allocation process.

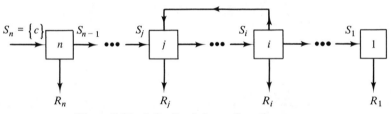

Figure 7.19. A feedback loop allocation process.

In either event, we have $R_j = R_j(s_j, d_j, d_i)$ for the present problem, and the directed arc connecting stages *i* and *j* indicates the dependence mentioned. The reader should not impute time to the flow of the main process; thus, there should be no difficulty in realizing that a downstream event can affect the state of affairs farther upstream, as in Figure 7.19.

The process of Figure 7.18 has been termed a *feedforward system* [4] and is seen to consist of a branch that first diverges, then converges into a normal serial process. The process of Figure 7.19 is the same as the feedforward system except that the direction of the flow in the branch is reversed, running opposite to that of the main process. Figure 7.19 depicts a *feedback* process.

Again attempting to keep our sample problems within simple allocation contexts, suppose we are designing some sort of system, and the decision at stage *i* is how much to spend on component *i*. The return $R_i(s_i, d_i)$ may be a reliability figure and the overall objective may be to maximize the product of these reliability measures, although here we are interested in the structure of the problem, not its solution. Let us imagine there are c dollars available and that there is one additional important aspect of the problem; namely, for one subsystem there is to be built a back-up system—perhaps not identical to the primary subsystem.

Viewing the system as a set of connected components, we find that it appears as follows, where components are the boxes, primes denoting the back-up system.

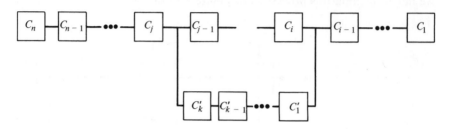

Here, determining *what* components to utilize is not a problem—the system will be constructed as shown. The problem is simply how much to spend on each component.

In our usual notation and symbology such a non-serial process would appear as in Figure 7.20.

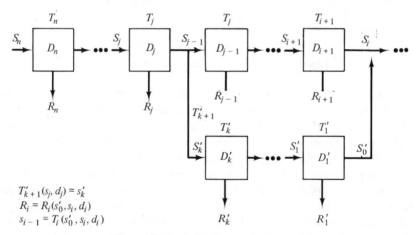

$$T'_{k+1}(s_j, d_j) = s'_k$$
$$R_i = R_i(s'_0, s_i, d_i)$$
$$s_{i-1} = T_i(s'_0, s_i, d_i)$$

Figure 7.20. A feedforward system.

Let the problem again be one of allocation. After stage j of a main process, the remaining resource must be allocated between the remaining stages of that process and a subprocess. Further, however, any of the resource remaining after the subprocess allocation may be made available to stages $\leq i$ of the main process.

It is best to divorce the concept of feedback system from considerations of time. In our serial processes, the higher numbered stages do occur first (although they are not treated first), and it is difficult to refrain from considering these as ordered in time. The branching processes and the feedforward process have also the look of time ordering, since the only gross structural change is the allowing of parallel processes. For the feedback system, however, there is a conceptual obstacle in attaching time points to stages. Stage j occurs prior to stage i, but the events at stage j are themselves influenced by the events at stage i.

Thus far, any decision process we have considered has been a one-time process, which is to say one pass through the sequence is all that the problem requires. For feedback systems of greatest familiarity, this is not true.

Consider, for example, that Figure 7.21 represents a chemical process, and

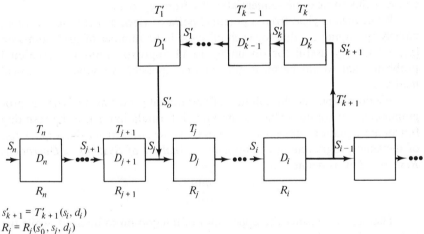

$$s'_{k+1} = T'_{k+1}(s_i, d_i)$$
$$R_j = R_j(s'_0, s_j, d_j)$$
$$s_{j-1} = T_j(s'_0, s_j, d_j)$$

Figure 7.21. A feedback system.

that after stage i, a certain proportion of the output, whatever it is, is made to undergo a different sequence of operations (stages) than that which runs on through, and is then *recycled* through the system beginning at stage j. The recycling may occur many times, so that observing the indicated process from start to finish entails more than one pass through the diagram. Such a feedback system as this, of course, fits well within a time framework—events during one pass affect events on the *next* pass. But, as was hopefully illustrated, there are one-pass problems which fit well within the feedback framework, and one should not find these paradoxical as a result of self-imposed time frameworks.

It must also be pointed out that dynamic programming is applicable to those processes that include more than a single pass.

7.9. A Brief Note on Infinite Stage Processes

Some serial decision processes have, in truth, an infinite number of stages; in fact, an uncountable infinity of them. For example, consider a variational problem of the kind discussed briefly in Section 2.6.

$$\min_y J(y) = \int_{x_1}^{x_2} F(x, y, y') \, dx$$

The answer here is a function $y(x)$ which is defined everywhere on the interval $[x_1, x_2]$.

As a serial decision process the problem may be viewed as selecting a value to associate with every point of the interval. The collection of these values determines a function defined over the interval, and, of course, we can consider this to be done sequentially, say from x_1 to x_2.

While meaningless from a practical point of view, a slight modification, namely specifying a value for each of some finite number of grid points on $[x_1, x_2]$ is very useful for obtaining useful approximations to variational problems that would be most difficult or impossible to solve by classical methods.

We do not pursue the solution of variational problems by dynamic programming, but we do include some sample formulation, since the resulting functional equations appear much like those resulting from the application of dynamic programming to problems in *Analysis of Systems in Operations Research.* Let us define

$$f(x_1) = \min_{\substack{y \\ \text{on } [x_1, x_2]}} J(y)$$

That is, we consider the upper limit of integration to be fixed, while the lower one is variable.

We know that

$$\int_{x_1}^{x_2} F(x, y, y') \, dx = \int_{x_1}^{x_1 + \Delta x} F(x, y, y') \, dx + \int_{x_1 + \Delta x}^{x_2} F(x, y, y') \, dx$$

so we may write

$$f(x_1) = \min_{\substack{y \\ \text{on } [x_1, x_1 + \Delta x]}} \left[\int_{x_1}^{x_1 + \Delta x} F(x, y, y') \, dx + f(x_1 + \Delta x) \right]$$

Thus results the general reasoning. There is a problem above, however, and that is that for a particular function $y_1, f(x_1 + \Delta x)$ must be restricted by the fact that y_1 assumes a specific value at $x_1 + \Delta x$. Furthermore, our previous examination of variational problems found us specifying both $y(x_1) = c_1$ and $y(x_2) = c_2$.

Therefore, it is reasonable to define

$$f(x_1, c_1) = \min_{\substack{y \\ \text{on } [x_1, x_2] \\ y(x_1) = c_1}} J(y)$$

and then write

$$f(x_1, c_1) = \min_{\substack{y \\ \text{on } [x_1, x_1 + \Delta x] \\ y(x_1) = c_1}} \left[\int_{x_1}^{x_1 + \Delta x} F(x, y, y') \, dx + f(x_1 + \Delta x, c_1(y)) \right]$$

where $c_1(y)$ indicates the "constant" value c_1 provided by a particular function y.

To avoid further cluttering the equation, we have omitted the constraint $y(x_2) = c_2$.

REFERENCES

[1] BELLMAN, R., *Adaptive Control Processes*, Princeton University Press, Princeton, N.J., 1961.

[2] BELLMAN, R., and S. DREYFUS, *Applied Dynamic Programming*, Princeton University Press, Princeton, N.J., 1962.

[3] EVERETT, H., "Generalized Lagrange Multiplier Method for Solving Problems of Optimum Allocation of Resource," *Operations Research*, Vol. 11, p. 399, 1963.

[4] NEMHAUSER, G., *Introduction to Dynamic Programming*, John Wiley & Sons, Inc., New York, 1966.

PROBLEMS

1. The reader has unquestionably utilized a dynamic-programming approach in solving certain problems previously. By this we mean a strategy that recognizes the desired end result, and, by working backwards, synthesizes a means to the end. Write down a number of problems amenable to the dynamic-programming approach and the means of "backward" solution. Think in terms of chess, card games, and certain puzzle-type problems, such as nim-type games, for example.

2. Formulate each of the following word problems as an n-stage serial decision process, and give detailed descriptions of all the necessary sets and mappings. Also, write the general recursion relation.
   (a) There exists a certain base and locations as in the following diagram:

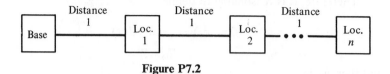

Figure P7.2

There is an object in one of the locations, and, at the outset $P\{\text{object in loc. } i\} = 1/n$. A searcher begins at the base and examines locations until the object is found. If in place i_k, and the searcher examines place i_q, the cost incurred is $|i_k - i_q| + t$, where t is a constant; that is, the cost of one step is the distance traversed plus a constant.

(1) There is no back-to-base cost after the object is found and

(2) If $n - 1$ searches have not yielded the object, the searcher still examines the remaining place.

Specify an order of examining locations so as to minimize the expected value of the search cost. (Can you solve this problem without dynamic programming?)

(b) An airplane has a cargo weight capacity of W pounds and a cargo space of V cubic feet. There are m kinds of objects it can carry, and each object of kind i weighs w_i pounds, takes v_i cubic feet, will sell for c_i dollars at the destination point, and there are $a_i =$ positive integer of these available as cargo, $i = 1, \ldots, m$. Find the number of objects of each kind (a non-negative integer) that will maximize total sales price of the cargo, while allowing the plane to get off the ground, and without causing overflow in the cargo hold.

(c) Return to part (a) and treat the problem identical to it except that initially

$$P\{\text{object is in place } i\} = \frac{2i}{n(n+1)}, \quad i = 1, \ldots, n$$

[Does this problem appear as easily solvable as (a) was?]

(d) Suppose that f is a unimodal function on an interval $[a, b]$ and a sequential search is to be utilized to locate a smaller interval in which x^*, the global maximum, lies. The first iteration has us evaluate f at two points, as does the second; then the search ends. Determine how to place the various evaluations so as to minimize the maximum length of the final interval. (After formulating the problem, can you solve it by dynamic programming?)

(e) What is the maximum number of distinct regions in the plane produced by n distinct lines in the plane? (Can you give a recursive expression for this number?)

(f) $a_1, a_2, \ldots, a_m$ are positive real numbers. Find nonnegative real numbers $\lambda_1, \lambda_2, \ldots, \lambda_m$ such that

$$\sum_{i=1}^{m} \lambda_i = 1$$

and

$$\prod_{i=1}^{m} a_i^{\lambda_i}$$

is a maximum.

3. Once the general problem of Section 7.4-iv has been properly formulated, a specific one of that class should be solved. With the following data, solve the reliability problem by dynamic programming.

First, suppose that $n = 4$, $c_1 = 2$, $c_2 = 5$, $c_3 = 1$, $c_4 = 3$, and $c = 18$. The probabilities will be defined in tabular fashion.

x_1	$p_1(x_1)$		x_2	$p_2(x_2)$
0	$\frac{1}{4}$		0	$\frac{1}{4}$
1	$\frac{1}{8}$		1	$\frac{1}{16}$
2	$\frac{1}{8}$		2	$\frac{1}{32}$
3	$\frac{1}{16}$		3	$\frac{1}{64}$
4	$\frac{1}{32}$		> 3	$\frac{1}{128}$
> 4	$\frac{1}{64}$			

x_3	$p_3(x_3)$		x_4	$p_4(x_4)$
0	$\frac{1}{2}$		0	$\frac{1}{4}$
1	$\frac{1}{4}$		1	$\frac{1}{16}$
2	$\frac{1}{8}$		2	$\frac{1}{32}$
3	$\frac{1}{8}$		3	$\frac{1}{64}$
4	$\frac{1}{32}$		4	$\frac{1}{64}$
> 4	$\frac{1}{64}$		> 4	$\frac{1}{128}$

4. Solve the following problem with a multiplicative objective function.

$$\max \prod_{j=1}^{n} x_j$$

subject to

$$\sum_{j=1}^{n} x_j = c > 0$$

$$x_j \geq 0, \quad j = 1, \ldots, n$$

5. For the following objective functions, determine whether or not recursive optimization is valid for each.
  (a) Maximize [maximum $\{R_1, R_2, \ldots, R_n\}$].
  (b) Maximize [$|R_1| + |R_2| + \cdots + |R_n|$].
  (c) Minimize [maximum $\{R_1, R_2, \ldots, R_n\}$].
  (d) Maximize [$R_5 + R_4 + R_3 R_2 + R_1$], $R_j \geq 0$, all j.
  (e) Minimize [$R_6 R_5 + R_4 R_3 + R_2 R_1$], $R_j \geq 0$, all j.
  (f) Maximize $\sum\limits_{j=1}^{n} (-1)^j R_j$.

  (g) Maximize $\sum\limits_{k=1}^{n} \dfrac{R_k}{c - \sum\limits_{i=1}^{k-1} R_i}$, $R_0 = 0$, c constant.

6. Solve by dynamic programming

$$\min \sum_{i=1}^{n} \sqrt{x_i}$$

subject to

$$\sum_{i=1}^{n} x_i = C$$

$$x_i \geq 0, \quad i = 1, \ldots, n$$

7. Define the transformations T_j and the sets S_j in Example 7.4 in a different way; then solve Example 7.4 with the new formulation.

8. Consider the following network whose undirected edges correspond to distances and whose vertices are cities. A traveler wishes to begin at city v_0, end at v_5,

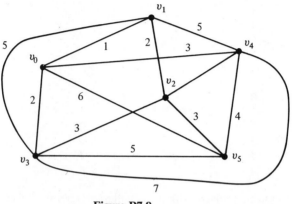

Figure P7.8

and pass through each city exactly once. Using dynamic programming find a shortest route for the traveler in terms of total distance traveled.

9. Find values of $p_j \geq 0$ that maximize the entropy function

$$H(p_1, \ldots, p_n) = -\sum_{j=1}^{n} p_j \log_2 p_j$$

subject to

$$\sum_{j=1}^{n} p_j = 1$$

10. Find values of the p_j satisfying the same constraints as in Problem 9 in order to *minimize* the entropy function; $0 \log_2 0$ is taken to be 0. (Is that justifiable?)

11. Recall Example 7.5 and the gentleman traveling through the cities. Suppose that we add to the problem four cities S_{41}, S_{42}, S_{43}, and S_{44}. Distances from these cities to those of the set S_3 are as follows:

		To			
		S_{31}	S_{32}	S_{33}	S_{34}
	S_{41}	2	3	1	3
From	S_{42}	4	1	1	2
	S_{43}	1	2	5	1
	S_{44}	1	6	1	3

The problem to solve here is to select the city from S_4, S_{41}, S_{42}, S_{43}, and S_{44} for which the shortest distance to S_0 is less than for any other city of that set. Solve the new problem, utilizing the results of the example to the fullest, in order to save computational effort.

12. Solve the search problem of Problem 2(a) for $n = 4$, and
   (a) for $t = 0$
   (b) for $t = 10$
   (c) for $t = 100$

13. Solve the search problem of Problem 2(c) for $n = 4$, where we assume it is *not* necessary to look in the last place if the preceding three searches were unfruitful and that there is a back-to-base cost of $t + i_4$. First, let $t = 0$; then solve again for $t = 10$.

14. Using the same network as in Problem 8, find a shortest distance route that terminates at v_2 and passes through each other city exactly once. (The route may begin at any city.)

15. Consider the classical traveling salesman problem. There are n towns within the territory of a traveling salesman, and the distances between pairs of these towns are known, where d_{ij} denotes the distance from town i to town j for all $i \neq j$. We do not require $d_{ij} = d_{ji}$.

The salesman's problem is to select a sequence of towns in order to pass through each town once and only once and to minimize the total distance traveled. (Popular narrative versions attribute other objectives to the traveling salesman, but here we are concerned only with distance.) Using dynamic programming, solve the traveling salesman problem for $n = 4$ towns utilizing the following array of distances.

$$D = \begin{pmatrix} 0 & 3 & 4 & 2 \\ 2 & 0 & 7 & 4 \\ 3 & 5 & 0 & 6 \\ 6 & 2 & 5 & 0 \end{pmatrix}$$

Discuss the amount of computational effort as a function of n, and decide whether or not dynamic programming appears to be a desirable method for this problem.

16. Obtain an approximate solution to

$$\min \sum_{i=1}^{4} \sqrt{x_i}$$

subject to

$$\sum_{i=1}^{4} x_i = 7, \quad x_i \geq 0$$

where at every stage, decision sets and state sets are approximated by discrete sets $\{0, .5, 1, 1.5, \ldots\}$. If a state is computed that does not exist in the grid allowed, choose the element of the discrete state set nearest the one computed. Comment on the accuracy of the approximately optimal solution obtained.

17. Consider the problem

$$\min \sum_{i=1}^{4} x_i^2 - 2a_i x_i + a_i^2$$

subject to

$$\sum_{i=1}^{4} x_i = 7$$

with $a_1 = 4$, $a_2 = 3$, $a_3 = 2$, and $a_4 = 4$.

(a) Obtain an approximation to the optimal solution by proceeding as in Problem 16. Find the best and second-best approximations.

(b) For each of the solutions obtained in part (a), construct intervals of unit length symmetric about optimizing decisions and states, use a partition of fineness twice that of part (a), and re-solve the problem over the restricted sets. What is now the best approximation to an optimal solution?

18. The linear-programming problem

$$\max x_1 + 2x_2 - x_3$$

subject to

$$3x_1 + 6x_2 + 4x_3 \leq 12$$
$$0 \leq x_1 \leq 3$$
$$0 \leq x_2 \leq 2$$
$$0 \leq x_3 \leq 2$$

may be solved geometrically, by the methods of linear programming, or by inspection.

With grids of fineness of $\frac{1}{2}$ unit, solve the problem by dynamic programming and compare that approximation to the optimal solution.

19. Using the one-at-a-time method, carry the problem

$$\min \sum_{i=1}^{4} x_i y_i$$

subject to

$$\sum_{i=1}^{4} x_i = 10$$

$$\sum_{i=1}^{4} y_i = 15$$

$$x_i, y_i \geq 0, \quad \text{for } i = 1, \ldots, 4$$

through three iterations, with $(4, 3, 2, 1)$ an initial guess for (x_1, x_2, x_3, x_4).

20. Solve Problem 19 with maximization the objective.

21. Define $x_0 = 0$, $x_{n+1} = 1$. Using dynamic programming, determine the values of $x_1, \ldots, x_n$ that minimize the

$$\max \{x_2 - x_0, x_3 - x_1, x_4 - x_2, \ldots, x_{n+1} - x_{n-1}\}$$

subject to

$$x_0 \leq x_1 \leq x_2 \cdots \leq x_n \leq x_{n+1}$$

22. Using grids of unit coarseness, approximate an optimal solution to

$$\max \tfrac{3}{2}x_1 + 4x_2$$

subject to

$$\tfrac{4}{3}x_1 - x_2 \geq \tfrac{8}{3}$$
$$-\tfrac{4}{7}x_1 - x_2 \geq \tfrac{32}{7}$$

GEOMETRIC PROGRAMMING

8.1. Introduction

In the present chapter we intend an introduction to one of the most recent techniques of mathematical optimization. The class of optimization problems to which these techniques are directly applicable is quite specialized, although certain important problems are typically of this nature. Furthermore, these problems are highly nonlinear, and their solution using any other technique is not as efficiently accomplished as by the method of geometric programming.

This state of affairs raises a sort of philosophical question: should investigators be concerned with the discovery of techniques applicable to general optimization problems or to the invention of specialized techniques for comparatively narrow classes of problems? (It must be pointed out, however, that by utilizing various approximation techniques and by invoking a degree of cleverness, the methods to be presented can be extended to more general problems.)

The discovery of the basic technique of geometric programming and the proof of the important and fundamental theorems is due to R. Duffin, C. Zener, and E. Peterson [1]. If the reader attempts to infer the nature of geo-

8

metric programming from its name, he will likely miss the mark entirely. The name would seem to derive from the technique's close relationship to (indeed, its dependency upon) a basic and powerful inequality, the name of which heads the following section. In Reference 1, however, the authors mention the strong geometric connections of the techniques as their motivation for so naming it.

8.2. Arithmetic Mean–Geometric Mean Inequality

i. *The Inequality*

In its full generality the arithmetic mean–geometric mean inequality states: If $x_1, \ldots, x_n$ are *any* n nonnegative numbers and if $\lambda_1, \ldots, \lambda_n$ have the property that

$$\sum_{j=1}^{n} \lambda_j = 1$$

and $\lambda_j > 0$ for all j ($j = 1, 2, \ldots, n$), then

$$\sum_{j=1}^{n} \lambda_j x_j \geq \prod_{j=1}^{n} x_j^{\lambda_j} \tag{8.1}$$

In the special case $\lambda_j = 1/n$, the left-hand side and the right-hand side of (8.1) are, by definition, the arithmetic mean and the geometric mean, respectively, of the numbers $x_1, \ldots, x_n$. Inequality (8.1) is then written

$$\frac{1}{n} \sum_{j=1}^{n} x_j \geq \prod_{j=1}^{n} x_j^{1/n} \tag{8.2}$$

and is known as the arithmetic mean–geometric mean inequality. (It is also the case that both the special and general forms hold as strict inequalities unless $x_1 = x_2 = \cdots = x_n$.)

ii. Establishing the Special Case

The proof of the special case can be accomplished easily and may be done in a variety of ways, a straightforward method being by induction. First, for any real numbers u_1, u_2,

$$(u_1 - u_2)^2 \geq 0$$

or

$$u_1^2 + u_2^2 \geq 2u_1 u_2 \tag{8.3}$$

Let

$$x_1 = u_1^2, \qquad x_2 = u_2^2$$

Then (8.3) becomes

$$x_1 + x_2 \geq 2x_1^{1/2} x_2^{1/2}$$

or

$$\tfrac{1}{2}x_1 + \tfrac{1}{2}x_2 \geq x_1^{1/2} x_2^{1/2} \tag{8.4}$$

and (8.4) is the special case (8.2) of inequality (8.1) for $n = 2$. (Clearly, the desired result is true for $n = 1$.)

If we replace, in (8.4), x_1 by $(x_1' + x_2')/2$ and x_2 by $(x_3' + x_4')/2$, and use (8.4) on the new variables we find

$$\frac{x_1' + x_2' + x_3' + x_4'}{4} \geq ((x_1' x_2')^{1/2} (x_3' x_4')^{1/2})^{1/2}$$
$$= (x_1' x_2' x_3' x_4')^{1/4}$$

which proves inequality (8.2) for $n = 4$. Continuing in this fashion, inequality (8.2) may be proved for $n = 1, 2, 4, 8, \ldots, 2^k, 2^{k+1}, \ldots$.

The induction may then be employed backward to obtain the result for the remaining n. Suppose that inequality (8.2) holds for some $n \geq 3$. If for $n = 3, 4, \ldots$, we specify

$$x_n = \frac{\sum_{j=1}^{n-1} x_j}{n - 1}$$

(8.2) becomes

$$\frac{x_1 + \cdots + x_{n-1} + \dfrac{x_1 + \cdots + x_{n-1}}{n - 1}}{n}$$

$$\geq (x_1 \cdots x_{n-1})^{1/n} \left(\frac{x_1 + \cdots + x_{n-1}}{n - 1} \right)^{1/n},$$

or

$$\frac{n(x_1 + \cdots + x_{n-1})}{n(n - 1)} \geq (x_1 \cdots x_{n-1})^{1/n} \left(\frac{x_1 + \cdots + x_{n-1}}{n - 1} \right)^{1/n} \qquad (8.5)$$

Dividing both sides of (8.5) by $[(x_1 + \cdots + x_{n-1})/(n - 1)]^{1/n}$, we obtain

$$\left(\frac{x_1 + \cdots + x_{n-1}}{n - 1} \right)^{(n-1)/n} \geq (x_1 \cdots x_{n-1})^{1/n} \qquad (8.6)$$

Raising both sides of (8.6) to the power $n/(n - 1)$, we find

$$\frac{x_1 + \cdots + x_{n-1}}{n - 1} \geq (x_1 \cdots x_{n-1})^{1/(n-1)}$$

as the desired result for $n - 1$. Thus, if inequality (8.2) holds for any $n = 2^k$ ($k = 2, 3, \ldots$), it holds true for any $n = 3, 4, \ldots, 2^k - 1$. Consequently it is true for *all* values of $n = 1, 2, \ldots$.

The inequality may be demonstrated by other techniques familiar to the reader of this book. Exercise 8.1 should be solved using the methods of Chapter 2.

Exercise 8.1

Solve the problem

$$\min \sum_{j=1}^{n} x_j$$

subject to

$$\prod_{j=1}^{n} x_j = 1$$

$$x_j \geq 0, \quad \text{for } j = 1, 2, \ldots, n$$

Hence prove inequality (8.2).

iii. *Using the Special Form in an Optimization Problem*

To see how the special form of the arithmetic mean–geometric mean inequality may be used in certain minimization problems, suppose that we have a given function f of the form

$$f = f(x_1, \ldots, x_n) = y_1 + y_2 + \cdots + y_m$$

where $\qquad y_j = y_j(x_1, \ldots, x_n) \geq 0, \quad j = 1, \ldots, m$

That is, each term y_j is itself a function of $x_1, \ldots, x_n$, and f is a simple sum of such terms. Suppose (and here we make a *very* strong assumption) that

$$\prod_{j=1}^{m} y_j^{1/m} = c \qquad (8.7)$$

where c is a constant. From (8.2) we know that

$$\sum_{j=1}^{m} \frac{1}{m} y_j \geq \prod_{j=1}^{m} y_j^{1/m}$$

so by virtue of (8.7), we obtain

$$\sum_{j=1}^{m} y_j = f(x_1, \ldots, x_n) \geq mc$$

for all vectors $(x_1, \ldots, x_n)$.

Thus we have an easily obtainable lower bound on f, and if f is to be minimized, then we know the optimal value can be no smaller than mc. It remains, of course, to discover whether f ever assumes that value, that is, whether there exists $(x_1^0, x_2^0, \ldots, x_n^0)$ such that

$$f(x_1^0, x_2^0, \ldots, x_n^0) = \sum_{j=1}^{m} y_j(x_1^0, \ldots, x_n^0) = mc$$

Example 8.1

Let

$$f(x) = \frac{1}{x^3} + 2x^2 + 4x, \quad x > 0$$

Here we have $m = 3$ and

$$y_1 = \frac{1}{x^3}, \qquad y_2 = 2x^2, \qquad y_3 = 4x$$

yielding

$$y_1^{1/3} y_2^{1/3} y_3^{1/3} = \left(\frac{1}{x^3}\right)^{1/3} (2x^2)^{1/3} (4x)^{1/3} = (8)^{1/3} = 2$$

Thus condition (8.7) is satisfied, and $f(x) \geq 6$, for $x > 0$.

Example 8.2

As a next example, consider the finding of a constant lower bound on a given function f, in the same fashion, except that now the inequality must be used in its most general form. Let

$$f(x) = x^3 + \frac{2}{x^2}$$

If we select equal weights and set

$$y_1 = x^3 \quad \text{and} \quad y_2 = \frac{2}{x^2}$$

we obtain

$$y_1^{1/2} y_2^{1/2} = (x^3)^{1/2} \left(\frac{2}{x^2}\right)^{1/2} = \sqrt{2x}$$

Thus condition (8.7) is not satisfied. Suppose, however, that we select arbitrary weights λ_1 and λ_2 such that

$$\lambda_1 + \lambda_2 = 1$$

We then have

$$y_1^{\lambda_1} y_2^{\lambda_2} = (x^3)^{\lambda_1} \left(\frac{2}{x^2}\right)^{\lambda_2} = x^{3\lambda_1 - 2\lambda_2} 2^{\lambda_2}$$

Clearly, by selecting

$$3\lambda_1 - 2\lambda_2 = 0$$

condition (8.7) is satisfied. Solving simultaneously the two linear equations in λ_1 and λ_2 as obtained, we find

$$\lambda_1 = \tfrac{2}{5}, \qquad \lambda_2 = \tfrac{3}{5}$$

Inequality (8.1) can then be written as

$$\lambda_1 y_1 + \lambda_2 y_2 \geq y_1^{\lambda_1} y_2^{\lambda_2}$$

or

$$\tfrac{2}{5} y_1 + \tfrac{3}{5} y_2 \geq y_1^{2/5} y_2^{3/5}$$

or

$$\frac{2}{5} x^3 + \frac{3}{5} \left(\frac{2}{x^2}\right) \geq (x^3)^{2/5} \left(\frac{2}{x^2}\right)^{3/5}$$

which is valid, of course, but we obtain a left-hand side that is *not* $f(x)$. The idea is to construct two (in this case) terms such that $f(x)$ *is* obtained as the left-hand side in the statement of the inequality *and* the constant term is obtained as the right-hand side.

This could have been accomplished if we had originally defined

$$z_1 = \frac{1}{\lambda_1} x^3 \quad \text{and} \quad z_2 = \frac{1}{\lambda_2} \frac{2}{x^2}$$

Inequality (8.1) for z_1 and z_2 would then become

$$\lambda_1 z_1 + \lambda_2 z_2 \geq z_1^{\lambda_1} z_2^{\lambda_2}$$

or

$$x^3 + \frac{2}{x^2} \geq \left(\frac{5}{2} x^3\right)^{2/5} \left(\frac{5}{3} \cdot \frac{2}{x^2}\right)^{3/5}$$

The left-hand side is clearly $f(x)$, and we obtain after reduction

$$f(x) \geq (\tfrac{5}{2})^{2/5} (\tfrac{10}{3})^{3/5}$$

or

$$f(x) \geq 5(\tfrac{2}{27})^{1/5}$$

8.3. Solving Unconstrained Geometric-Programming Problems

i. *General Inequality Form*

Let $Y_1, Y_2, \ldots, Y_m$ be m nonnegative numbers; then we may write inequality (8.1) as

$$\lambda_1 Y_1 + \cdots + \lambda_m Y_m \geq Y_1^{\lambda_1} \cdots Y_m^{\lambda_m}$$

with

$$\sum_{j=1}^{m} \lambda_j = 1 \quad \text{and} \quad \lambda_j > 0, \quad \text{for all } j$$

If we now set $y_j = \lambda_j Y_j$, then

$$y_1 + y_2 + \cdots + y_m \geq \left(\frac{y_1}{\lambda_1}\right)^{\lambda_1} \left(\frac{y_2}{\lambda_2}\right)^{\lambda_2} \cdots \left(\frac{y_m}{\lambda_m}\right)^{\lambda_m} \qquad (8.8)$$

where we have specified the Y_j so as to obtain for the function $f = f(x_1, \ldots, x_m)$

$$f = \sum_{j=1}^{m} y_j$$

and the λ_j cause the right-hand side of (8.8) to be a constant; that is, the power of x_i is zero in the right-hand side of (8.8), for all $i = 1, \ldots, n$.

ii. *The Class of Functions—Posynomials*

The general function f we wish to consider is composed of a sum of m terms y_j ($j = 1, \ldots, m$) such that

$$y_j = c_j x_1^{a_{j1}} x_2^{a_{j2}} \cdots x_n^{a_{jn}}$$

where $c_j > 0$; and for $k = 1, 2, \ldots, n$, a_{jk} is a real number and x_k is a variable that assumes only *positive* values. [Notice the slight difference; terms need only be nonnegative to satisfy the requirements of inequality (8.1). Such a function y_j has been dubbed "posynomial" by Wilde [2], since it is a multinomial that is positive for all admissible values of its arguments.]

iii. *Basic Problem*

Consider, then, the posynomial

$$f(x_1, \ldots, x_n) = \sum_{j=1}^{m} c_j x_1^{a_{j1}} \cdots x_n^{a_{jn}} = \sum_{j=1}^{m} c_j p_j$$

Applying inequality (8.8), we obtain

$$\sum_{j=1}^{m} \lambda_j c_j p_j \geq \prod_{j=1}^{m} \left(\frac{c_j p_j}{\lambda_j} \right)^{\lambda_j}$$

or

$$\sum_{j=1}^{m} \lambda_j c_j x_1^{a_{j1}} \cdots x_n^{a_{jn}} \geq \prod_{j=1}^{m} \left(\frac{1}{\lambda_j} c_j x_1^{a_{j1}} \cdots x_n^{a_{jn}} \right)^{\lambda_j}$$

$$= \left(\frac{c_1}{\lambda_1} \right)^{\lambda_1} \cdots \left(\frac{c_m}{\lambda_m} \right)^{\lambda_m} x_1^{(a_{11}\lambda_1 + \cdots + a_{m1}\lambda_m)} \cdots x_n^{(a_{1n}\lambda_1 + \cdots + a_{mn}\lambda_m)}$$

An immediate question, then, is whether it is possible to choose $\lambda_1, \ldots, \lambda_m$ to satisfy the conditions

$$
\begin{aligned}
a_{11}\lambda_1 + a_{21}\lambda_2 + \cdots + a_{m1}\lambda_m &= 0 \\
a_{12}\lambda_1 + a_{22}\lambda_2 + \cdots + a_{m2}\lambda_m &= 0 \\
&\;\;\vdots \\
a_{1n}\lambda_1 + a_{2n}\lambda_2 + \cdots + a_{mn}\lambda_m &= 0 \\
\lambda_1 + \lambda_2 + \cdots + \lambda_m &= 1 \\
\lambda_1, \lambda_2, \ldots, \lambda_m &> 0
\end{aligned}
\tag{8.9}
$$

The fact that the a_{ij} appear in "transposed" form in the array of coefficients (8.9) results from the manner we have elected to index them, and should not be a source of confusion.

In attempting to locate the lower bound b, questions of existence and uniqueness of solutions to the system (8.9) must be answered. Furthermore, there remains the important task of locating, if they exist, positive values of $x_1, \ldots, x_n$, say $x_1^0, x_2^0, \ldots, x_n^0$, such that $f(x_1^0, x_2^0, \ldots, x_n^0) = b$; that is, such that f actually achieves the lower bound calculated. These questions will be taken up later.

From what we have said, the reader may infer that in the course of problem solving one obtains an optimal objective function value prior to obtaining an optimal solution. This is indeed the case, as we shall see.

Example 8.3

Consider the problem of selecting x_1, x_2, x_3 so as to minimize

$$f(x_1, x_2, x_3) = \frac{4}{x_1 x_2 x_3} + 2(2x_2 x_3 + 2x_1 x_3) + 2(4x_1 x_2)$$

$$= \frac{4}{x_1 x_2 x_3} + 8x_1 x_2 + 4x_2 x_3 + 4x_1 x_3$$

Examining

$$(4x_1^{-1} x_2^{-1} x_3^{-1})^{\lambda_1}(8x_1 x_2)^{\lambda_2}(4x_2 x_3)^{\lambda_3}(4x_1 x_3)^{\lambda_4}$$

it is clear that we require

$$\lambda_2 = \lambda_3 = \lambda_4 = \tfrac{1}{2}\lambda_1$$

and with

$$\sum_{i=1}^{4} \lambda_i = 1, \quad \lambda_i > 0$$

it must be that

$$\lambda_1 = \tfrac{2}{5}, \qquad \lambda_2 = \lambda_3 = \lambda_4 = \tfrac{1}{5}$$

But, clearly,

$$\frac{2}{5}\left(\frac{4}{x_1 x_2 x_3}\right) + \frac{1}{5}(8x_1 x_2) + \frac{1}{5}(4x_2 x_3) + \frac{1}{5}(4x_1 x_3) \neq f(x_1, x_2, x_3)$$

We can write

$$f(x_1, x_2, x_3) = \frac{2}{5}\left(\frac{5}{2}\frac{4}{x_1 x_2 x_3}\right) + \frac{1}{5}(5 \cdot 8x_1 x_2) + \frac{1}{5}(5 \cdot 4x_2 x_3) + \frac{1}{5}(5 \cdot 4x_1 x_3)$$

$$\geq \left(\frac{5}{2}\frac{4}{x_1 x_2 x_3}\right)^{2/5}(5 \cdot 8x_1 x_2)^{1/5}(5 \cdot 4x_2 x_3)^{1/5}(5 \cdot 4x_1 x_3)^{1/5}$$

$$= (10)^{2/5}(40)^{1/5}(20)^{1/5}(20)^{1/5}$$

$$= (1{,}600{,}000)^{1 \cdot 5}$$

The reader is invited to show that there are positive values x_1, x_2, x_3 which produce that value for f, that is, which will satisfy

$$\frac{4}{x_1 x_2 x_3} = 10^{2/5}$$

$$8x_1 x_2 = 40^{1/5}$$

$$4x_2 x_3 = 20^{1/5}$$

$$4x_1 x_3 = 20^{1/5}$$

Example 8.4

A man decides to build a fallout shelter. It will be rectangular in design, will have its ceiling 5 feet below ground level and will itself be 6.5 feet high. Its length will be l, its width w. A partition will be constructed so that there will be two rooms, one more heavily fortified than the other. The partition will be constructed of a sturdy material, which will also be used to insulate floor, ceiling, and other walls of the primary room. This material costs $3.00 per square foot. Floor, ceiling, and other walls of the other room created by the partition will be covered with a material costing $1.00 per square foot. The heavy insulator is two feet thick, the other has negligible thickness.

It costs $.25 per cubic foot to remove dirt; $.05 per cubic foot to dump dirt back on top the finished structure; and $.30 per cubic foot to dispose of remaining dirt.

The remaining cost relates to the opportunities for survival; the volume of the fortified room is the most critical, its cost component being $10/l_1 w$ while the other room contributes cost $1/l_2 w$. (These expressions indicate that the smaller the volume, the higher the cost, which is to say a decreased chance of survival results.) l_1, l_2 are the useful lengths of the subrooms constructed.

Thus, the various costs components are

> dirt removal
>
> dirt replacement
>
> dirt hauling
>
> finishing primary room
>
> finishing secondary room
>
> cost of risk

Write the entire function to be minimized, the total cost of the shelter.

iv. *Obtaining a Linear Dual Problem*

Our posynomial function f is clearly differentiable for all values of x_1, $\ldots$, x_n that are positive. Therefore, we may apply the usual necessary conditions of differential calculus to f. The result is of theoretical importance. Let

$(x_1^0, x_2^0, \ldots, x_n^0) = \mathbf{x}^0$ be a global minimum. Then

$$\frac{\partial f(\mathbf{x}^0)}{\partial x_i} = 0, \quad i = 1, \ldots, n$$

and, consequently,

$$x_i^0 \frac{\partial f(\mathbf{x}^0)}{\partial x_i} = 0, \quad i = 1, \ldots, n$$

Since

$$f(\mathbf{x}) = \sum_{j=1}^{m} y_j(\mathbf{x}) = \sum_{j=1}^{m} c_j x_1^{a_{j1}} x_2^{a_{j2}} \cdots x_n^{a_{jn}}$$

we obtain, for $i = 1, \ldots, n$,

$$\sum_{j=1}^{m} x_i^0 \frac{\partial y_j(\mathbf{x}^0)}{\partial x_i} = \sum_{j=1}^{m} x_i^0 a_{ji} c_j (x_1^0)^{a_{j1}} (x_2^0)^{a_{j2}} \cdots (x_i^0)^{a_{ji}-1} \cdots (x_n^0)^{a_{jn}}$$

$$= \sum_{j=1}^{m} x_i^0 a_{ji} y_j(\mathbf{x}^0)(x_i^0)^{-1} \qquad (8.10)$$

$$= \sum_{j=1}^{m} a_{ji} y_j(\mathbf{x}^0) = 0$$

   Observing (8.10), we see that the $y_j(\mathbf{x}^0)$ come quite close to satisfying the system (8.9). They do, in fact, except for the final equation of (8.9), which requires that the weights sum to unity. Since $f(\mathbf{x}^0) > 0$, dividing both sides of all equations in (8.10) by $f(\mathbf{x}^0)$, we obtain

$$\frac{1}{f(\mathbf{x}^0)} \sum_{j=1}^{m} a_{ji} y_j(\mathbf{x}^0) = 0$$

Thus, we may define

$$\lambda_j^0 = \frac{y_j(\mathbf{x}^0)}{f(\mathbf{x}^0)}, \quad j = 1, \ldots, m \qquad (8.11)$$

which do satisfy the system (8.9). Since we have

$$\sum_{j=1}^{m} \lambda_j^0 = 1$$

we may write $f(\mathbf{x}^0)$ as

$$f(\mathbf{x}^0) = f(\mathbf{x}^0)^{\lambda_1^0} f(\mathbf{x}^0)^{\lambda_2^0} \cdots f(\mathbf{x}^0)^{\lambda_m^0}$$

and using (8.11) we obtain

$$f(\mathbf{x}^0) = \left(\frac{y_1(\mathbf{x}^0)}{\lambda_1^0}\right)^{\lambda_1^0} \left(\frac{y_2(\mathbf{x}^0)}{\lambda_1^0}\right)^{\lambda_2^0} \cdots \left(\frac{y_m(\mathbf{x}^0)}{\lambda_m^0}\right)^{\lambda_m^0} \qquad (8.12)$$

It may be helpful at this point to utilize a new notation for the function on the right side of (8.12). Let

$$g(\lambda_1, \ldots, \lambda_m) = \left(\frac{y_1(\mathbf{x}^0)}{\lambda_1}\right)^{\lambda_1} \cdots \left(\frac{y_m(\mathbf{x}^0)}{\lambda_m}\right)^{\lambda_m}$$

From inequality (8.1)

$$f(\mathbf{x}^0) \geq g(\lambda_1, \ldots, \lambda_m)$$

But in (8.12) we have

$$f(\mathbf{x}^0) = g(\boldsymbol{\lambda}^0)$$

Thus $(\lambda_1^0, \ldots, \lambda_m^0)$ solves the "dual" problem

$$\max g(\lambda_1, \ldots, \lambda_m)$$

subject to (8.9).
    We have then

Theorem 8.1

If $(x_1^0, x_2^0, \ldots, x_n^0) = \mathbf{x}^0$ is the global minimum of f, $f(\mathbf{x}^0) > 0$, then the values of $\lambda_j^0, j = 1, \ldots, m$, defined by

$$\lambda_j^0 = \frac{y_j(\mathbf{x}^0)}{f(\mathbf{x}^0)}$$

solve the dual problem

$$\max g(\lambda_1, \ldots, \lambda_m) = \left(\frac{y_1(\mathbf{x}^0)}{\lambda_1}\right)^{\lambda_1} \cdots \left(\frac{y_m(\mathbf{x}^0)}{\lambda_m}\right)^{\lambda_m}$$

subject to (8.9), and we have $f(\mathbf{x}^0) = g(\boldsymbol{\lambda}^0)$.

We have written g as a function of $\mathbf{x}^0$, also, but it should be clear that if the system (8.9) is satisfied, then all powers of $x_1, \ldots, x_n$ are zero, and g is not a function of $x_1, \ldots, x_n$.
    Now suppose that $(\lambda_1^0, \ldots, \lambda_m^0)$ maximizes g subject to (8.9). We have seen that for minimum f we have

$$f(\mathbf{x}^0) = g(\boldsymbol{\lambda}^0)$$

However, as seen in Section 8.2-i, the equality condition in the arithmetic mean–geometric mean inequality holds, if and only if the terms are all equal. Thus in our case the equality condition in $f(\mathbf{x}^0) \geq g(\lambda_1, \ldots, \lambda_m)$ holds if and only if

$$\frac{y_1(\mathbf{x}^0)}{\lambda_1^0} = \frac{y_2(\mathbf{x}^0)}{\lambda_2^0} = \cdots = \frac{y_m(\mathbf{x}^0)}{\lambda_m^0} = c_0$$

We thus have

$$g(\lambda^0) = f(\mathbf{x}^0) = \sum_{j=1}^{m} y_j(\mathbf{x}^0)$$

$$= \sum_{j=1}^{m} \frac{y_j(\mathbf{x}^0)}{\lambda_j^0} \lambda_j^0$$

$$= \sum_{j=1}^{m} c_0 \lambda_j^0 = c_0$$

Hence
$$g(\lambda^0) = \frac{y_j(\mathbf{x}^0)}{\lambda_j^0}, \quad \text{for all } j$$

Thus knowledge of the λ_j^0 and $g(\lambda^0)$ *provides a system of equations from which the optimal values* $x_1^0, x_2^0, \ldots, x_n^0$ *may be determined.* This system is

$$g(\lambda^0)\lambda_j^0 = y_j(\mathbf{x}^0) = c_j x_1^{a_{j1}} x_2^{a_{j2}} \cdots x_n^{a_{jn}}, \quad j = 1, \ldots, m \tag{8.13}$$

System (8.13) shows how an optimal "dual" solution provides an optimal solution to the given "primal" problem; but nonlinear systems such as (8.13) are not so easily solved in general. The posynomial nature of $y_1, \ldots, y_m$, however, allows the transformation of (8.13) into an equivalent system of linear equations. Taking natural logarithms of each member of system (8.13), we obtain the system

$$\ln\left(g(\lambda^0)\lambda_j^0\right) = \ln c_j + \sum_{i=1}^{n} a_{ji} \ln x_i, \quad j = 1, \ldots, m \tag{8.14}$$

Effecting the substitution

$$\theta_i = \ln x_i \quad \text{and} \quad k_j^0 = \ln\left(g(\lambda^0)\lambda_j^0\right) - \ln c_j$$

we obtain a system of the form

$$\sum_{i=1}^{n} a_{ji}\theta_i = k_j^0, \quad j = 1, \ldots, m \tag{8.15}$$

Consequently, one is led to inquire whether (8.15) has a solution, and, if so, whether it has a unique solution; for if not, then the corresponding x's are not uniquely determined. To investigate such questions and observe the implications, it is desirable to treat more general problems that include the unconstrained problems as special cases; that is, we begin examining constrained geometric-programming problems.

8.4. General Constrained Problem

i. *General Problem with One Constraint*

Constraints will be assumed to be of the form "posynomial form" ≤ 1, or

$$h_k(x_1, \ldots, x_n) = \sum_{l=1}^{p} w_l(x_1, \ldots, x_n) \leq 1$$

where each term w_l is of the same form as our previous terms y_j.

Consider a problem with a single inequality constraint:

$$\min f(\mathbf{x}) = \sum_{j=1}^{m} y_j(\mathbf{x})$$

subject to

$$h_1(\mathbf{x}) = \sum_{l=1}^{p} w_l(\mathbf{x}) \leq 1$$

$$x_i > 0, \quad i = 1, \ldots, n$$

ii. *Extending the Previous Methods*

One possible approach to solving the given problem might be the following extension of previous methods. We may write

$$f(\mathbf{x}) \geq \left(\frac{y_1}{\lambda_1}\right)^{\lambda_1}\left(\frac{y_2}{\lambda_2}\right)^{\lambda_2} \cdots \left(\frac{y_m}{\lambda_m}\right)^{\lambda_m} \tag{8.16}$$

where the positive λ_j sum to unity. We also have, writing w_k for $w_k(\mathbf{x})$,

$$1 \geq h_1(\mathbf{x}) \geq \left(\frac{w_1}{\alpha_1}\right)^{\alpha_1}\left(\frac{w_2}{\alpha_2}\right)^{\alpha_2} \cdots \left(\frac{w_p}{\alpha_p}\right)^{\alpha_p} \tag{8.17}$$

where the positive α_k sum to unity.

Multiplying inequality (8.16) by inequality (8.17), we have

$$f(\mathbf{x}) \geq \left(\frac{y_1}{\lambda_1}\right)^{\lambda_1} \cdots \left(\frac{y_m}{\lambda_m}\right)^{\lambda_m}\left(\frac{w_1}{\alpha_1}\right)^{\alpha_1} \cdots \left(\frac{w_p}{\alpha_p}\right)^{\alpha_p} \tag{8.18}$$

with

$$\sum_{j=1}^{m} \lambda_j = \sum_{k=1}^{p} \alpha_k = 1$$

Then it seems we should be able to choose a set of positive λ's and a set of positive α's, so that the right-hand side of (8.18) becomes the greatest lower bound of $f(\mathbf{x})$ subject to the constraint $h_1(\mathbf{x}) \leq 1$. But will this work? The answer is no; because in order to apply our previous "duality" theory, it is necessary that the weights sum to unity, which they *cannot* if (8.17) and (8.18) are to hold for all values of their variables. Thus the need for being able to work with more general weights arises.

iii. *Generalizing the Inequality—Obtaining a Workable Approach*

It is possible to write the arithmetic mean–geometric mean inequality (8.1) in a form where the weights need not sum to unity. Let $\gamma_1, \gamma_2, \ldots, \gamma_m$ be an arbitrary set of positive weights with

$$\Gamma = \gamma_1 + \gamma_2 + \cdots + \gamma_m$$

As before let

$$f(\mathbf{x}) = y_1(\mathbf{x}) + \cdots + y_m(\mathbf{x})$$

Define

$$\lambda_j = \frac{\gamma_j}{\Gamma}, \quad j = 1, \ldots, m \tag{8.19}$$

so that

$$\sum_{j=1}^{m} \lambda_j = 1$$

Define

$$y_j = \gamma_j Y_j, \quad j = 1, \ldots, m \tag{8.20}$$

so that from (8.1)

$$\lambda_1 Y_1 + \lambda_2 Y_2 + \cdots + \lambda_m Y_m \geq \left(\frac{y_1}{\gamma_1}\right)^{\lambda_1} \cdots \left(\frac{y_m}{\gamma_m}\right)^{\lambda_m} \tag{8.21}$$

which is the usual form of the inequality with weights that sum to unity, and with terms $Y_1, Y_2, \ldots, Y_m$. Substituting (8.19) and then (8.20) into (8.21), we obtain first

$$\frac{\gamma_1}{\Gamma} Y_1 + \frac{\gamma_2}{\Gamma} Y_2 + \cdots + \frac{\gamma_m}{\Gamma} Y_m \geq \left(\frac{y_1}{\gamma_1}\right)^{\gamma_1/\Gamma} \cdots \left(\frac{y_m}{\gamma_m}\right)^{\gamma_m/\Gamma}$$

and then

$$y_1 + y_2 + \cdots + y_m \geq \Gamma \left[\left(\frac{y_1}{\gamma_1}\right)^{\gamma_1/\Gamma} \cdots \left(\frac{y_m}{\gamma_m}\right)^{\gamma_m/\Gamma} \right] \tag{8.22}$$

To eliminate fractional powers from (8.22), both sides may be raised to the power Γ, yielding

$$(y_1 + y_2 + \cdots + y_m)^{\Gamma} \geq \Gamma^{\Gamma} \left[\left(\frac{y_1}{\gamma_1}\right)^{\gamma_1} \cdots \left(\frac{y_m}{\gamma_m}\right)^{\gamma_m} \right] \tag{8.23}$$

which is the arithmetic mean–geometric mean inequality in terms of the arbitrary positive weights $\gamma_1, \gamma_2, \ldots, \gamma_m$.

Applying (8.23) to both the objective function and the constraining inequality of the problem defined in Section 8.4-i, we obtain

$$(f(\mathbf{x}))^\Gamma \geq \Gamma^\Gamma \left[\left(\frac{y_1}{\gamma_1} \right)^{\gamma_1} \cdots \left(\frac{y_m}{\gamma_m} \right)^{\gamma_m} \right] \tag{8.24}$$

and

$$1 \geq (w_1 + \cdots + w_p)^\beta \geq \beta^\beta \left[\left(\frac{w_1}{\beta_1} \right)^{\beta_1} \cdots \left(\frac{w_p}{\beta_p} \right)^{\beta_p} \right] \tag{8.25}$$

where the γ_j are arbitrary positive weights that sum to Γ, and the β_l are arbitrary positive weights that sum to β.

Multiplying (8.24) with (8.25), we obtain

$$(f(\mathbf{x}))^\Gamma \geq \Gamma^\Gamma \beta^\beta \left[\left(\frac{y_1}{\gamma_1} \right)^{\gamma_1} \cdots \left(\frac{y_m}{\gamma_m} \right)^{\gamma_m} \left(\frac{w_1}{\beta_1} \right)^{\beta_1} \cdots \left(\frac{w_p}{\beta_p} \right)^{\beta_p} \right] \tag{8.26}$$

Inequality (8.26) may be simplified a little to obtain a relationship involving f, rather than a power thereof, by setting $\Gamma = 1$, which is legitimate, since all weights in (8.26) are arbitrary (but positive). Thus (8.26) becomes

$$f(\mathbf{x}) \geq \beta^\beta \left[\left(\frac{y_1}{\gamma_1} \right)^{\gamma_1} \cdots \left(\frac{y_m}{\gamma_m} \right)^{\gamma_m} \left(\frac{w_1}{\beta_1} \right)^{\beta_1} \cdots \left(\frac{w_p}{\beta_p} \right)^{\beta_p} \right] \tag{8.27}$$

with

$$\gamma_1 + \gamma_2 + \cdots + \gamma_m = 1$$

Now we can apply our previous procedures to (8.27) to transform the righthand side into a greatest lower bound constant for f as constrained by h_1.

iv. *Obtaining the Dual of the General Constrained Problem*

Considering more than a single posynomial constraint leads in a similar way to the formulation of a general dual problem. The general constrained geometric-programming problem is

$$\min f(\mathbf{x}) = \sum_{j=1}^{m} y_j(\mathbf{x})$$

subject to

$$h_i(\mathbf{x}) = \sum_{j=p_{i-1}+1}^{p_i} y_j(\mathbf{x}) \leq 1, \quad i = 1, \ldots, k \tag{8.28}$$

$$(x_1, \ldots, x_n) = \mathbf{x} > 0$$

where

$$y_j(\mathbf{x}) = c_j x_1^{a_{j1}} x_2^{a_{j2}} x_3^{a_{j3}} \cdots x_n^{a_{jn}}, \quad j = 1, \ldots, p$$

and

$$c_j > 0, \quad j = 1, \ldots, p$$

with

$$p_0 = m, \qquad p_k = p, \qquad p_i > p_{i-1}$$

It is convenient for purposes of writing a dual problem to label all polynomial terms y followed by an appropriate subscript in the statement of the general problem.

The reader should understand the significance of the several integers appearing in problem (8.28). We have n variables, k constraints, m polynomial terms in the objective function f, and a total of p posynomial terms in the entire problem, objective function and constraints combined.

Problem (8.29) is then taken to be the dual of (8.28).

$$\max g(\lambda_1, \lambda_2, \ldots, \lambda_p) = \max g(\lambda)$$

$$= \prod_{i=1}^{p} \left(\frac{c_i}{\lambda_i}\right)^{\lambda_i} \prod_{r=1}^{k} (\lambda_{p_{r-1}+1} + \cdots + \lambda_{p_r})^{(\lambda_{p_{r-1}+1} + \cdots + \lambda_{p_r})}$$

subject to

$$\lambda_1, \ldots, \lambda_p \geq 0 \tag{8.29}$$

$$\sum_{i=1}^{m} \lambda_i = 1$$

$$\sum_{i=1}^{p} a_{ij}\lambda_i = 0, \quad j = 1, \ldots, n$$

Example 8.5

Consider the following primal problem:

$$\min \{2x_1^{3/2}x_2^{-3}x_3^{1/2}x_4^{-3/4} + \tfrac{1}{2}x_1^{-2}x_2^2x_3^{-1/2}x_4^{-1/2} + 8x_1^{\pi/2}x_2^{-3/8}x_3^{3/2}x_4^{-4/5}\}$$

subject to

$$\tfrac{3}{8}x_1^2 x_2^{-2} x_3^{1/5} x_4^{-3/7} + 45x_1^{-2/7}x_2^{4/5}x_3^3 x_4^{-38} \leq 1 \tag{8.30}$$

$$2x_1^{-3/4}x_2^{1/2}x_3^{-4}x_4^{2/3} \leq 1$$

$$\tfrac{2}{3}x_1^{-1}x_2^{2/9}x_3^{-3/7}x_4^{-4/5} + 12x_1^{10}x_2^{-9}x_3^{1/8}x_4^{-3/11} + \tfrac{7}{3}x_1^{-2}x_2^{-3}x_3^4 x_4^{-1/2} \leq 1$$

$$x_1, x_2, x_3, x_4 > 0$$

According to (8.29), the dual of (8.30) will be

$$\max g(\lambda_1, \ldots, \lambda_9) = \left(\frac{2}{\lambda_1}\right)^{\lambda_1}\left(\frac{1/2}{\lambda_2}\right)^{\lambda_2} \cdots \left(\frac{45}{\lambda_5}\right)^{\lambda_5} \cdots \left(\frac{7/3}{\lambda_9}\right)^{\lambda_9}(\lambda_4 + \lambda_5)^{\lambda_4 + \lambda_5}$$

$$\times \lambda_6^{\lambda_6}(\lambda_7 + \lambda_8 + \lambda_9)^{\lambda_7 + \lambda_8 + \lambda_9}$$

subject to

$$\lambda_1 + \lambda_2 + \lambda_3 = 1$$

$$\tfrac{3}{2}\lambda_1 - 2\lambda_2 + \tfrac{\pi}{2}\lambda_3 + 2\lambda_4 - \tfrac{2}{7}\lambda_5 - \tfrac{3}{4}\lambda_6 - \lambda_7 + 10\lambda_8 - 2\lambda_9 = 0$$

$$-3\lambda_1 + 2\lambda_2 - \tfrac{3}{8}\lambda_3 - 2\lambda_4 + \tfrac{4}{5}\lambda_5 + \tfrac{1}{4}\lambda_6 + \tfrac{2}{3}\lambda_7 - 9\lambda_8 - 3\lambda_9 = 0$$

$$\tfrac{1}{2}\lambda_1 - \tfrac{1}{2}\lambda_2 + \tfrac{3}{2}\lambda_3 + \tfrac{1}{4}\lambda_4 + 3\lambda_5 - 4\lambda_6 - \tfrac{3}{7}\lambda_7 + \tfrac{1}{8}\lambda_8 + 4\lambda_9 = 0$$

$$-\tfrac{3}{4}\lambda_1 - \tfrac{1}{2}\lambda_2 - \tfrac{4}{5}\lambda_3 - \tfrac{3}{7}\lambda_4 - 38\lambda_5 + \tfrac{3}{2}\lambda_6 - \tfrac{4}{5}\lambda_7 - \tfrac{3}{11}\lambda_8 - \tfrac{1}{2}\lambda_9 = 0$$

$$\lambda_1, \ldots, \lambda_9 \geq 0$$

Example 8.6

The problem

$$\min \{5x_1^{-2}x_2^{3/5}x_3^{-1/8}x_4^{1/3}x_5^{7} + \tfrac{1}{2}x_1^{-4}x_2^{3/2}x_3^{-6/7}x_4^{4/3}x_5^{-3}\}$$

subject to

$$\tfrac{3}{2}x_1^2x_2^4x_3^{-3}x_4^{-1/10}x_5^{7/10} + 8x_1^{-4/5}x_2^{3/2}x_3^{-6/5}x_4^{-7/3}x_5^{4} \leq 1$$

$$4x_1^{-4}x_2^{3/5}x_3^8x_4^2x_5^{-1} + \tfrac{5}{3}x_1^{-3}x_2^{-6/5}x_3^{3/8}x_4^{2/9}x_5^{-7/11} \leq 1$$

$$x_1, \ldots, x_5 > 0$$

has the dual

$$\max \left(\tfrac{5}{\lambda_1}\right)^{\lambda_1}\left(\tfrac{1/2}{\lambda_2}\right)^{\lambda_2} \cdots \left(\tfrac{5/3}{\lambda_6}\right)^{\lambda_6}(\lambda_3 + \lambda_4)^{\lambda_3+\lambda_4}(\lambda_5 + \lambda_6)^{\lambda_5+\lambda_6}$$

subject to

$$\lambda_1 + \lambda_2 = 1$$

$$-2\lambda_1 - 4\lambda_2 + 2\lambda_3 - \tfrac{4}{5}\lambda_4 - 4\lambda_5 - 3\lambda_6 = 0$$

$$\tfrac{3}{5}\lambda_1 + \tfrac{3}{2}\lambda_2 + 4\lambda_3 + \tfrac{3}{2}\lambda_4 + \tfrac{3}{5}\lambda_5 - \tfrac{6}{5}\lambda_6 = 0$$

$$-\tfrac{1}{8}\lambda_1 - \tfrac{6}{7}\lambda_2 - 3\lambda_3 - \tfrac{6}{7}\lambda_4 + 8\lambda_5 + \tfrac{3}{8}\lambda_6 = 0 \qquad (8.31)$$

$$\tfrac{1}{3}\lambda_1 + \tfrac{4}{3}\lambda_2 - \tfrac{1}{10}\lambda_3 - \tfrac{7}{3}\lambda_4 + 2\lambda_5 + \tfrac{2}{3}\lambda_6 = 0$$

$$7\lambda_1 - 3\lambda_2 + \tfrac{7}{10}\lambda_3 + 4\lambda_4 - \lambda_5 - \tfrac{7}{11}\lambda_6 = 0$$

$$\lambda_1, \ldots, \lambda_6 \geq 0$$

Regarding the system of equations (8.31), suppose that the coefficient matrix has full rank, 6. Then there is *at most* one solution to the dual constraints. On the other hand, if the matrix does not have full rank, then the system of equations (8.31) has an infinite number of solutions, as may the set of *all solutions to the dual constraints in the problem.* It is important to appreciate that although the primal problem is nonlinear throughout, the dual constraints

are *always* just a set of linear equations along with nonnegativity requirements.

v. General Results

a. THEORETICAL RESULTS

Two theorems describe some of the relationships between the primal
problem and the dual problem. These are proved in Reference 1 and will only
be stated here.

Theorem 8.2

If the primal problem (8.28) has a feasible solution and if there is a vector
$\lambda^0 = (\lambda_1^0, \ldots, \lambda_p^0) > 0$ that is feasible for the dual problem, then the primal
problem has an optimal solution.

Theorem 8.2 gives a condition that is *sufficient for the existence* of a global
minimum in the primal problem.

Theorem 8.3

If the primal problem (8.28) has an optimal solution, and if there is at
least one point x^0 satisfying

$$h_i(x^0) < 1, \quad i = 1, \ldots, k$$

then
  i. The dual problem (8.29) has an optimal solution.
  ii. For the respective optimal solutions x^*, λ^*, we have $f(x^*) = g(\lambda^*)$.
  iii. If λ^* is an optimal solution for the dual problem, then *every* optimal
solution x^* for the primal problem satisfies

$$c_i x_1^{a_{i1}} x_2^{a_{i2}} \cdots x_n^{a_{in}} = \lambda_i^* g(\lambda^*) \qquad i = 1, \ldots, m$$

$$= \frac{\lambda_i^*}{\sum_{j=p_{r-1}+1}^{p_r} \lambda_j^*} \tag{8.32}$$

for all $p_{r-1} + 1 \leq i \leq p_r, r = 1, 2, \ldots, k$, if $\sum_{j=p_{r-1}+1}^{p_r} \lambda_j^* > 0$.

Thus, for r fixed, expression (8.32) contributes $p_r - p_{r-1}$ equations if the
corresponding λ_j^* sum is positive. Otherwise, (8.32) contributes no equations.

b. OBSTACLES TO IMPLEMENTATION

Our strategy for solving the primal problem might then proceed as follows:

1. With the hypotheses of Theorems 8.3 satisfied, we first solve the dual problem, since part i of Theorem 8.3 assures the existence of an optimal solution to the dual.
2. Having obtained an optimal dual solution, determine the optimal primal objective function value using part ii of Theorem 8.3.
3. Use part iii of Theorem 8.3 to find an optimal solution $\mathbf{x}^*$ for the primal.

It is not always the case that the foregoing strategy is easily implemented. Neglecting the problem of verifying whether Theorem 8.3 is actually applicable, the dual problem does not appear to be particularly trivial. And, even if it is solved, part iii of Theorem 8.3 does not say that *any* set of x's satisfying equations (8.32) is an optimal primal solution, but, rather, that an *optimal* set of x's will satisfy them.

There are at any rate, fortuitous instances in which the proposed strategy is quite simply effected, so let us see when such might be the case and why.

vi. *Special Cases*

First, the equality constraints in the dual consist of a system of $n + 1$ inhomogeneous linear equations in the p variables $\lambda_1, \ldots, \lambda_p$. We wish to assume that $n < p$; that is, in the primal problem there are *fewer variables than total terms*.

For the moment, suppose also that the matrix of coefficients of the dual equality constraints has full rank, that is, $n + 1$. In other words, the constraining dual equations are linearly independent. This in turn implies that the matrix of real exponents with the elements a_{ij} has full rank, n. If under the assumption of independence we should have $n + 1 = p$, then the system of linear equations in the dual has a *unique* solution $(\lambda_1^0, \ldots, \lambda_p^0)$. If it is the case that $\lambda_j^0 \geq 0$, $j = 1, \ldots, p$, then $\boldsymbol{\lambda}^0$ is a (unique) optimal solution to the dual problem; otherwise, of course, the dual has no feasible solution.

Should we also have our $\boldsymbol{\lambda}^0 > \mathbf{0}$ and the knowledge that the primal has a feasible solution, then Theorem 8.2 assures that the primal has an optimal solution, one of the necessities for applying Theorem 8.3. Suppose that the other hypothesis of Theorem 8.3 is satisfied. Then, part ii of Theorem 8.3 gives the optimal primal objective function value.

Furthermore, since we have $\boldsymbol{\lambda}^0 > \mathbf{0}$, part iii of Theorem 8.3 describes a system of p equations in n variables, that is, $n + 1$ equations in n variables. We observe that each of the right-hand sides of the equations generated by part iii of Theorem 8.3 is positive, and each coefficient c_i is positive; thus one may take natural logarithms of the ith equation, obtaining

$$\ln c_i + a_{i1} \ln x_1 + a_{i2} \ln x_2 + \cdots + a_{in} \ln x_n = \ln b_i$$

where b_i is a positive real number. Making the substitution $z_j = \ln x_j$, we obtain a system of linear equations:

$$\sum_{j=1}^{n} a_{ij} z_j = d_i, \quad i = 1, \ldots, p \tag{8.33}$$

But since the matrix of coefficients in (8.33) has rank n, exactly one of the equations in (8.33) must be redundant. Hence the system (8.33) has a unique solution, $z_1^*, \ldots, z_n^*$. The optimizing x_j^* are then found by

$$x_j^* = e^{z_j^*}$$

and we have $x_j^* > 0, j = 1, \ldots, n$.

Let us illustrate the problem-solving procedure under these ideal circumstances.

Example 8.7

Consider the problem

$$\min f(x_1, x_2, x_3) = 3x_2 x_3 + \tfrac{1}{9}x_1$$

subject to

$$\tfrac{1}{3}x_1^{1/2}x_2^{-2}x_3^{1/2} + 3x_1^{-1}x_2 x_3^{-2} \leq 1$$

The dual is

$$\max g(\lambda_1, \lambda_2, \lambda_3, \lambda_4) = \left(\frac{3}{\lambda_1}\right)^{\lambda_1}\left(\frac{1/9}{\lambda_2}\right)^{\lambda_2}\left(\frac{1/3}{\lambda_3}\right)^{\lambda_3}\left(\frac{3}{\lambda_4}\right)^{\lambda_4}(\lambda_3 + \lambda_4)^{\lambda_3 + \lambda_4}$$

subject to

$$
\begin{aligned}
\lambda_1 + \lambda_2 &&&= 1 \\
\lambda_2 + \tfrac{1}{2}\lambda_3 &- \lambda_4 &&= 0 \\
\lambda_1 \quad\quad &- 2\lambda_3 + \lambda_4 &&= 0 \\
\lambda_1 \quad\quad &+ \tfrac{1}{2}\lambda_3 - 2\lambda_4 &&= 0
\end{aligned}
\tag{8.34}
$$

The array of coefficients

$$
\begin{pmatrix}
1 & 1 & 0 & 0 \\
0 & 1 & \tfrac{1}{2} & -1 \\
1 & 0 & -2 & 1 \\
1 & 0 & \tfrac{1}{2} & -2
\end{pmatrix}
$$

has the inverse

$$\begin{pmatrix} \frac{7}{9} & -\frac{7}{9} & -\frac{1}{9} & \frac{1}{3} \\ \frac{2}{9} & \frac{7}{9} & \frac{1}{9} & -\frac{1}{3} \\ \frac{2}{3} & -\frac{2}{3} & -\frac{2}{3} & 0 \\ \frac{5}{9} & -\frac{5}{9} & -\frac{2}{9} & -\frac{1}{3} \end{pmatrix}$$

whence

$$\lambda_1 = \tfrac{7}{9}, \quad \lambda_2 = \tfrac{2}{9}, \quad \lambda_3 = \tfrac{2}{3}, \quad \lambda_4 = \tfrac{5}{9}$$

Thus we have a situation where the dual has a unique solution, and, furthermore, one in which $\lambda^* > 0$.

It is also simple to verify that the primal has a feasible solution for which the constraining inequality is strictly satisfied; $x_1 = 1$, $x_2 = 3$, $x_3 = 4$ is effective, for example.

Therefore, Theorem 8.3 is entirely applicable. The optimal primal objective function value is

$$(\tfrac{27}{7})^{7/9}(\tfrac{1}{2})^{2/9}(\tfrac{1}{2})^{2/3}(\tfrac{27}{5})^{5/9}(\tfrac{11}{9})^{11/9} = g(\lambda^*)$$

and the minimizing x's are found as the (unique) solution to the system of equations

$$3x_2 x_3 = \tfrac{7}{9}g(\lambda^*)$$
$$\tfrac{1}{9}x_1 = \tfrac{2}{9}g(\lambda^*)$$
$$\tfrac{1}{3}x_1^{1/2}x_2^{-2}x_3^{1/2} = \tfrac{6}{11}$$
$$3x_1^{-1}x_2 x_3^{-2} = \tfrac{5}{11}$$

or

$$\ln x_2 + \ln x_3 = \ln\tfrac{7}{9} + \ln g(\lambda^*) - \ln 3$$
$$\ln x_1 = \ln\tfrac{2}{9} + \ln g(\lambda^*) - \ln\tfrac{1}{9}$$
$$\tfrac{1}{2}\ln x_1 - 2\ln x_2 + \tfrac{1}{2}\ln x_3 = \ln\tfrac{6}{11} - \ln\tfrac{1}{3}$$
$$-\ln x_1 + \ln x_2 - 2\ln x_3 = \ln\tfrac{5}{11} - \ln 3$$

Substituting $z_i = \ln x_i$, and writing d_i ($i = 1, 2, 3, 4$) for the constant quantities appearing in the right-hand side, we obtain

$$z_2 + z_3 = d_1$$
$$z_1 \qquad\qquad = d_2$$
$$\tfrac{1}{2}z_1 - 2z_2 + \tfrac{1}{2}z_3 = d_3$$
$$-z_1 + z_2 - 2z_3 = d_4$$

and then discover the minimizing x's from $e^{z_i} = x_i$, $i = 1, 2, 3$.

Example 8.8

Suppose that in the primal the functions f and h_1 are each defined by two posynomials, and let the problem be

$$\min c_1 x_2 + c_2 x_1 x_3^{1/2}$$

subject to

$$c_3 x_2^{-2} x_3^{-1} + c_4 x_2 x_3^{2/3} \leq 1 \tag{8.35}$$

$$x_1, x_2, x_3 > 0$$

We observe that this problem has no minimum, for x_1 is unconstrained. Thus, for any feasible x_2 and x_3, x_1 may be made to approach zero and

$$c_1 x_2 + c_2 x_1 x_3^{1/2} \longrightarrow c_1 x_2$$

However, there is no minimum for $x_1 > 0$. Let us next apply our geometric programming theory to the problem. The system of dual constraining equations has the coefficient matrix

$$\begin{pmatrix} 1 & 1 & 0 & 0 \\ 0 & 1 & 0 & 0 \\ 1 & 0 & -2 & 1 \\ 0 & \frac{1}{2} & -1 & \frac{2}{3} \end{pmatrix} \tag{8.36}$$

The inverse of (8.36) is

$$\begin{pmatrix} 1 & -1 & 0 & 0 \\ 0 & 1 & 0 & 0 \\ 2 & -\frac{7}{2} & -2 & 3 \\ 3 & -6 & -3 & 6 \end{pmatrix}$$

and, thus, the unique optimal solution to the dual is

$$\lambda_1^* = 1, \qquad \lambda_2^* = 0, \qquad \lambda_3^* = 2, \qquad \lambda_4^* = 3$$

Problems arise immediately, however, when we attempt to obtain an optimal primal solution. Assuming Theorem 8.3 is applicable, the fact that $\lambda_2^* = 0$ implies that a primal posynomial must assume the value zero, which it cannot, so we verify that the primal has no minimum.

It should be noticed that when one or more dual variables are zero, the dual objective function possesses terms like

$$\left(\frac{c_i}{0}\right)^0$$

or perhaps $(0 + 0 + \cdots + 0)^{0+0+\cdots 0}$. Both such terms are taken to be unity, since, for example,

$$\lim_{x \to 0} x^x = 1$$

which the reader may easily verify by the methods of elementary calculus.

Example 8.9

Let the primal problem be

$$\min c_1 x_1^{1/2} + c_2 x_1^{-1/2}$$

subject to

$$c_3 x_1 x_2^{-1} \le 1 \tag{8.37}$$

$$x_1, x_2 > 0$$

The dual constraining equations have the coefficient matrix

$$\begin{pmatrix} 1 & 1 & 0 \\ \frac{1}{2} & -\frac{1}{2} & 1 \\ 0 & 0 & -1 \end{pmatrix}$$

and the inverse is

$$\begin{pmatrix} \frac{1}{2} & 1 & 1 \\ \frac{1}{2} & -1 & -1 \\ 0 & 0 & -1 \end{pmatrix}$$

and the unique dual solution is

$$\lambda_1^* = \lambda_2^* = \tfrac{1}{2}, \qquad \lambda_3^* = 0$$

As opposed to the previous example, where the dual variable associated with an objective function posynomial was zero, we have at present a zero value for a dual variable associated with a term in a primal constraint, and the implications of the two differ.

Assuming Theorem 8.3 to be valid, its part iii creates the following system of equations:

$$c_1 x_1^{1/2} = \tfrac{1}{2} g(\lambda^*)$$
$$c_2 x_1^{-1/2} = \tfrac{1}{2} g(\lambda^*) \tag{8.38}$$

From the way it has been written, the system of equations (8.38) may appear to have no solution, or at least to depend heavily upon c_1 and c_2 for the existence of a solution. However, using

$$g(\lambda^*) = \left(\frac{c_1}{1/2}\right)^{1/2}\left(\frac{c_2}{1/2}\right)^{1/2}$$

(8.38) becomes

$$c_1 x_1^{1/2} = \tfrac{1}{2}(2c_1)^{1/2}(2c_2)^{1/2}$$
$$c_2 x_1^{-1/2} = \tfrac{1}{2}(2c_1)^{1/2}(2c_2)^{1/2}$$

The first equation gives

$$x_1 = \frac{c_2}{c_1}$$

and the second equation gives

$$x_1^{-1} = \frac{c_1}{c_2}$$

so system (8.38) *is* consistent and has an infinite number of solution pairs (x_1, x_2).

We may apply classical methods to problem (8.37) and interpret all these events. First, consider the unconstrained minimization over $x_1 > 0$ of

$$f(x_1) = c_1 x_1^{1/2} + c_2 x_1^{-1/2}, \quad \text{where } c_1, c_2 > 0$$
$$f'(x_1) = \tfrac{1}{2}c_1 x_1^{-1/2} - \tfrac{1}{2}c_2 x_1^{-3/2}$$

With $f'(x_1) = 0$, we have

$$c_1 x_1^{-1/2} - c_2 x_1^{-3/2} = 0$$

or

$$c_1 x_1^{-1/2}\left[1 - \frac{c_2}{c_1}x_1^{-1}\right] = 0 \qquad (8.39)$$

The solution is $x_1 = c_2/c_1$. To show that this value minimizes $f(x_1)$, we show $f''(c_2/c_1) > 0$:

$$f''(x_1) = -\frac{1}{4}c_1 x_1^{-3/2} + \frac{3}{4}c_2 x_1^{-5/2}$$

and

$$f''\left(\frac{c_2}{c_1}\right) = -\frac{1}{4}c_1\left(\frac{c_2}{c_1}\right)^{-3/2} + \frac{3}{4}c_2\left(\frac{c_2}{c_1}\right)^{-5/2}$$

$$= \frac{1}{2}\frac{c_1^2}{c_2}\left(\frac{c_1}{c_2}\right)^{1/2} > 0, \quad \text{for all } c_1, c_2 > 0$$

So $x_1 = c_2/c_1$ is a relative minimum and is, in fact, a *global* minimum, since $f(x_1) \to \infty$ as $x_1 \to \infty$.

Our primal problem (8.37) does in fact have an optimal solution, for with $x_1 = c_2/c_1$, x_2 may be chosen sufficiently large that the constraint is satisfied, regardless of the value of c_3, and, of course, the value of x_2 does not affect the objective function value. Clearly, too, the primal constraint is satisfied as a strict inequality for some values of (x_1, x_2).

By independent means, therefore, we have shown that the hypotheses of Theorem 8.3 are satisfied by problem (8.37). We have, in fact, solved the problem without recourse to the dual methods of geometric programming. This provides us with an opportunity to verify Theorem 8.3, and we point out that the sufficient condition, Theorem 8.2, is inapplicable for the problem (8.37).

First, the dual does have an optimal solution, as we have seen. Also, the optimal objective function values for the primal and dual problems are identical, since for the primal that value is

$$\frac{c_2^{1/2}}{c_1^{-1/2}} + \frac{c_2^{1/2}}{c_1^{-1/2}} = 2c_1^{1/2}c_2^{1/2}$$

Finally, it was the case that the optimal values of x_1 and x_2 satisfied the system of equations in part iii of Theorem 8.3. The problem of course, caused by $\lambda_3^* = 0$, was that the system gave *no* value for x_2^*. Thus even a unique solution to the dual constraints does not guarantee that an optimal primal solution is immediate.

vii. *Degree of Difficulty*

Still assuming that our $n + 1$ dual constraints are independent (and consistent), suppose that we have $n + 1 < p$. Then there are an infinite number of solutions to the dual constraining equations, and perhaps as many solutions satisfying $\lambda_j \geq 0, j = 1, \ldots, p$.

In such a case, we know (Chapter 1) that if $p - (n + 1)$ variables are arbitrarily specified, then the remaining $n + 1$ are uniquely determined. The number $p - n - 1$ has therefore been called [2] the "degree of difficulty" of solving the dual problem (In general, of course, if the system of dual equations has rank $k \leq n + 1$, then the degree of difficulty will be $p - k$.), since the trick is to find the proper values for those $p - n - 1$ independent variables. Cases where a unique solution arises have zero degree of difficulty. Let us consider a problem in which we obtain a nonzero degree of difficulty.

Example 8.10

Let the primal problem be

$$\min x_3^{-1} + x_1 x_2^{-1}$$

subject to

$$\tfrac{1}{2}x_1^{-1}x_2^{-1/2} + \tfrac{1}{2}x_1^{1/2}x_2 + \tfrac{1}{3}x_1^2 x_3 \leq 1$$

$$x_1, x_2, x_3 > 0$$

(8.40)

The dual equality constraints become

$$\begin{aligned}
\lambda_1 + \lambda_2 \qquad\qquad\qquad\qquad &= 1 \\
\lambda_2 - \lambda_3 + \tfrac{1}{2}\lambda_4 + 2\lambda_5 &= 0 \\
-\lambda_2 - \tfrac{1}{2}\lambda_3 + \lambda_4 \qquad\qquad &= 0 \\
-\lambda_1 \qquad\qquad\qquad + \lambda_5 &= 0
\end{aligned} \tag{8.41}$$

From Chapter 1 we know that the system (8.41) is solvable. Employing complete elimination, as far as it can be carried, on the matrix

$$\begin{pmatrix}
1 & 1 & 0 & 0 & 0 & 1 \\
0 & 1 & -1 & \tfrac{1}{2} & 2 & 0 \\
0 & -1 & -\tfrac{1}{2} & 1 & 0 & 0 \\
-1 & 0 & 0 & 0 & 1 & 0
\end{pmatrix}$$

we obtain

$$\begin{pmatrix}
1 & 0 & 0 & 0 & -1 & 0 \\
0 & 1 & 0 & 0 & 1 & 1 \\
0 & 0 & 1 & 0 & -\tfrac{2}{3} & 2 \\
0 & 0 & 0 & 1 & \tfrac{2}{3} & 2
\end{pmatrix}$$

All solutions to (8.41) then have the form

$$\begin{aligned}
\lambda_4 &= 2 - \tfrac{2}{3}\lambda_5 \\
\lambda_3 &= 2 + \tfrac{2}{3}\lambda_5 \\
\lambda_2 &= 1 - \lambda_5 \\
\lambda_1 &= \lambda_5
\end{aligned} \tag{8.42}$$

where λ_5 is arbitrary.

From the system of equations (8.42), it is easily verified that a solution to (8.41) with $\lambda_1, \ldots, \lambda_5 \geq 0$ occurs if and only if $0 \leq \lambda_5 \leq 1$. Then, the objective function for the dual may be expressed in terms of a single parameter $\lambda = \lambda_5$.

$$\begin{aligned}
g(\lambda_1, \lambda_2, \lambda_3, \lambda_4, \lambda_5) &= \tilde{g}(\lambda) \\
&= \left(\frac{1}{\lambda}\right)^{\lambda}\left(\frac{1}{1-\lambda}\right)^{1-\lambda}\left(\frac{\tfrac{1}{2}}{2+\tfrac{2}{3}\lambda}\right)^{2+(2/3)\lambda}\left(\frac{\tfrac{1}{2}}{2-\tfrac{2}{3}\lambda}\right)^{2-(2/3)\lambda}\left(\frac{1}{3}\right)^{\lambda}(4+\lambda)^{4+\lambda}
\end{aligned} \tag{8.43}$$

Now we ask whether or not Theorem 8.3 applies. First, the dual constraints have an infinity of positive solutions. Furthermore, the primal problem has a feasible solution, which satisfies the primal constraint as a strict inequality; $x_1 = 2.25$, $x_2 = 1$, $x_3 = .001$ is effective. Thus Theorem 8.3 is applicable.

Evaluating $\tilde{g}(\lambda)$ for $\lambda = 0, \tfrac{1}{3}, \tfrac{1}{2}$, and 1, $\tilde{g}$ does seem to have a property of increasing up to a point and then decreasing beyond. Thus we might inquire whether $\tilde{g}$ is concave, in which case solution of the dual would entail the maximization of a concave function over a convex set, a case we know to be relatively simple to handle (Chapter 2). Unfortunately, the general dual function of a single parameter say $g(\lambda_k)$ need not be concave over the range of λ_k that provides nonnegativity for all dual variables.

However, in the present example, consider

$$\ln \tilde{g}(\lambda) = \lambda \ln\left(\frac{c_1}{\lambda}\right) + (1 - \lambda) \ln\left(\frac{c_2}{1 - \lambda}\right) + \left(2 + \frac{2}{3}\lambda\right) \ln\left(\frac{c_3}{2 + \frac{2}{3}\lambda}\right)$$

$$+ \left(2 - \frac{2}{3}\lambda\right) \ln\left(\frac{c_4}{2 - \frac{2}{3}\lambda}\right) + \lambda \ln\left(\frac{c_5}{\lambda}\right) \tag{8.44}$$

$$+ (4 + \lambda) \ln(4 + \lambda)$$

The reader will verify that for $0 \leq \lambda \leq 1$

$$\frac{d^2}{d\lambda^2}(\ln \tilde{g}(\lambda)) = -\frac{1}{\lambda} - \frac{1}{1 - \lambda} - \frac{\frac{4}{9}}{2 + \frac{2}{3}\lambda} - \frac{\frac{4}{9}}{2 - \frac{2}{3}\lambda} - \frac{1}{\lambda}$$

$$+ \frac{1}{4 + \lambda} < 0 \tag{8.45}$$

which implies that $\ln \tilde{g}(\lambda)$ is indeed strictly concave over $0 \leq \lambda \leq 1$. Thus the problem of maximizing $\ln \tilde{g}(\lambda)$ over the same constraints as in the dual problem *is* a concave programming problem; however, we need to maximize $\tilde{g}(\lambda)$, not $\ln \tilde{g}(\lambda)$.

A property of the logarithm, however, comes to the fore. We know that if $x_1 > x_2$ and $x_1, x_2 \geq 0$, then $\ln x_1 > \ln x_2$. Thus, if λ^* maximizes $\ln \tilde{g}(\lambda)$, then certainly λ^* maximizes $\tilde{g}(\lambda)$.

Returning to equation (8.44), let us attempt to maximize $\ln \tilde{g}(\lambda)$. Set

$$\frac{d}{d\lambda}(\ln \tilde{g}(\lambda)) = \ln\left(\frac{1}{\lambda}\right) - \ln\left(\frac{1}{1 - \lambda}\right) + \frac{2}{3} \ln\left(\frac{1}{4 + \frac{4}{3}\lambda}\right)$$

$$- \frac{2}{3} \ln\left(\frac{1}{4 - \frac{4}{3}\lambda}\right) + \ln\left(\frac{1}{3\lambda}\right) + \ln(4 + \lambda) = 0 \tag{8.46}$$

Rewriting (8.46), we have

$$-\ln \lambda + \ln(1 - \lambda) - \frac{2}{3} \ln(4 + \frac{4}{3}\lambda) + \frac{2}{3} \ln(4 - \frac{4}{3}\lambda)$$

$$- \ln(3\lambda) + \ln(4 + \lambda) = 0 \tag{8.47}$$

By virtue of (8.45), $\ln \tilde{g}(\lambda)$ is unimodal, and an optimizing value of λ could be found, therefore, by a sequential search method of Chapter 6 (this problem is found at the end of the chapter). Another possibility is to seek a point which satisfies (8.47). Now, (8.47) is a transcendental equation and its solution will require an approximation technique, such as Newton's method. It should be noted that the *derivative* of $\ln \tilde{g}(\lambda)$ behaves in a nice fashion, owing to $\tilde{g}(\lambda)$'s strict concavity; that is, as λ increases from 0 to 1, the derivative is positive up to a point, say λ^0, where it is zero, and thereafter remains negative (see Figure 8.1).

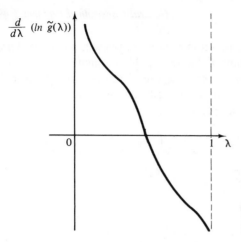

$\frac{d}{d\lambda} (\ln \tilde{g}(\lambda))$

0   1 λ

Figure 8.1. Solution to Example 8.10.

If one evaluates the function $f(\lambda) = (d/d\lambda)(\ln \tilde{g}(\lambda))$ at the interval mid-point, $\frac{1}{2}$, then the zero lies in $(0, \frac{1}{2})$ or $(\frac{1}{2}, 1)$, accordingly as $f(\frac{1}{2}) < 0$ or $f(\frac{1}{2}) > 0$. Thus a new interval is obtained, and it is one half the length of the original. Knowing the signs of f at the end points, the fact that a zero is *somewhere* in the interval, and that f behaves monotonically allows that midpoint evaluation to halve the interval. Thus accuracy to any desired degree is possible—subject to roundoff error, of course. The problems ask that the optimization of $\ln \tilde{g}(\lambda)$ be completed by finding a zero of its derivative. The reader should also attempt to solve the primal problem (8.40).

In the general case where the dual problem has degree of difficulty $\delta > 0$, we shall find $p - \delta$ of the dual variables, say $\lambda_1, \ldots, \lambda_{p-\delta}$, uniquely determined by δ of the remaining variables, say $\lambda_{p-\delta+1}, \ldots, \lambda_p$, which may vary over some region of the δ-dimensional Euclidean vector space. In that case the dual objective function may be written as a function of the form $\tilde{g}(\lambda_{p-\delta+1}, \ldots, \lambda_p)$, and it may be shown that $\ln \tilde{g}(\lambda_{p-\delta+1}, \ldots, \lambda_p)$ is itself concave over the region of interest.

REFERENCES

[1] DUFFIN, R., E. PETERSON, and C. ZENER, *Geometric Programming*, John Wiley & Sons, Inc., New York, 1967.

[2] WILDE, D., and C. BEIGHTLER, *Foundations of Optimization*, Prentice-Hall, Inc., Englewood Cliffs, N.J., 1967.

ADDITIONAL SOURCES

[1] DUFFIN, R., "Cost Minimization Problems Treated by Geometric Means," *Operations Research*, Vol. 10, p. 668, 1962.

[2] ZENER, C., "A Further Mathematical Aid in Optimizing Engineering Designs," *Proceedings of the National Academy of Science*, Vol. 48, p. 518, 1962.

[3] ZENER, C., "A Mathematical Aid in Optimizing Engineering Designs," *Proceedings of the National Academy of Science*, Vol. 47, p. 537, 1961.

PROBLEMS

1. Apply the arithmetic mean–geometric mean inequality to obtain $f(x) \geq b$ (b a constant) for the following functions:

(a) $f(x) = \dfrac{1}{2x^2} + x^{-3} + 2x^{3/2}$

(b) $f(x) = \dfrac{1}{2}x^{1/3} + \dfrac{2}{x^2} + 3x^3$

(c) $f(x) = 5x^{5/2} + \dfrac{1}{x^{1/3}} + x^{3/2} + (4x)^{-2}$

(d) $f(x) = 2x^{-1/2} + x^{5/3} + \frac{2}{3}x^{3/5}$

(e) $f(x) = 1 + x + \dfrac{1}{x} + \dfrac{1}{x^2}$

(f) $f(x) = \dfrac{3}{4}x^{-3/4} + (2x)^4 + \left(\dfrac{1}{x}\right)^3$

(g) $f(x) = \frac{2}{9}x^3 + \frac{1}{9}x^2 + \frac{2}{9}x^{-3}$

(h) $f(x) = (2x)^{4/3} + \left(\dfrac{1}{x}\right)^3 + \dfrac{3}{4}x$

(i) When and only when is the special form of the inequality effective?

2. Write posynomial expressions for four or five real-world quantities.

3. Some information regarding system (8.9) and solutions to it is immediately available.
   (a) Write a condition on the a_{ji} that is necessary for the existence of a solution to (8.9).
   (b) With regard to m, n, when is there no solution to the system?
   (c) In terms of *numbers* of solutions to (8.9), state all the possibilities, and for each determine the coefficient array.

4. There are some famous theorems (useful in linear programming theory, incidentally) that are related to system (8.9).
   (a) The key theorem (A. W. Tucker, 1956). The two systems of equations, $Ax = 0$, $x \geq 0$, and $A^T u \geq 0$, have solutions x^*, u^* such that

$$x^* + A^T u^* > 0$$

   What has this to do, if anything, with solutions to (8.9)?
   (b) The Farkas lemma (1902), which may be obtained as an easy corollary of the key theorem (proof?). If for every solution u to $A^T u \geq 0$ we have $u^T b \geq 0$, then $Ax = b$, $x \geq 0$ has a solution.

5. Show that if the $(n + 1)$st equality is eliminated from (8.9), and the remaining system has a solution, then (8.9) has a solution.

6. Show that if the reduced system in Problem 8.5 has a solution it has an infinity of them. Does this imply that (8.9) will have an infinity of solutions?

7. Write the dual problem for each of the following:
   (a) $\min 4x_1^{-1}x_2^{3/2}x_3^{5/4}x_4^2x_5^{-2} + \frac{3}{5}x_1^2x_2x_3x_4x_5 + 10x_1^2x_2^{-2}x_3^2x_5^{-2} + 5x_3x_4x_5$
   subject to

$$12x_1^{3/5}x_2^{7/8}x_4^{-10}x_5^{4/3} + \frac{3}{8}x_1^{-2/5}x_3^3x_4^{-1/2} \leq 1$$

$$6x_1^2x_2^{-4/5}x_3^{3/8}x_4^5x_5^{-3} \le 1$$

$$x_1, \ldots, x_5 > 0$$

(b) $\qquad$ min $\frac{3}{4}x_1^{-6}x_2^{-3}x_3^4x_4^{7/8} + 4x_1^{7/5}x_2^{-3}x_3^{-4}x_4^{-1}$

subject to

$$7x_1^2x_2^{3/8}x_3^{8/3}x_4^{-2} + 4x_1^3x_3^{-3}x_4^{5/2} + 6x_2^6x_3^{-4} \le 1$$

$$\frac{3}{2}x_1^{-10}x_2^{3/5}x_3^{7/2}x_4^{-4/5} + 8x_2^{4/9}x_4^{3/9} \le 1$$

$$x_1, \ldots, x_4 > 0$$

(c) $\qquad$ min $\frac{5}{7}x_1^2x_2^{-1}x_3^3 + 4x_1^{-2}x_2^2x_3^{-3/2}$

subject to

$$3x_1^2x_2^{-7}x_3^4 + 7x_1^3x_2^{-2}x_3^1 + x_1^8x_2^5x_3^{-2} \le 1$$

$$\frac{7}{8}x_1^{-3}x_2^{-6}x_3^4 + 2x_1^7x_2^{-2}x_3^{-3} + 5x_1^{-1}x_2^{-1}x_3^{-1} \le 1$$

$$6x_1^{-1}x_2^{-1}x_3^{10} + 14x_1^5x_2^2x_3^{-2} \le 1$$

$$x_1, x_2, x_3 > 0$$

(d) min $7x_1^{-1}x_2^2x_3^3x_4^8 + 14x_1^{-1}x_2^{-4}x_3^{-3}x_4^7 + 3x_1^{7/2}x_2^{10}x_3^{-5}x_4^{-3} + x_1^{-6}x_2^{-4}x_3^2x_4^{10}$

subject to

$$3x_1^{-3/2}x_2^{-6/5}x_3^4x_4^2 + x_1^1x_2^{-2}x_3^3x_4^{-3} + x_2^4x_3^2x_4^{-6} \le 1$$

$$x_1, \ldots, x_4 > 0$$

(e) min $10x_1^{-2}x_2^3x_3^{7/2}x_4^{6/7}x_5^{7/6} + 7x_1^5x_2^{-2}x_3^{-2}x_4^{-1}x_5^{-1}$

subject to

$$\frac{7}{5}x_1^{-2}x_2^3x_3^4x_4^{-1}x_5^{-7/2} \le 1$$

$$\frac{3}{8}x_1^{-5/2}x_2^{4/3}x_3^3x_5 \le 1$$

$$\frac{2}{9}x_2^{6/7}x_3^{-4}x_4^{-2} \le 1$$

$$\frac{1}{10}x_1^9x_3^{10}x_5^{-11} \le 1$$

$$x_1, \ldots, x_5 > 0$$

8. In discussing linear-programming problems, unrestricted variables were easily accommodated by certain substitutions. Would the identical substitution make any noticeable change in a geometric-programming problem with an unrestricted variable? Investigate and discuss.

9. For each of the following problems, determine whether or not Theorem 8.2 is applicable, that is, whether or not its hypotheses are satisfied.

(a) $\qquad$ min $x_1^{-3}x_2 + x_1^{3/2}x_2^{-1}$

subject to

$$x_1^2x_2^{-1} + \frac{1}{2}x_1^{-2}x_2^3 \le 1$$

$$x_1, x_2 > 0$$

(b) $\qquad$ min $x_1^{-2}x_2^{1/2}x_3^{-1} + 2x_1^2x_2^{-1}x_3^{-1/2}$

subject to

$$x_1x_2x_3^{-1} \le 1$$

$$2x_1^{-1}x_2^{-1}x_3 + \frac{1}{4}x_1^2x_2^{-2}x_3^2 \le 1$$

$$x_1, x_2, x_3 > 0$$

(c)
$$\min x_1^{3/2}x_2^{-1}x_3^{-2}x_4^{-3/2} + \tfrac{4}{3}x_1^{-2}x_2x_3^{-1}x_4^{1/3}$$
subject to
$$4x_1^2x_2^{-2/3}x_3^3x_4^{-1/2} + x_1^{-1}x_2^2x_4^3 \leq 1$$
$$\tfrac{1}{2}x_1^2x_2^{-1/2}x_3^{-2}x_4 + 2x_2^{-1}x_3^{-1/2}x_4^2 \leq 1$$
$$x_1^{-3/2}x_2^{-1/2}x_3^2x_4^{3/2} \leq 1$$
$$x_1, \ldots, x_4 > 0$$

(d)
$$\min x_1^{-1/2}x_2^2x_3^{-3/2} + \tfrac{1}{2}x_1^{-1}x_2^{-3}x_3^{3/2}$$
subject to
$$2x_1^2x_2^{-3}x_3 + 3x_1^{-1}x_2^2x_3 \leq 1$$
$$x_1^{-1}x_2^2x_3^{-2} \leq 1$$
$$3x_1^{-1}x_2^{-2}x_3^2 + \tfrac{1}{2}x_1^2x_2^{-1}x_3^{-2} \leq 1$$
$$x_1, x_2, x_3 > 0$$

10. (a) Show that $g(\lambda_1, \ldots, \lambda_5)$, whose arguments are expressed in (8.42), is a concave function of $\lambda_1, \ldots, \lambda_4$.
 (b) Using the results of Chapter 2, on the relationship between quadratic forms and concave functions, show *how* one might prove that the general dual function is concave. Write the matrix of interest for the dual function of Example 8.7.

11. For each of the following problems determine whether or not Theorem 8.3 is applicable.
(a)
$$\min x_1^{-1}x_2^{3/2} + x_1^2x_2^{-1/2}$$
subject to
$$5x_1^{-1}x_2^{-1} + \tfrac{1}{3}x_1^{5/2} \leq 1$$
$$x_1, x_2 > 0$$

(b)
$$\min x_1^{-1}x_2^{3/2} + x_1^2x_2^{-1/2}$$
subject to
$$6x_1^{3/2}x_2^{-1/2} + \tfrac{1}{4}x_1^{4/3}x_2^{-1/2} \leq 1$$
$$x_1x_2^{-1} + x_1^{-3/2}x_2^{3/2} \leq 1$$
$$x_1, x_2 > 0$$

(c)
$$\min x_1^{-2}x_2^{3/2}x_3^{1/3} + 4x_1^3x_2^{1/2}x_3^{-3} + \tfrac{2}{7}x_1^7x_3^2$$
subject to
$$\tfrac{1}{3}x_1^2x_2^{-2}x_3 + \tfrac{1}{3}x_1x_2^{-1}x_3^{-1} \leq 1$$
$$x_1^2x_2^3x_3^{-1} \leq 1$$
$$x_1, x_2, x_3 > 0$$

(d)
$$\min \tfrac{3}{4}x_1^{1/2}x_2^2x_3^{-1} + \tfrac{1}{4}x_1^{-1}x_2^{-2}x_3 + \tfrac{1}{4}x_2^{-2}x_3^{-2}$$
subject to
$$\tfrac{1}{3}x_1^{-1}x_2^{-1}x_3^{3/2} + \tfrac{1}{2}x_1^{2/3}x_2^{-2/5}x_3^2 \leq 1$$
$$x_1x_2x_3^2 \leq 1$$
$$x_1, x_2, x_3 > 0$$